Iron Age and Roman salt production and the medieval town of Droitwich

Excavations at the Old Bowling Green and Friar Street

Iron Age and Roman salt production and the medieval town of Droitwich

Excavations at the Old Bowling Green and Friar Street

Edited by Simon Woodiwiss

Hereford and Worcester County Council

Contributions by: James Bond BA FSA MIFA, Father Jerome Bertram MA FSA, Susan Colledge BSc, Anne Crone, Brenda Dickinson BA, Christopher Gregson NDD ATC Dip Cons, James Greig, Mark Hassall, Julian Henderson BA MA PhD, Jennifer Hillam BSc FSA MIFA, Hilary Howard, Justin Hughes BA AIFA, Alan Hunt, Derek Hurst, Steven Juggins, Alison Locker BSc, Louise Monk BA, Donald Mackreth, Gerry McDonnell BTech PhD MIFA, Ruth Morgan PhD, Elaine Morris, Michael Nellist Bsc LDS RCS(Eng), Robert Otlet, Allan Peacey, Helen Rees BA, Fiona Roe MA MLitt, Alan Saville BA FSA MIFA, Wilfred Seaby MA FSA, Michael Sekulla, Robin Symonds PhD, Jane Timby BA PhD, Roberta Tomber, David Williams PhD FSA, Roger Williams BSc, Simon Woodiwiss BA AIFA

Illustrations by:
Anne Crone and Carolyn Hunt MAAIS

1992

CBA Research Report 81

Council for British Archaeology

Published 1992 by the Council for British Archaeology
112 Kennington Road, London SE11 6RE

British Library Cataloguing in Publication Data
A catalogue record for this book is available from the British Library

ISBN 1 872414 24 9

Typeset in house by Julie Gardiner and Lisa Pickering

Printed by Alden Press, Oxford

Contents

Sections in parentheses refer to fiche

Illustrations

All illustrations by Carolyn Hunt except Figs 36-8, 47, 49: Nos 5-9, 50 and 52-4 by Anne Crone

General discussion

Fiche

Old Bowling Green

Plates

Tables

Fiche

Summary

This is the first of a proposed three volume series, reporting on six excavations carried out in Droitwich between 1967 and 1985. The town has been a centre for the large-scale production of salt from the Iron Age until the early part of this century. This industry was based on the brine springs of exceptional purity and strength that naturally occurred in the area of the town.

This volume commences with a general introduction which considers some of the geological, technological, and economic aspects of the industry.

The main evidence for salt production from these excavations dates to the late Iron Age. It consists of large tanks lined with clay and revetted by stakes and probably used primarily for the storage of brine. These tanks were also associated with hearths and vast quantities of briquetage. The size and arrangement of these structures indicate a well organised and large-scale industry. The term *briquetage* is here used to refer to a coarse sandy or organic pottery fabric, almost exclusively in the form of vessels, probably used for the draining of salt crystals. A high water table led to the preservation of a range of wooden artefacts recovered from the tanks and again probably used in the production of salt.

At the Old Bowling Green there is some evidence for the continuity of use of the tanks into the early Roman period, though for how long is uncertain. After the disuse of the structures associated with salt production, occupation in the Friar Street area becomes more domestic in nature. Industrial activity at the Old Bowling Green continued in the 3rd or 4th centuries. The nature of this industry could not be conclusively determined, though the primary production of salt or the processing of animal products as a secondary process are equally probable.

The post-Roman phases at the Old Bowling Green did not produce much useful evidence of activity. At Friar Street, however, the later deposits were better preserved. After an initial phase of agricultural activity the area assumed an industrial character (later Saxo-Norman), possibly associated with tanning or horn-working. Friar Street itself may also have been in existence at this time. From the 12th century there follows a fairly complex development of domestic enclosures and buildings.

The reports on pottery and other ceramic materials are the first substantial ones to be published in the area, covering a virtually continuous sequence from the late Iron Age to the later medieval periods. Many of the objects described in other specialist reports (wood, lead, inscriptions, etc) are unique and may relate directly to the salt industry. Unfortunately it is often impossible to establish this connection with any degree of certainty.

In the concluding general discusson two major themes are examined: Iron Age and Roman salt production, and the development of the Saxon and medieval town. For the historic period discussion of the excavations is integrated with an outline of the documentary and topographical evidence. These themes form the major aspects of two excavations at the Old Bowling Green and Friar Street.

The combination of a primary industrial resource which was exploited over many centuries and well preserved archaeological deposits, gives Droitwich considerable potential for research on the development of European industry.

Résumé

Ceci est le premier d'une série de trois volumes proposés qui rend compte sur six fouilles effectuées à Droitwich entre 1967 et 1985. La ville fut centre pour la production de sel de l'âge de Fer jusqu'au début du vingtième siècle. Cette industrie était fondée sur les sources de sel d'une pureté et intensité exceptionelles qui existaient naturellement autour de la ville.

Ce volume commence avec une introduction générale qui traite quelques-uns des aspects géologiques, technologiques et économiques de l'industrie.

L'évidence principale pour la production de sel de ces fouilles vient de vers la fin de l'âge de Fer. Elle consiste de grandes cuves chemisées d'argile et soutenues à l'aide de poteaux qui furent utilisées probablement pour stoker l'eau salée. Ces cuves furent aussi associées à des foyers et à d'énormes quantités de briquetage. La grandeur et l'arrangement de ces structures indiquent une industrie bien organisée et à grande échelle. Le terme briquetage s'emploie ici pour indiquer un matériel de poterie de gros sable ou d'une substance organique, presque entièrement en forme de vaisseaux, utilisés en toute probabilité pour le drainage de crystales de sel. A cause d'un haut niveau hydrostatique un assortiment d'objets fabriqués en bois furent préservés et retrouvés dans les cuves, eux aussi étant probablement utilisés pour produire le sel.

À l'Old Bowling Green (le Terrain de Boules sur Gazon) il y a aussi de l'évidence pour l'usage des cuves jusqu'au début de l'ère romaine mais pendant combien de temps n'est pas certain. Après la désuétude des structures associées à la production de sel, le quartier de Friar Street devient plus résidentiel. L'activité industrielle à l'Old Bowling Green continua pendant le troisième ou quatrième siècle. On ne pouvait pas distinguer de façon concluante la nature de cette industrie bien que la production de base de sel ou le traitement de produits animaux soient également probables.

Les périodes post-romaines à l'Old Bowling Green n'ont pas fourni beaucoup d'évidence d'activité. À Friar Street, cependant, les gisements de cette époque étaient mieux préservés. Après une première période d'activité agricole, le quartier prit un charactère industriel (vers la fin de l'ère saxo-normande) éventuellement associé avec le tannàge ou l'artisanat de cornes. Friar Street elle-même existait possiblement à cette époque. À partir du douzième siècle se suivaient d'une manière assez complexe des enclos de famille et des bâtiments.

Les rapports sur la poterie et d'autres matières céramiques sont les premiers publiés dans ce domaine, englobant une séquence presque continue de la fin de l'âge de Fer vers la fin du moyen âge. Beaucoup des objets décrits dans d'autres rapports spécialisés (bois, plomb, graffiti etc) sont uniques et il se peut qu'ils soient directement liés à l'industrie de sel. Malheureusement, il est souvent difficile d'établir ce lien avec certitude.

Dans la discussion générale qui conclue, deux thèmes majeurs sont examinés: l'âge de Fer et la production romaine de sel, et le développement de la ville saxonne et médiévale. En ce qui concerne la période historique la discussion de la fouille s'intègre à un résumé de l'évidence documentaire et topographique. Ces thèmes forment les principaux aspects de deux fouilles à l'Old Bowling Green et Friar Street.

La conbinaison d'une ressource primaire industrielle qui fut exploitée pendant de nombreux siècles et des gisements archéologiques bien préservés accorde à Droitwich beaucoup de possibilités pour des recherches sur le développement de l'industrie européenne.

Zusammenfassung

Dieser ist der erste einer vorgeschlagenen Reihe von drei Bändern, die über sechs Ausgrabungen zwischen 1967 und 1985 in Droitwich berichten. Die Stadt war Zentrum für die Salzherstellung in großem Masse von der Eisenzeit bis zu früh in diesem Jahrhundert. Diese Industrie wurde auf die außerordentlich reinen und starken Salzwasserquellen gegründet, die im Stadtgebiet ganz natürlich vorkamen.

Dieser Band beginnt mit einer allgemeinen Einleitung, die einige der geologischen, technologischen und wirtschaftlichen Hinsichte betrachtet.

Das wichtigste Anzeichen von Salzherstellung von diesen Ausgrabungen kommt von der späten Eisenzeit. Es besteht aus großen mit Lehm ausgekleideten und von Stangen bekleideten Bassins. Diese Bassins wurden auch mit Herden und großen Quantitäten «Briquetage» benutzt. Die Bassins wurden wahrscheinlich meistens für das Lagern von Salzwasser benutzt. Die Größe und die Anordnung dieser Strukturen deuten auf eine gut organisierte und große Industrie hin. Das Wort «Briquetage» bedeutet hier eine große sandige oder organische Töpfererde, fast immer für Behälter benutzt, worin Salzkristalle wahrscheinlich abgetrocknet wurden. Ein hoher Grundwasserspiegel hat geholfen, eine Menge holzerner Werkzeuge in den Bassins in gutem Zustand zu erhalten. Diese Werkzeuge wurden auch wahrscheinlich für die Salzherstellung benutzt.

Am Old Bowling Green [Alten Rasen zum Bowlingspiel] gibt es einige Spuren, daß diese Bassins bis zur Frührömerzeit benutzt wurden, aber für wie lange ist noch unklar. Nachdem diese Salzherstellungsstrukturen nicht mehr gebraucht wurden, wurde das Leben im Gebiet von Friar Street häuslicher. Industrielle Aktivität am Old Bowling Green fuhr im dritten oder vierten Jahrhundert fort. Die Art dieser Industrie war nicht beweiskräftig zu entscheiden, aber die Ursalzherstellung oder die Verarbeitung von Tierprodukte als zweiter Vorgang sind ebenso wahrscheinlich.

Die Nachrömerstufen am Old Bowling Green statteten nicht viel nützliche Aktivitätsanzeichen aus. In Friar Street jedoch wurden die späteren Ablagerungen besser erhalten. Nach einer ersten Stufe von landwirtschaftlichen Tätigkeit nahm das Gebiet eine industrielle Natur an (später Sachse-normannisch), die vielleicht etwas mit dem Lohgerben oder dem Hornwerken zu tun hatte. Friar Street selbst mag wohl in dieser Zeit existiert haben. Vom zwölften Jahrhundert folgte eine ziemlich komplizierte Entwicklung von häuslichen Einfriedigungen und Gebäude.

Die Berichte über Töpferwaren und andere keramische Materialien sind die ersten beträchtlichen veröffertlichten Berichte über diese Gegend, und sie fassen eine fast ununterbrochene Reihenfolge um, von der späten Eisenzeit bis zum späten Mittelalter. Viele der in anderen Spezialistberichten beschriebenen Gegenstände (Holz, Blei, Einschreibungen usw) sind einzig, und sind vielleicht direkt mit der Salzindustrie verbunden. Unglücklicherweise ist es oft unmöglich, diese Verbindung mit irgendeiner Sicherheit festzustellen.

Während der abschließenden allgemeinen Diskussion werden zwei Themen untersucht: die Salzherstellung von der Eisenzeit und der Römerzeit, und die Entwicklung von der Sachsestadt und der Mittelalterlichestadt. Für die historische Periode wird eine Diskussion der Ausgrabungen mit einem Abriß der Beweisstücke und der Topographie verbunden. Diese Themen stellen die größten Aspekte von zwei Ausgrabungen am Old Bowling Green und in Friar Street dar.

Die Verknüpfung eines ursprünglichen industriellen Hitfsmittels, das über viele Jahrhunderte ausgenutzt wurde, und gut erhaltene archäologische Ablagerungen statten Droitwich mit beträchtlichen Forschungsmöglichkeiten aus, auf die Entwicklung von europaischer Industrie.

Acknowledgements

The projects reported in this volume were supported by the Department of the Environment, Droitwich Town Council, Hereford and Worcester County Council, Historic Buildings and Monuments Commission for England, Manpower Services Commission and Wychavon District Council.

Anne Crone would like to thank Jennifer Hillam (Department of Archaeology and Prehistory University of Sheffield), the staff of the Physics Department, Worcester Technical College, and the staff of the County Analyst's Department (Hereford and Worcester County Council).

Jennifer Hillam is grateful to Ruth Morgan for the original work on the Friar Street timbers and for help in preparing the report, and to the Department of the Environment for financing the work.

Alan Hunt would like to thank Philip Barker, Clifford Denham, Tony Fleming, James Greig, Tony Hartshorne, Justin Hughes, Elaine Morris, Philip Rahtz, Geoffrey Showell, Ian Walker, Hereford and Worcester Fire Service, and all the contributors for their assistance at Friar Street. James Bond initiated the original survey of towns in Worcestershire which was the main catalyst of all subsequent work in Droitwich.

Derek Hurst gratefully acknowledges the particular assistance of the following in the preparation the Friar Street pottery report: C Cane, J G Hurst, H Leech, M Mellor, J Rutter, Dr D P S Peacock (for confirmation of the *amphorae* fabric identifications), and D Walker (Worcester Technical College). Especial thanks are also due to the Bibliographical Section of the Libraries Department (Hereford and Worcester County Council) for their indispensable service in supplying many and diverse publications.

Michael Nellist gratefully acknowledges the assistance of Dr E Corbett (Department of Oral Pathology, Dental School), Mr T F Spence (Department of Anatomy, Medical School), Mrs S Waring (Photographic Unit, Department of Oral Pathology, Dental School), all of the University of Birmingham, and Dr D K M Toye and Miss D Kelly (Department of Oral Surgery and Oral Medicine), of the Dental Hospital, Birmingham.

Helen Rees wishes to acknowledge with gratitude the help and advice concerning all aspects of the post-excavation work given by A Anderson, D Barker, P Booth, B Dickinson, C Goudge, K Hartley, M Hassall, H Howard, J Keeley, D F Mackreth, E Morris, J R Perrin, J Price, the late H Russell, J Sawle, R Symonds, J Timby, R Tomber, A Vince, Dr G Webster, Dr D Williams, and Dr C J Young. P Hefford managed the finds shed with efficiency and dispatch. A Crone, C Hunt, W Riddick, R Styles, B Willoughby, and N Wood drew the pottery. C Hunt prepared the illustrations and J D Hurst suggested many changes to the original text to make it compatible with the revised phasing.

Simon Woodiwiss wishes to thank Alan Hunt for initiating the current programme of archaeological work in Droitwich, which was subsequently guided by the members of the Droitwich Archaeology Committee, especially former chairmen, Lawrence Barfield, Geoffrey Showell and Anne Walton. Over the years many officers of the constituent local authorities have offered and given assistance to the two excavations reported here, especially Colin Garrett, Tony Hartshorne, Ken Jennings, and Richard Taylor. Successive Inspectors of Ancient Monuments, namely Tony Fleming, Michael Parker-Pearson, and Anthony Streeten have given valuable assistance and advice over the years.

All the contributors to this volume have shown great understanding of the difficulties and delays in the publication programme. David Freezer supervised the excavation at Friar Street. The preliminary work by John Sawle and Heather James on the excavated data from the Old Bowling Green contributed greatly to the stratigraphic analysis. Conservation of finds was carried out by Martin Read (HBMCE's regional conservator) of Bristol City Museum and Joanna Williams and Alexander Trone of Birmingham City Museum; unless otherwise acknowledged. Hilary White commented on the inlaid floor tile from the Old Bowling Green. David Jordan of the Ancient Monuments Laboratory kindly calibrated the radiocarbon dates. Derek Hurst is especially to be thanked for his patient criticism and advice. Thanks are also due to the CBA's referee, to Julie Gardiner, Christine Pietrowski, and Lisa Pickering and to the Ancient Monuments Laboratory for their much needed observations. Jeremy Owen kindly translated the summaries. Elspeth Entwhistle, Amanda Woodiwiss and Christine Pitt typed the manuscripts. Adrian Tindall found time at a very busy period to correct the worst excesses of grammar and punctuation.

My work on this volume is dedicated to the memory of my father who brought back many treasures from his travels; and to my mother for introducing me to archaeology.

Abbreviations

SR	Specialist reference
S	Structure
P	Phase
HWCM	Hereford and Worcester County Monument, the succeeding number refers to the County Sites and Monuments Record primary reference number
HWRO	Hereford and Worcester Records Office, the succeeding number refers to the accession number

Numbers in parentheses, without prefixes, refer to context numbers.

Project staff

R Armitage, D Barratt, A Bell, D Berwick, L Bourne, D Britton, N Cafferty, S Clarke, M Coker, A Colgan, A Crone, R Croton, F Dancyger, G Dipple, D Dyson, M Edwards, D Eveson, D Freezer, N Green, V Groves, D Hales, D Halls, D Harris, P Hefford, P Hillier, J Hughes, A Hunt, C Hunt, R Hunt, J D Hurst, H James, R Kemp, P Kurn, M Lind, H Lubin, A Marling, T Mason, M Meadows, C Moor, H Mytum, R Oram, C Pitt, W Pitt, A Price, E A Price, R Price, N Pritchard, H Rees, W Riddick, P Rose, H Russell, J Sawle, M Schofield, M Shaw, P Simkins, A Simmons, E Smith, F Smith, D Staniforth, M Stevens, R Styles, D Taylor, A Thomas, L Utchinson, M Wallace, A Ward, J Ward, K White, M Williams, B Willoughby, D Windell, N Wood, S Woodiwiss and G Wright.

1 Introduction *Simon Woodiwiss*

This is the first of a proposed three volume series, reporting on six excavations carried out in Droitwich between 1967 and 1985. All were partly funded by the Department of the Environment and latterly by the Historic Buildings and Monuments Commission for England. The excavations relate in varying degrees to different aspects and periods of the town's development, of which Iron Age and Roman salt production and the medieval town are dealt with here. The second volume will report on excavations at the Roman fort (Dodderhill, HWCM 603), villa (Bays Meadow, HWCM 678) and settlement (Hanbury Street, HWCM 681), and the final volume will report on a continuous series of deposits relating to salt production from the Iron Age to the post-medieval periods (Upwich, HWCM 4575).

Geology and topography

The town of Droitwich, in the County of Hereford and Worcester, is located on the River Salwarpe, a tributary of the Severn, amongst the gently-undulating topography of the English Midlands (Fig 1). Droitwich is situated in the middle of a band of Mercian Mudstone (Keuper Marl and Sandstone) extending from Cumberland and County Durham in the north to Devon in the south. Within the marl are saliferous beds deposited during the Triassic period, probably when the area was covered by a shallow sea (Northolt and Highley 1973, 4). Although the saliferous beds are not continuous from north to south, Poole and Williams (1981, 3) believe the Worcestershire salt field to be a marginal extension of the main Cheshire field. The full extent of the local saliferous beds is imperfectly known and they are expected to extend over a larger area than so far established. For example, it is thought that the field extends at a greater depth in the Worcester basin towards Evesham (Northolt and Highley 1973, 7). A general account of the local geology is given in Mitchell *et al* (1961).

Under present climatic conditions salt will not outcrop, and at Droitwich the saliferous beds are covered by breccias formed by collapse of the surrounding rock (Poole and Williams 1981, 4). Water percolating through the beds takes up salt in solution until it forms saturated brine and is forced to the surface under artesian pressure. The resulting springs are today restricted to the low river flood plain in the area of the town. This may be expected to have been the main location of brine springs in the past, though salt derived from the saliferous beds has possibly had localised effects outside this area. For example, the pools identified at Upton Warren (HWCM 4151, five kilometres to the north-east of Droitwich) and dated to 42,000 years bp, were mildly brackish, which was attributed to the saliferous beds of the area (Coope *et al* 1961, 398).

Both of the excavated areas reported in this volume were sited on Mercian Mudstone (Fig 2). The bedrock and its heavy clay soil would give poor natural drainage and artificial drainage is to be expected. Friar Street (HWCM 605) lies just to the north of a small area of gravel, the third Severn terrace. The gravel would alleviate the otherwise heavy clays of the area and may have been preferentially selected for cultivation. The Old Bowling Green (HWCM 600) lay to the west of a former meander of the river Salwarpe. The river was probably diverted to its present course in this area in the mid 19th century, when the Droitwich Junction Canal was constructed. The meander is shown on a 17th century map (reproduced in Crickmore 1984b, fig 5) and presumably determined the course of the ecclesiastical parish boundary between St Andrew's and St Augustine's. The course of the meander, reconstructed from the parish boundary, is shown in Figure 2. It is possible that the meander existed from the Roman period or earlier, though not necessarily following precisely the same course.

The 'open pan' method of salt production

In order to produce salt crystals from brine, the water content must be evaporated. This may be achieved either by solar evaporation or through direct heating. There is no mention in early documents of solar evaporation being used in Droitwich, and the equipment and features mentioned (furnaces and pans) indicate evaporation by direct heating. Solar evaporation cannot be totally discounted when considering the function of archaeological features and artefacts. However, the inland situation of Droitwich (away from sea breezes), and the prevailing climate, suggest that direct heating would have been the most feasible method. Throughout the historical period, until the adoption of the 'vacuum' method in the 20th century, the 'open pan' method, requiring a simple technology, was used. The method itself changed little throughout its period of use, variation occurring in such details as the increase in boiling pan size, with the advent of iron pans in the 17th century (Berry 1975, 78). Perhaps the best contemporary account of the method is that submitted by Thomas Rastell in 1678 to the Royal Society. The manual by Georgius Agricola on European industries of the 16th century (Hoover

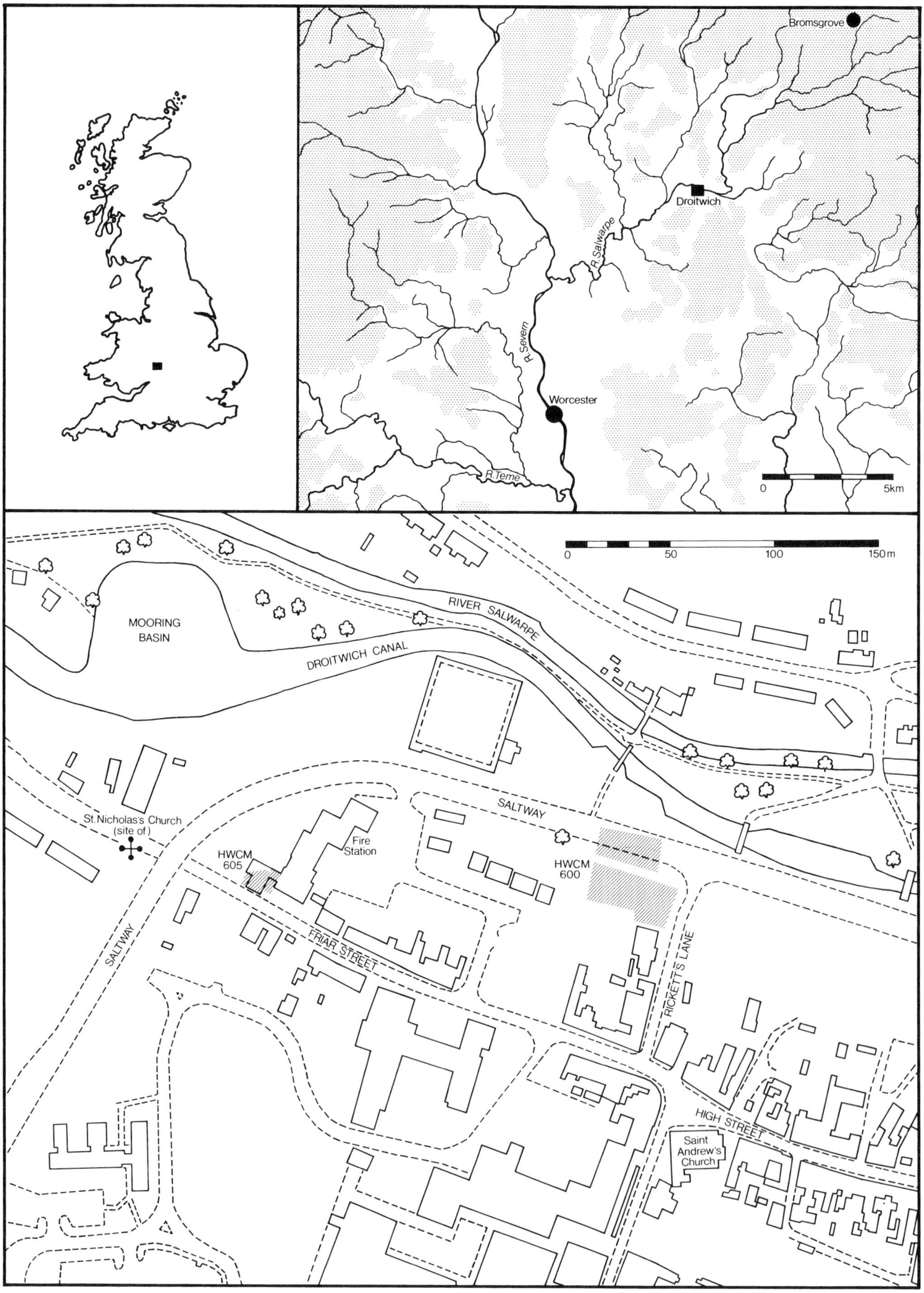

Figure 1 Location of excavations

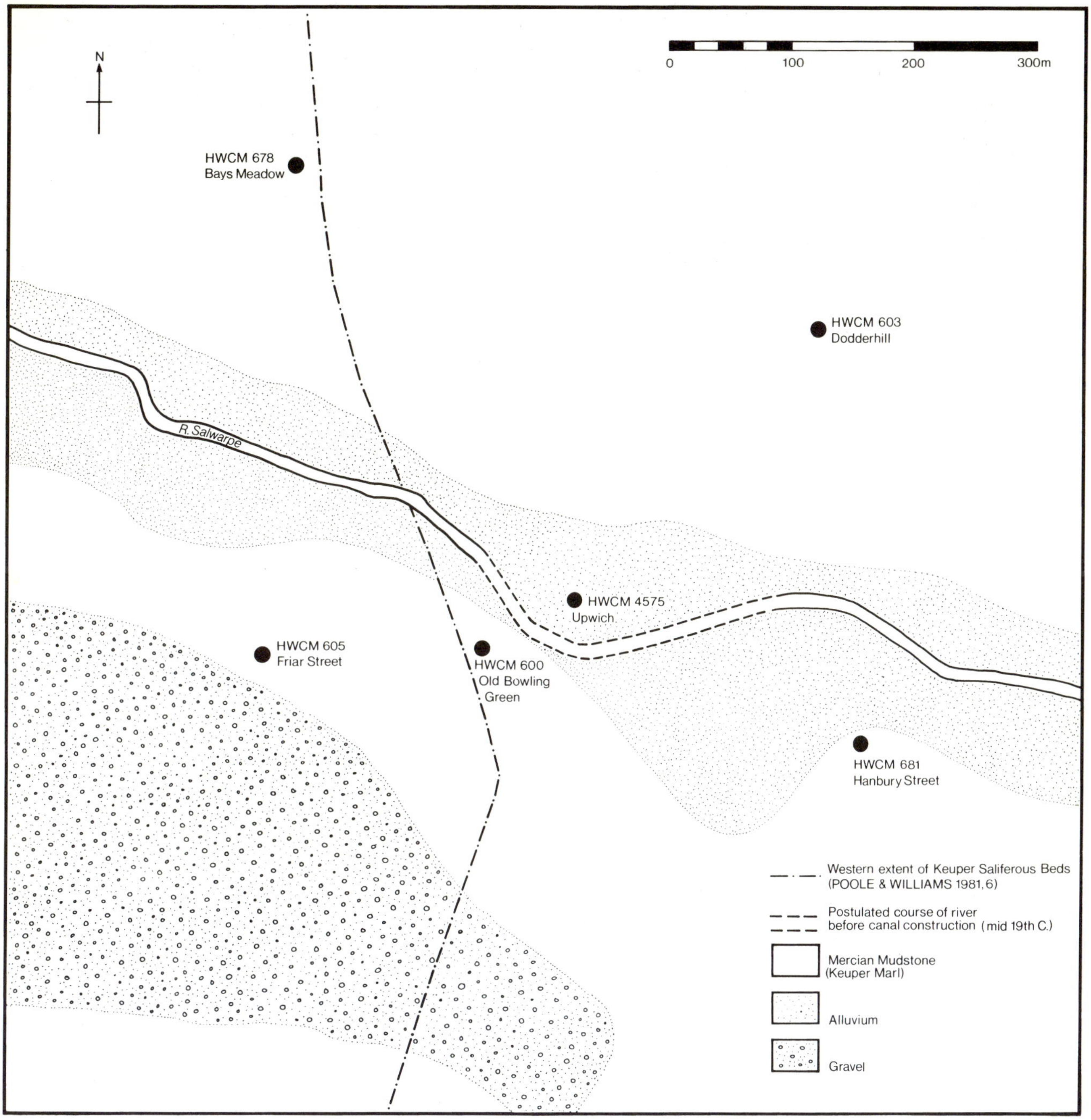

Figure 2 Geology, showing former course of River Salwarpe and western extent of Keuper Saliferous Beds

and Hoover 1912) also has useful illustrations and descriptions of the salt making process.

The process and terminology of salt production by the open pan method is summarised in Figure 3, together with the various energy inputs required. Both the processes and inputs change in detail throughout the development of the salt industry in Droitwich.

Extraction was from natural brine springs, perhaps within pits, which may have been lined to control contamination by fresh ground water and material from the sides of the pit. Brine was presumably originally extracted using buckets or other containers. Later, pumps were used, and documents of the late 16th century mention reciprocating pumps in the brine pits (Berry 1975, 77). The motive power for extraction by the pumps is uncertain, however the Salwarpe may have provided one source.

It would be necessary to settle any silts contaminating the brine prior to boiling, especially where the brine had been taken from poorly-lined, or unlined pits. Both Rastel (1678, 1062) and Agricola (Hoover and Hoover 1912, 548) mention

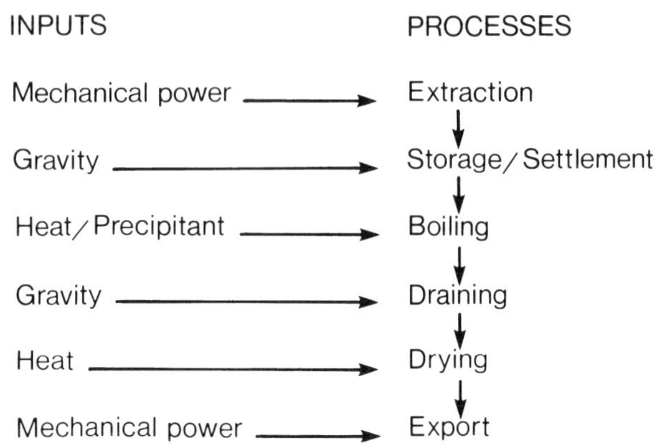

Figure 3 The process of salt production by the 'open pan' method

the use of 'tuns' or tubs close to the furnaces, in which brine was stored.

Salt houses, furnaces and wood for fuel are mentioned in a number of Saxon documents (eg Finberg 1972, nos 197, 206 and 282). Additionally, lead pans are mentioned in the Domesday Book (Thorn and Thorn 1982, nos 1, 1a:5). Similar structures and implements are described by Rastel (1678, 1062) in the late 17th century, though coal was then a recently adopted fuel. Changes in scale of the furnaces and pans came as a result of a number of developments, such as the use of coal, the breaking of the borough's monopoly on salt production, an increase in demand and the adoption of iron pans (Berry 1975, 78). To produce different sized salt crystals the heating of the pans was controlled, and Rastel (1678, 1063) mentions the use of resin to aid the formation of fine crystals.

The crystals were removed from the pan before the water had completely evaporated. It is suspected that the presence of liquid had a cooling effect on the lead pans used in the medieval and earlier periods, since lead was a metal whose low melting point may otherwise have been reached (John Price pers comm). The form and porous nature of the coarse ceramic vessels, in a fabric known as briquetage, indicates that drying and moulding of salt crystals was their probable function (Rees 1986, 51). Conical wicker baskets were used for this purpose in the medieval period (Berry 1975, 78). Agricola (Hoover and Hoover 1912, 552–3) and Rastel (1678, 1062) describe in detail the implements used in the production of salt. The crystals may have been dried naturally in the open air or artificially, perhaps by heat ducted from the boiling furnaces. The usual model for a 19th or 20th century open pan salt works had the stoking area, furnaces and pan at one end of the building and a drying room, heated through ducts from the furnaces, at the other. Having been dried, the crystals would have adhered to each other and could be turned out of their containers as cakes which could be easily handled.

The economic potential of Droitwich

The indirect exploitation of Droitwich salt may extend as far back as those periods when hunting and gathering was the main mode of subsistence. Hunters may have been drawn to the area by the presence of herbivores attracted by salt as a mineral supplement (John Price pers comm). Though direct ethnographic parallels have not been found, a modern parallel may be seen in the provision of 'salt licks' for farm animals. Nomads in Niger collect salt specifically for animal consumption (Gouletquer 1975, 47). The earliest dates for the production of salt lie in the 6th and 5th centuries BC when briquetage, petrologically identified as coming from Droitwich and thought to be used in the production of salt, is found on several sites distant from the source (Morris 1985, 346).

It appears that in a normal diet man has no physiological need for salt as a mineral supplement (Carter 1975, 13) and demand is determined by its use as a condiment, as well as for food processing (salting of meat) and other industrial processes (eg tanning, soap making, chemicals, etc). Its use as a condiment serves as a reminder that the exploitation of salt at Droitwich was not driven solely by economic forces, but that social factors would also have had an important role. Its use as currency or as an exchange standard, for example in the payment of the early Roman army and more recently in parts of Africa (Alexander 1975, 82), demonstrates the potential complexity of its social and economic functions.

Droitwich brine is saturated (or nearly so) at about 25% sodium chloride, and as such will produce about 0.294kg (0.65lb) of salt from 1l (0.22gal) of brine. By way of comparison, the same amount of sea water will produce only about 0.031kg (0.07lb). The economic advantages are immediately apparent. Contamination from fresh ground water must be taken into account and this would have affected the strength of the brine, especially when drawn from natural springs or from shallow, poorly-lined pits. Droitwich brine also contains very few impurities and does not require further refinement.

Another major resource necessary for the production of salt by the open pan method is timber for fuel. This must have been plentiful in the area, at least initially. The clays of the English Midlands are traditionally regarded as having been heavily forested throughout the prehistoric period (Evans 1975, 130). The scale of salt production appears to have been fairly stable and Berry (1975, 78) has calculated that it rose by only a third from the late 11th to the mid 17th century. Subsequently, with the importation of coal as fuel, production almost doubled in around 50 years. The importance of fuel is indicated by Saxon and later documents. These, when mentioning the ownership of salt and salt making property in Droitwich, may include provision for access to, or supply of, fuel (eg Finberg 1972, no 282). Shortages of timber for fuel are noted in the mid 16th century (Berry 1975, 78) and may be expected to have occurred in earlier periods (Hooke 1981, 149). Maintenance of a continuous and adequate supply of timber would have required a high degree of woodland management.

Before the construction of the canal reduced the flow of water, the Salwarpe was navigable (Whitley 1923, 1) between Droitwich and the Severn, eight kilometres distant. The topography would have presented no physical impediments to easy travel in any direction. The present system of major roads serving Droitwich is based on the Roman roads, to Worcester in the south, Wall via Metchley in the north and Alcester in the east. In addition a road connected Droitwich to the fort at Greensforge and there is slight evidence of a road leading westward from the town (HWCM 7510). Whitley (1923), Houghton (1929–30) and more recently Hooke (1981, 306), have demonstrated that Droitwich was the focus of a large number of routes in the Saxon and medieval periods.

Indirect evidence from several sources has been used to indicate the distribution of Droitwich salt in various periods. This includes the distribution of briquetage from Iron Age sites (Morris 1985), the mention of manors associated with salt from Droitwich (Darby and Terrett 1954, 251–6), and the saltways mentioned above. As with any indicators of distribution, care must be taken with interpretation, as the biases acting on the evidence may obscure the real distribution. This is especially so for salt, which itself leaves no physical trace. It must always be remembered that the distribution of briquetage and Domesday manors need not directly reflect the distribution of Droitwich salt. Perhaps the most dramatic indication of Droitwich's economic importance was its rating at £100 for the annual fee-farm in the early 13th century. This is the same as that levied on Newcastle-upon-Tyne, a major European port, and far exceeds that of Worcester (rated at £30) the main town of the county (Berry 1957, 40).

Droitwich is fairly central to the land mass of Britain. Figure 4 shows all the known salt sources and the areas around them nearest to each source. The figure indicates the potential hinterland for each source, all the other variables affecting distribution being equal. All the other variables, of course, constantly vary through time, as well as the individual circumstances of each source. The strength of brine, availability of fuel, demand, ease of transport, political constraints and existence of competing sources are a few of the variables that must be considered in examining the development of the salt industry.

Previous work and research objectives

Thomas Habington, in the 17th century, was the first to have considered the antiquity of Droitwich, 'These sprynge(s) of salt, havinge byn I thinke heere from the tyme of Noe's flud' (Amphlett 1895, 466). He also demonstrated from Saxon documents that salt production existed at this period (*ibid*, 469). Nash used Habington's documentary references extensively but also noted the possibility of the town being of Roman date (Nash 1781–2, 302). Jabez Allies discussed various finds of Roman date (coins, pottery, mosaics, etc), confirming archaeologically Nash's speculation and himself speculating that salt was produced in pre-Roman times (Allies 1852, 102). He also noted two important sites, one of Roman date at Bays Meadow (HWCM 678), the other a medieval tile kiln at Witton (*ibid*, 102.; HWCM 660). The former, a villa, was later extensively excavated (Hodgkinson 1925–6; Gelling 1957). Later excavation in the town was confined to small-scale trenches establishing the presence of an early Roman fort at Dodderhill (St Joseph 1938; Whitehouse 1962.; HWCM 603). Between 1967 and 1977 excavation at Bays Meadow continued as part of the training programme for the Department of Ancient History and Archaeology, Birmingham University.

In the mid 1960s Droitwich was designated a development area and from this time the town has expanded continuously. The main types of development were in housing construction surrounding the town centre and the enlargement of existing facilities (shopping centres, service buildings, recreation amenities) in the centre. Knowledge of the town's archaeology was largely confined to sites of Roman date outside the town centre. The potential existence of a major ancient industrial centre was only suggested by the wealth of documentary evidence and had yet to be archaeologically demonstrated.

In 1972 Alan Hunt, on behalf of Worcestershire County Museum, carried out a preliminary survey of the archaeological potential and implications of development in the smaller historic towns of Worcestershire. Of these towns, Droitwich appeared to have the greatest potential significance and was accordingly given the highest priority in the planning and funding of rescue excavations (Alan Hunt pers comm). Rescue excavation as an initial response to development at Friar Street (HWCM 605) demonstrated the existence of deeply-stratified archaeological deposits dating from the Iron Age and probably relating to the production of salt. In response to the threat posed by widespread development the post of Droitwich Archaeological Officer was established in 1976. At the same time the Droitwich Archaeology Committee was formed to advise on the coordination of resources.

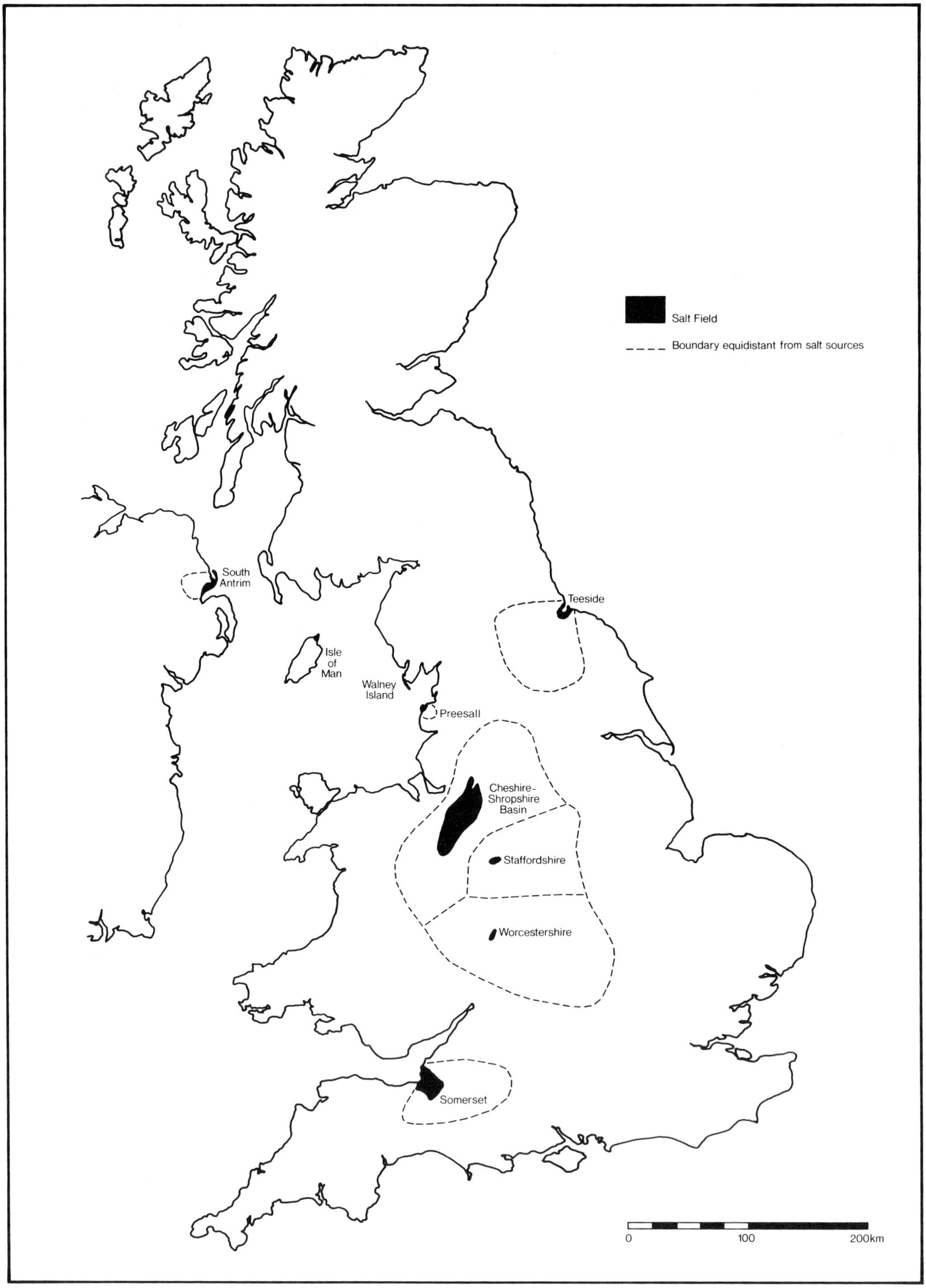

Figure 4 Known salt sources (after Northolt and Highley 1973, fig 2) and their potential hinterlands

Apart from the known archaeological sites of Bays Meadow, Dodderhill and Friar Street, three other areas were identified as potentially having deposits relating to industrial (Old Bowling Green and Vines Park) and domestic (Hanbury Street/Queen Street) aspects of the town's topography (Freezer 1977, 17).

Note on the stratigraphic analysis

A full description of the method of stratigraphic analysis is given elsewhere (1:B3–8). As roughly a decade had passed since the excavations were carried out the field record was not suited to the rigorous methods of analysis employed on more recent excavations. The method of stratigraphic analysis adopted here reflects the nature of the field record. The field record was first subjected to a process of rationalisation to make it internally consistent for the purposes of analysis. The term 'structure' is primarily used as a generic term for groups of related contexts and does not necessarily imply the presence of buildings. In addition, elements of 'construction', 'use' and 'disuse' of each structure were identified where possible. This aided the organisation of the phase descriptions, and the qualification of information derived from finds, especially in indicating residuality.

The publication and archive conform as far as possible to the recommendations of the Frere Committee (DoE 1975) and Cunliffe Committee (DoE 1982). The printed volume contains discussion of the results of analysis where this has a direct bearing on the major aspects of the excavations, or is of importance to specialists. The section of the publication contained on fiche presents more detailed description and discussion of the results. The archive contains all supporting data on which the above reports are based. The full archive (including finds) is held by the Archaeology Section, Hereford and Worcester County Council. A copy of the archive is also held by the National Archaeological Record.

'Phase 6' was removed at a late stage in the analysis which explains its absence in the following. 'Pers comm' references to Hurst, Roe, Tomber, Wills, Williams, Ford and Rees refer to a report on excavations at Beckford, Worcestershire, which is in preparation.

2 Old Bowling Green (HWCM 600), the excavation *Simon Woodiwiss*

The excavation was located at National Grid Reference SO 8992 6350 (Fig 1) and an initial season of work was carried out in the autumn of 1977. A second season commenced in March 1978 and continued until November 1978, and a third season commenced in January 1979 and continued until May 1979. The first season's work was directed by David Freezer, Droitwich Archaeological Officer, and the second and third seasons by John Sawle, Archaeological Excavations Officer.

In 1975/6 recording of a sewer pipe trench along the west side of the Bowling Green indicated the existence of substantial archaeological deposits. These deposits contained well-preserved wood, in waterlogged conditions, and fragments of briquetage. Similar deposits had already been excavated at Friar Street (HWCM 605) and were thought to be associated with the production of salt (Hunt 1975, 40).

Two development proposals threatened to disturb these deposits: the construction of the Inner Relief Road (now the Saltway) and a sewer on the northern part of the excavated area, and a proposed Salvation Army hall to the south. The ground conditions encountered during the excavation subsequently led to the latter proposal being withdrawn. This part of the site is now occupied by a car park belonging to Wychavon District Council. Due to difficult ground conditions (this area of the Salwarpe Valley has a high water table of 27 to 28m OD and is liable to subsidence), major ground disturbance is usually necessary to establish firm foundations.

The high water table and heavy clay soil derived from the marl bedrock made working conditions difficult. Resources did not permit overall artificial lowering of the local water table, water being removed from the site by pump or by hand. A significant proportion of the excavation was carried out during the wetter seasons of the year.

The following discussion of each phase is a summary of the results of the stratigraphic analysis, with the main conclusions of the other specialist reports added where appropriate. It includes only those structures which have a functional significance or are major components of the site. This naturally leads to accentuation of the constructional and use elements at the expense of those associated with disuse. A more comprehensive description of each phase is given in fiche (1:B9–E1), as are detailed drawings of the major structures (4:A1–E12).

Phase 1: Possible late Bronze Age to Iron Age

Buried soils (S61, S180, S212 and S215)

The components of this phase were characterised by layers of dark soil, often with pebbles, overlying natural marl. Briquetage and charcoal were recorded as minor components and charcoal from one layer (S180) provided a radiocarbon date of 1045–838 Cal BC (Table M17, 1:E10).

The dark coloration, pebbles and clay component of the layers of this phase support their interpretation as buried soils. The clay would have been derived from the underlying rock (Mercian Mudstone) and the coloration from a high humic content. The radiocarbon date was from residual material, indicated by the presence of Iron Age pottery, but suggests activity on the site during the late Bronze Age.

Discussion

The single sherd of earlier prehistoric pottery from this phase, the three residual sherds from later phases (Chapter 3), and the residual flints from phase 2 contexts (Chapter 9) complement the radiocarbon date in implying earlier activity on or near the site. As only a small proportion of the site was excavated to this level there is no evidence to indicate the nature and extent of this earlier activity.

Phase 2: Late Iron Age (Fig 5)

This phase was characterised by layers with briquetage as a minor component, interspersed with the construction and disuse of small pits and postholes. Only a few structures however had enough constructional detail to allow conclusive discussion of their function. All the structures of this phase were located in the eastern half of the excavation.

A brine tank and pit (S2 and S14)

Structure 2 was cut into the clay soil of phase 1 and the underlying Mercian Mudstone. It consisted of a rectangular pit (3.70 x 1.20 x 0.40m) with sloping sides and a wattle lining. Structure 14 was a much smaller circular pit, with a partially burnt clay lining.

Figure 5 Phase 2 features

The characteristics of structures 2, cutting into the poorly permeable Mercian Mudstone, and 14, with its clay lining, indicate the storage of liquid. The former was however considerably larger than the latter, indicating a different function for each.

In considering structure 2, the function of similar structures from other phases (S1, S3, S4, S5, S6, S7, S8, S9 and S10) may also be discussed. The bases of these structures were generally cut into marl, their more permeable sides lined with clay and revetted with wattle or stakes to form a deep tank. A tank with steep sides facilitates the filling of vessels such as buckets with liquid, as these are able to be fully immersed. The construction of the tank therefore indicates that ease of access was important to their proper function.

The suggestion that the tanks were for the direct collection of brine from the ground may be dismissed, as their base of Mercian Mudstone is not very permeable and showed no sign that springs had been present. That the tanks were used in the evaporation of the water content of brine by solar heating may be similarly dismissed. Features used in solar evaporation may be expected to be shallow with a large surface area, so that saturation and precipitation of salt crystals would take place in the shortest possible time. The ratio of volume to surface area of the tanks would be too great to allow efficient solar evaporation. The association of solar evaporation with similar, though unrevetted, clay lined pits from coastal salt production sites in Essex has also been dismissed (Rodwell 1979, 136).

Storage and settling would have been necessary if the source of brine were distant, if the brine held a substantial amount of silt in suspension, or if production methods required the close proximity of brine to another part of the salt making process (eg boiling). The most likely place for contemporary springs would be in the permeable gravels of the valley bottom. Numerous springs in the gravels were noted during a recent excavation at Upwich (HWCM 4575), just to the north-east of the Old Bowling Green. Without adequate lining, brine derived from pits or springs would have had a high silt content, impairing the final product and requiring settling prior to boiling. The boiling of brine requires an easily accessible supply to top up the pans. For instance, if the pans were allowed to dry out during boiling and the salt adhered to the pan bottom, overheating and melting of the pan would result. The form and structural details indicate that the primary function of these tanks, of this and later phases, was the storage and/or settling of brine, as suggested by Rees (1986, 51).

Possible hearth (S182)

One poorly-defined pit (0.74 x 0.64m), part of structure 182, contained a fill almost entirely composed of charcoal which the excavator suggested could be *in situ* fuel residue from a hearth. In the absence of obvious alternatives (eg domestic structures) it seems safest to associate this feature with the boiling of brine (see discussion of similar structures in phase 3, see below).

Layers of industrial waste (S71, S72, S81, S82, S87, S89, S131, S181 and S183)

The layers of this phase characteristically contained briquetage. In only four of the 31 layers did briquetage not form a component and six layers were almost totally composed of briquetage. In view of the interpretation of briquetage vessels as part of the salt production process (Rees 1986, 51), it would seem likely that these layers were formed from waste during salt making. It was also noted during excavation that these layers overlay very wet ground and this waste material may have been deliberately deposited to produce a better working surface.

Discussion

With the presence of more activity the range of pottery fabrics and forms also increased slightly. In explanation Rees (Chapter 3) points to general developments in communication and increasing specialisation in pottery in the later Iron Age, in addition to the development of Droitwich as a salt producing centre. This must however be viewed with reference to the relatively small sample sizes of phases 1 and 3. The presence of two saddle querns (Chapter 9), a bone needle (Chapter 17) and the characteristic food debris of the animal bone assemblage (Chapter 15) indicate that the area was not used exclusively for the production of salt.

Phase 3: Late Iron Age (Fig 6)

This phase was characterised by large structures of two different types. Eight very large clay-and stake-lined pits were used in conjunction with three hearths.

Brine tanks (S3, S4, S5, S6, S7, S8, S9 and S10)

Structures 3, 4, 5 and 6 formed a row and were orientated in the same direction. Structure 3 had two periods of reconstruction, the last one occurring in phase 4. Many of the original construction elements had been removed by later reconstruction, with only the base of the elongated pit remaining. A line of stakeholes ran along the southern side of the pit. The first reconstruction element consisted of the northern and eastern sides of a pit, of which the other sides had subsequently been removed. Neither the clay lining nor stake revetment, present in other similar structures, had survived. Structures 4 and 5 were partially destroyed by later activity; both however had a stake revetment, though no clay lining was recorded for structure 5. A radiocarbon date of 358–91 Cal BC was obtained from the stake revetment of structure 4 and another of 101–72 Cal BC from wood found within the sandy clay fill of structure 5 (Table M17, 1:E10). The latter at least is probably close to its actual date of disuse, as unlike the other brine tanks, the primary fills did not contain Roman finds. It may be that both were constructed much earlier than the others, though structure 4 also appeared to continue in use into later phases. Structure 6 (Fig 7) was the most complete pit in this alignment (4.62 x 2.10 x 1.40m) and had a clay lining (*c* 0.20m thick) revetted by 58 stakes. Most of these stakes, which were shaped from planks with one end cut for driving into the ground, were placed with their greatest width parallel to the sides of the pit. Several were however placed at right-angles to this at intervals along the sides of the tank and with similar stakes along the opposite side of the tank. A silt from behind the stake revetment may have accumulated during use of the structure. A dendrochronological *terminus post quem* of AD 19 was obtained for the stake revetment (Table M16, 1:E8).

The long axes of structures 7 and 8 were parallel to each other. These were the two largest of this type of structure (5.50 x 2.74 x 1.00 and 6.95 x 3.44 x 1.29m respectively). Both had a clay lining between 0.10 and 0.30m thick. Again, many of the stakes were shaped from planks. In structure 8 (Fig 8), four pieces of wood were laid along the bottom edge of the stake revetment and were themselves held in place by stakes. The stakes of structure 7 gave a radiocarbon date of 151 Cal BC–Cal AD 66

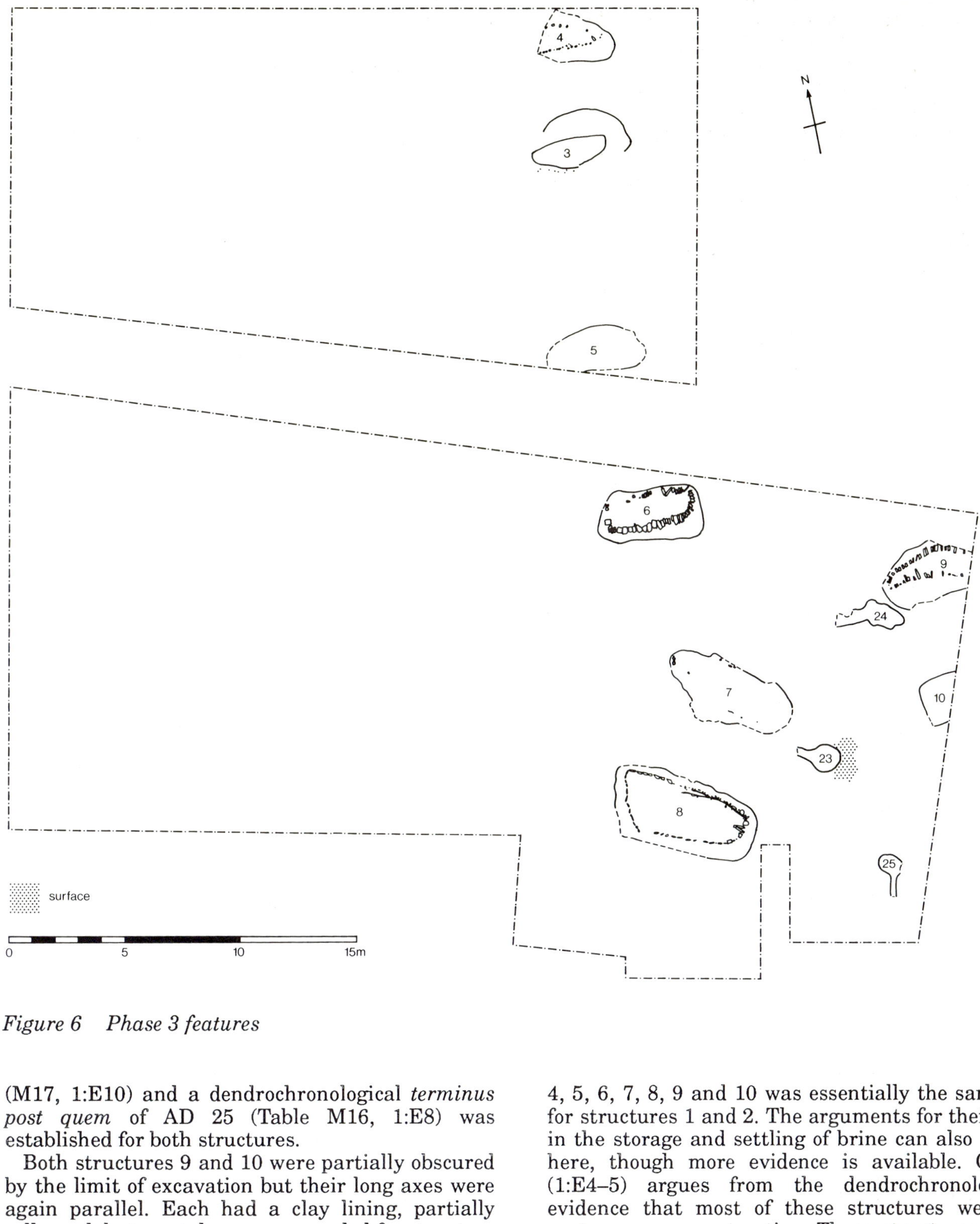

Figure 6 Phase 3 features

(M17, 1:E10) and a dendrochronological *terminus post quem* of AD 25 (Table M16, 1:E8) was established for both structures.

Both structures 9 and 10 were partially obscured by the limit of excavation but their long axes were again parallel. Each had a clay lining, partially collapsed, but no stakes were recorded for structure 10. A dendrochronological *terminus post quem* of AD 17 was obtained from the stakes of structure 9 (Table M16, 1:E8).

In all instances where the stake revetment had survived the tops were decayed and impressions in the clay lining existed to a higher level.

Though the sides were revetted with stakes, as opposed to wattle, the construction of structures 3, 4, 5, 6, 7, 8, 9 and 10 was essentially the same as for structures 1 and 2. The arguments for their use in the storage and settling of brine can also apply here, though more evidence is available. Crone (1:E4–5) argues from the dendrochronological evidence that most of these structures were of contemporary construction. These structures were completely excavated and the clay lining extended only over the more permeable sides of the pits, where these cut through archaeological deposits. The clay did not extend over the less permeable Mercian Mudstone at the base of the pits. The impressions above the highest surviving level of the stakes indicate that they were originally much longer, possibly extending to the original ground

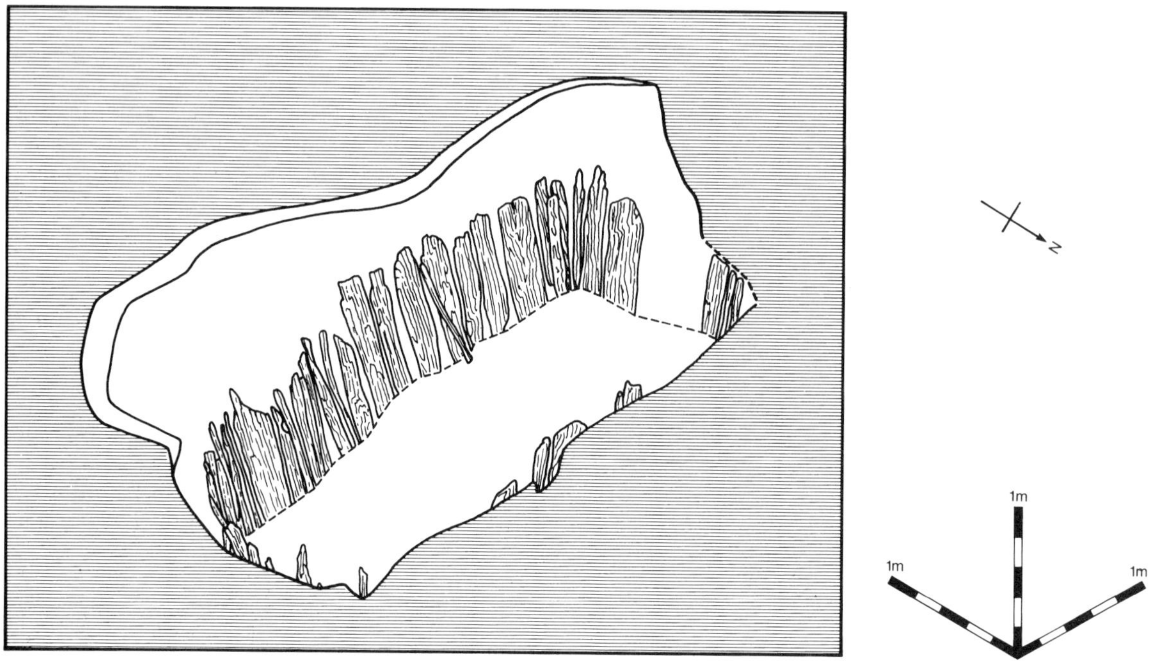

Figure 7 *Brine tank (S6 construction)*

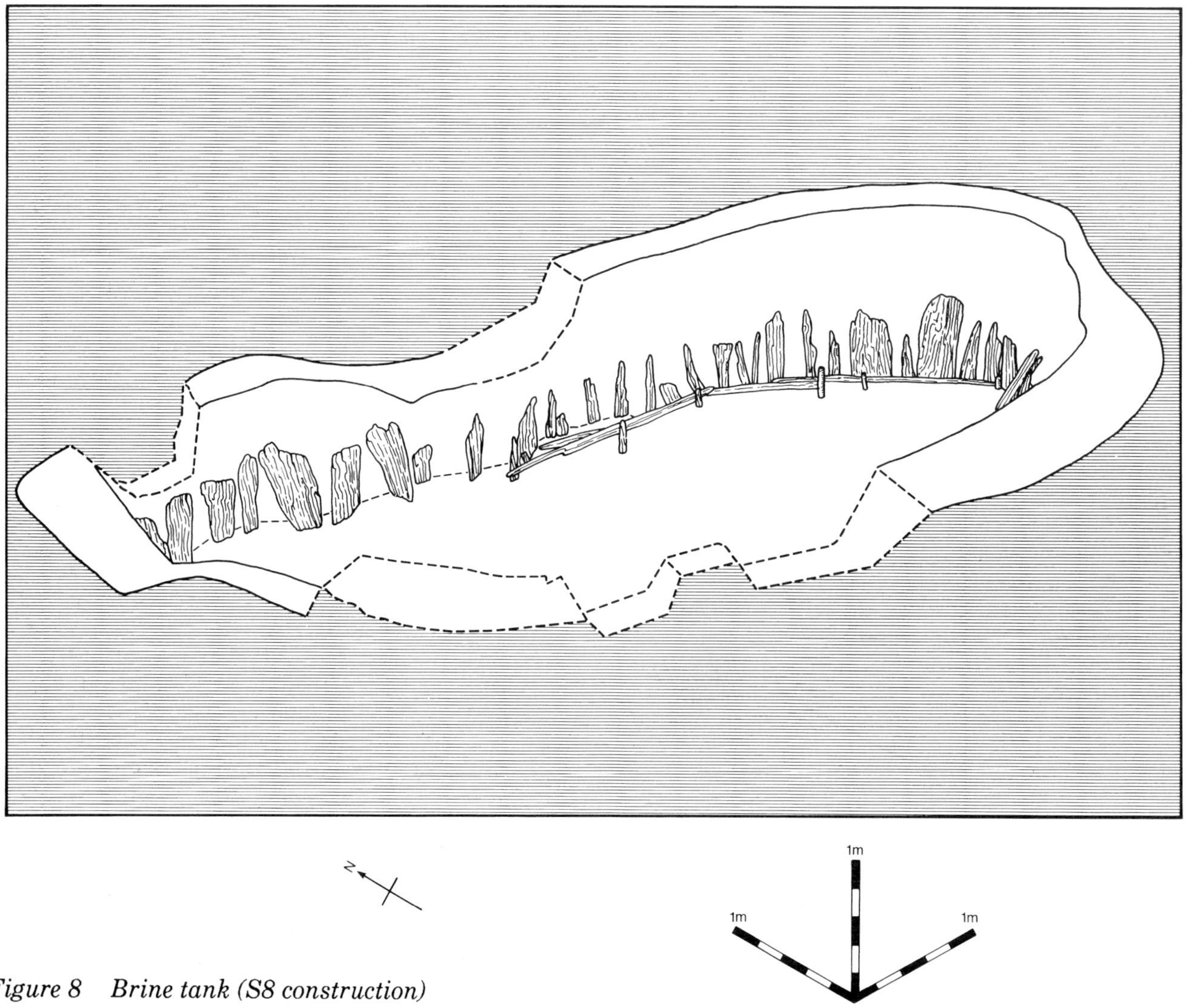

Figure 8 *Brine tank (S8 construction)*

surface. The arrangement of some of the stakes of structure 6, with their greatest width at right angles to the lining and with complementary stakes on the opposite side, suggests a facility for internal division. Only structure 6 had deposits that had possibly resulted from settling, a dark grey silt from behind the stakes revetting the clay sides. All of these structures, except structure 3, whose final reconstruction occurred in this phase, exhibited only one period of construction. The structures would have contained very large quantities of liquid, and estimates of the capacities have been made for the most complete examples: 6,950l (S6), 7,000l (S7), 14,000l (S8) and 5,300l (S9). Structure 8 was significantly larger and was estimated to hold twice as much as any of the others. If 1l (0.22 gal) of brine produced 0.294kg (0.65lb) of salt (Chapter 1) the largest and smallest tanks would have contained enough brine for 4.116 tonnes (4.05 ton) and 1.558 tonnes (1.53 ton) of salt respectively. These figures, though estimates, indicate that production was on a large scale.

Hearths (S23, S24 and S25)

All of these structures consisted of a linear cut extending from a pit. The structures varied in depth and degree of symmetry, structure 24 being particularly irregular. The long axis of structures 23 and 24 were parallel and a pebble surface existed on the eastern edge of the former. The only fill in this phase, that of structure 25, was composed of sand and charcoal.

The form of structures 23, 24 and 25 was obviously suggestive of hearths, the linear cut representing a flue. The fill of structure 25 contained much charcoal, not enough to indicate its use as a fuel but indicating burning. Like structure 5, the fill of structure 25 had an Iron Age *terminus post quem*, also probably falling into disuse in this phase. No evidence of any superstructure existed, save for a pebble surface at the eastern end of structure 23. This suggests that whatever was being heated needed at least periodic attention. Again the easiest association to make is with the production of salt, and here specifically with the boiling of brine. Discussion of receptacles (or pans) that may have been used for boiling is undertaken later (see below). The boiling of brine requires frequent agitation of the liquid to ensure the formation of evenly sized crystals and prevent them from adhering to the base of the pan, perhaps explaining the pebble surface of structure 23.

Discussion

The spatial arrangement of the structures of this phase suggests some degree of organisation. The north to south arrangement of the brine tanks may reflect the existence of a feature to the east of the excavation, possibly the meander in the River Salwarpe (Fig 2). The concentration of features in the eastern half of the excavated area may also be the result of clustering around some focal point, presumably an area of brine extraction. The structures appear to be paired, structure 3 with 4, 5 with 6, 7 with 8, and 9 with 10. Rather more dubious is an association of structures 24 and 23 with 7 and 8 respectively. If the pairing is genuine, it may be due to a specific characteristic of the salt making process. For instance, settling may occur in one tank whilst clear brine was being drawn for boiling from the other. Alternatively, and possibly additionally, an explanation may be sought in the administration of salt production. It may, for example, be speculated that ownership of a restricted number of tanks only was allowed.

The apparent absence of layers, such as those of phase 2, is probably a result of the stratigraphic analysis rather than a genuine absence of deposits. A number of groups of layers were uncertainly phased (S85, S86, S91 and S106) and could have been deposited in phase 3 but lacked conclusive stratigraphic relationships.

The paucity of layers and disuse fills has effectively reduced the amount of artefactual evidence that can be used in discussion. For instance, though this is the only phase in which ovicaprid and small ungulate bones outnumber ox and large ungulate, this is drawn from a very small sample and is probably not representative (Chapter 15). Much of the material incorporated in the disuse fills (P4) of structures constructed in phase 3 may be from equipment used in salt production (see below).

Phase 4: Mid 1st to early 2nd century AD (Fig 9)

Two of the brine tanks, which may have continued in use throughout this phase, lay to the north of a leat. Layers containing significant amounts of briquetage continued to be deposited. A number of small pits and postholes were also constructed.

Brine tanks, hearths and a leat (S3, S4, S6, S7, S8, S10, S23, S24 and S30)

The final reconstruction of structure 3 (Fig 10) consisted of a pit (4.74 x 3.26 x 0.85m) with a clay lining (*c* 0.10 to 0.20m thick) and a stake revetment. The phasing of this structure however was in some doubt as it relied on only two sherds of Roman pottery. The disuse of structures 3, 6, 7, 8, 10, 23, 24 and possibly 4 has been assigned to this phase, and that of structures 1 and possibly 4 and 9, does not happen until at least phase 5. The fill of structure 24 contained much charcoal and briquetage, including two substantially complete and unabraded (though distorted) vessels. Many pieces of wood from the disuse fills of the brine tanks were worked. The assemblages in structures 6 (31 pieces, Fig 11, Pl 1), 7 (59 pieces, Fig 12) and 8 (66 pieces, Fig 13) were mostly aligned in the

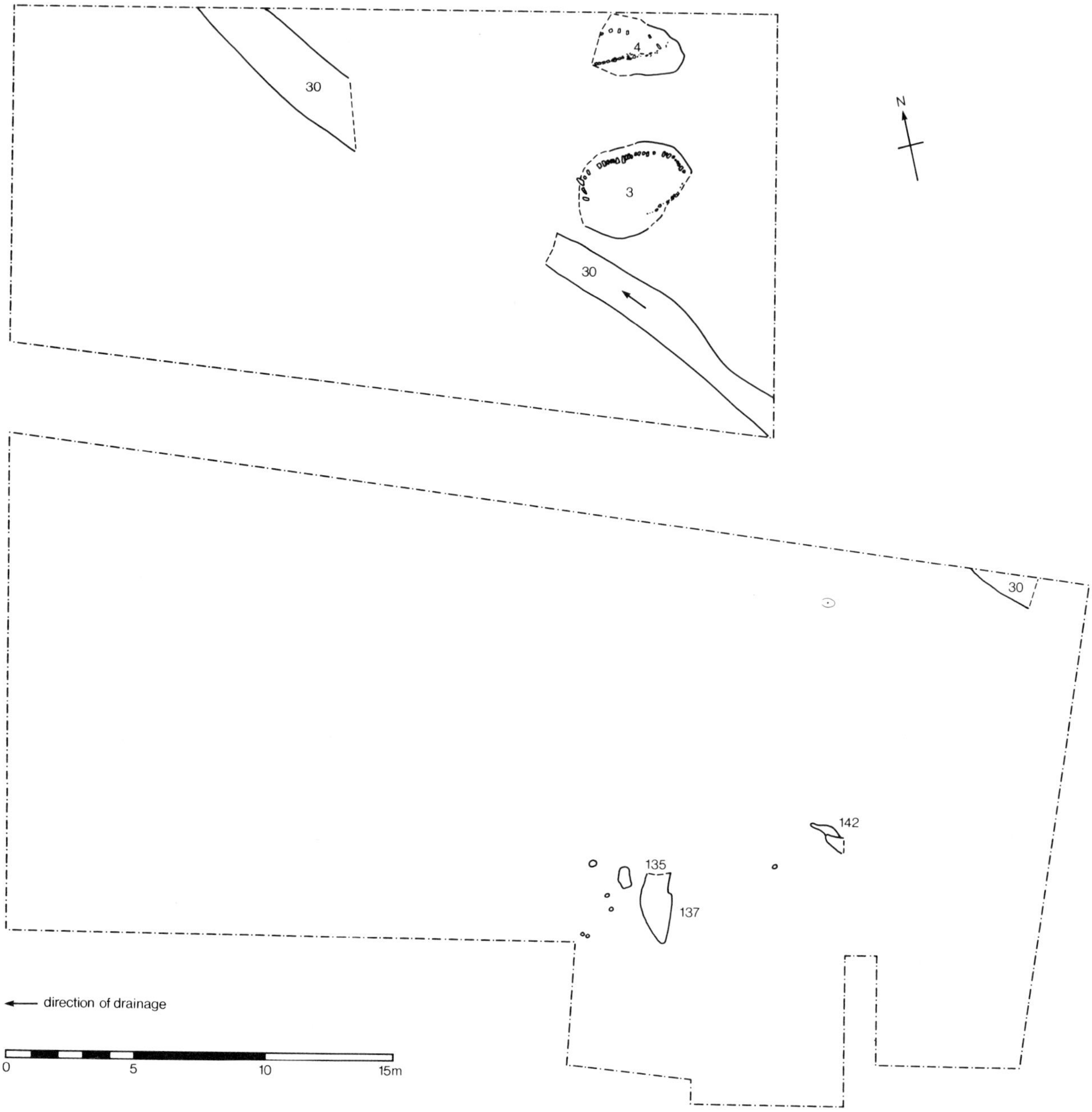

Figure 9 Phase 4 features

same direction and grouped together. Much smaller assemblages were recovered from the fills of the other structures. Wooden artefacts deposited in structures 6, 7 and 8 may have been related to the structures' use as they were part of the primary disuse fills (see below).

A linear cut (S30) was orientated north-west to south-east, with a gradient down to the north-west of 1 in 20. Its depth ranged from 1.00 to 1.50m. Reconstruction occurred four times along the same alignment. Related to these reconstructions were constructional fills composed of wattle and clay. A number of the disuse elements also contained redeposited wattle and clay. A radiocarbon date of 388–168 Cal BC was obtained for a fill (Table M16, 1:E10).

The final form of structure 3 is estimated to have contained about 3,000l of liquid. Just to the south, the linear cut had a constant gradient and a clay and wattle lining, indicating that it transported a controlled flow of water. It probably formed a leat between the upstream bend in the Salwarpe (Fig 2), to some point downstream to the north-west of the excavation.

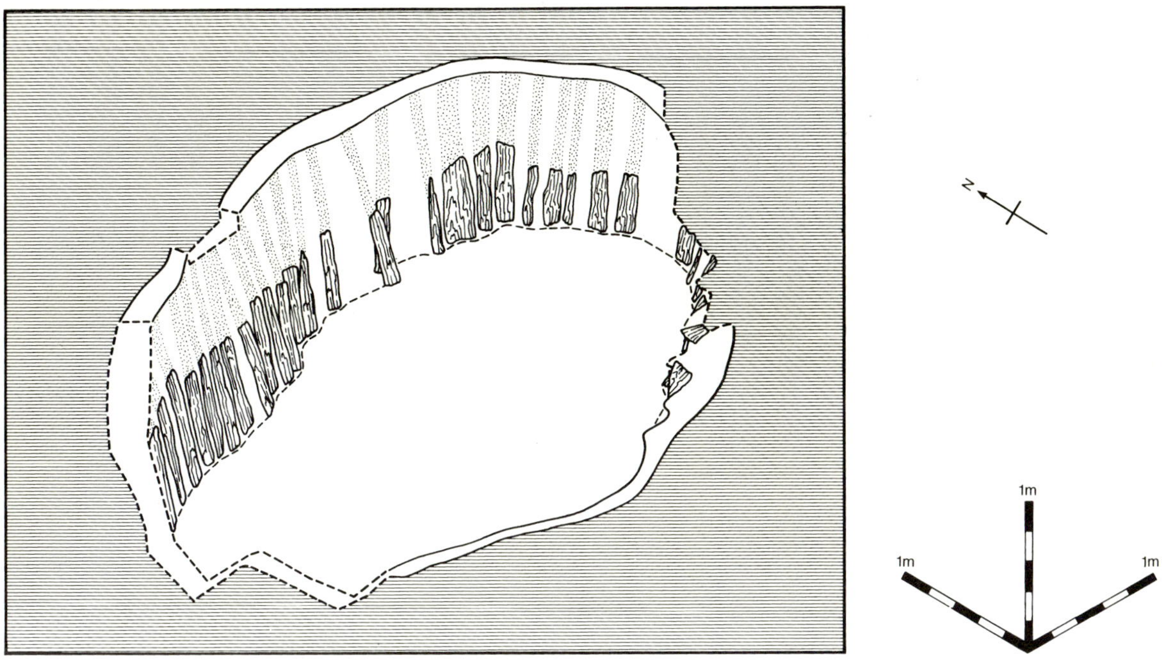

Figure 10 Brine tank (S3 final construction)

Layers of industrial waste (S63, S93, S134, S136 and S143)

The layers of phases 2 and 4 were very similar in character. Forty-one of the total of 52 layers in phase 4 contained briquetage and five were almost completely composed of it. The fragmentary nature and large quantity of briquetage, especially compared with the layers of succeeding phases, indicated deposition of waste material. Deposition continued in the eastern area of the excavation but also occurred in the west.

Discussion

Both the layers and the brine tanks strongly suggest that the methods, and perhaps even the organisation, of salt production continued with little change into the Roman period, though for how long is uncertain. To suggest the contraction, or even interruption, of salt production from excavation of a single, restricted area is clearly inadvisable. Nevertheless the possibility may be suggested.

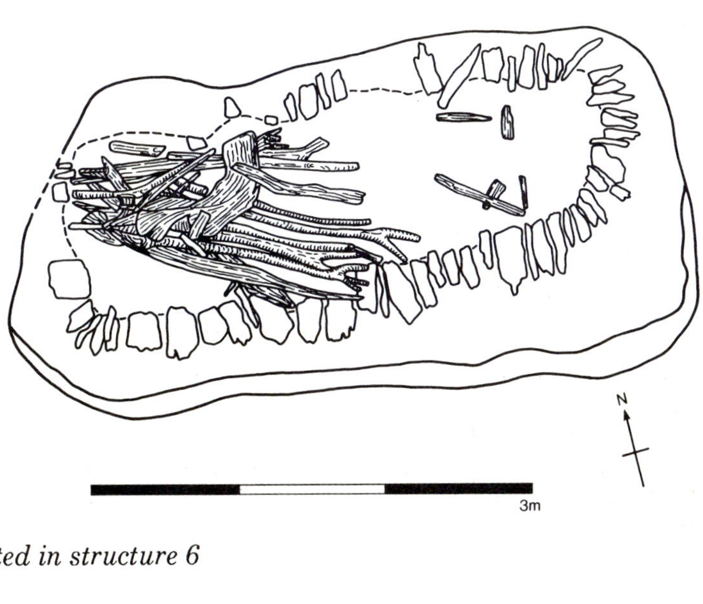

Figure 11 Wood deposited in structure 6

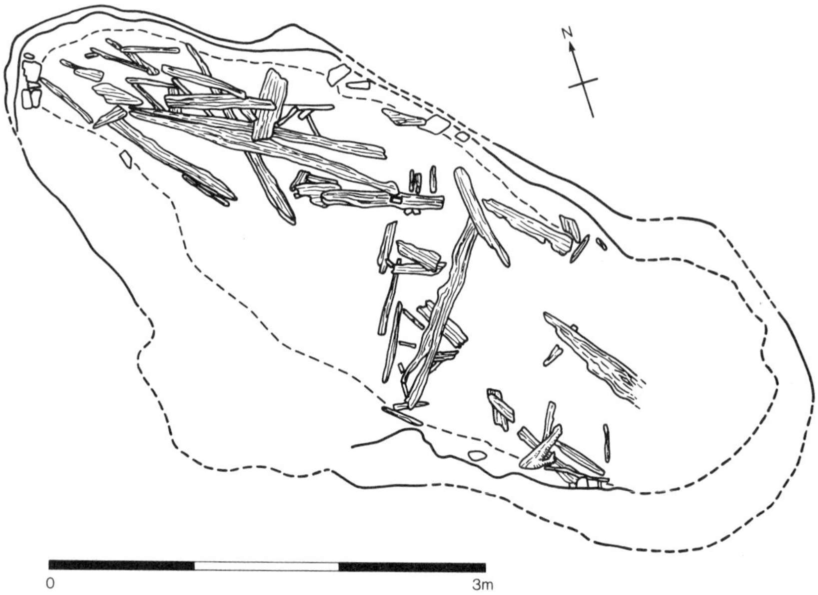

Figure 12 Wood deposited in structure 7

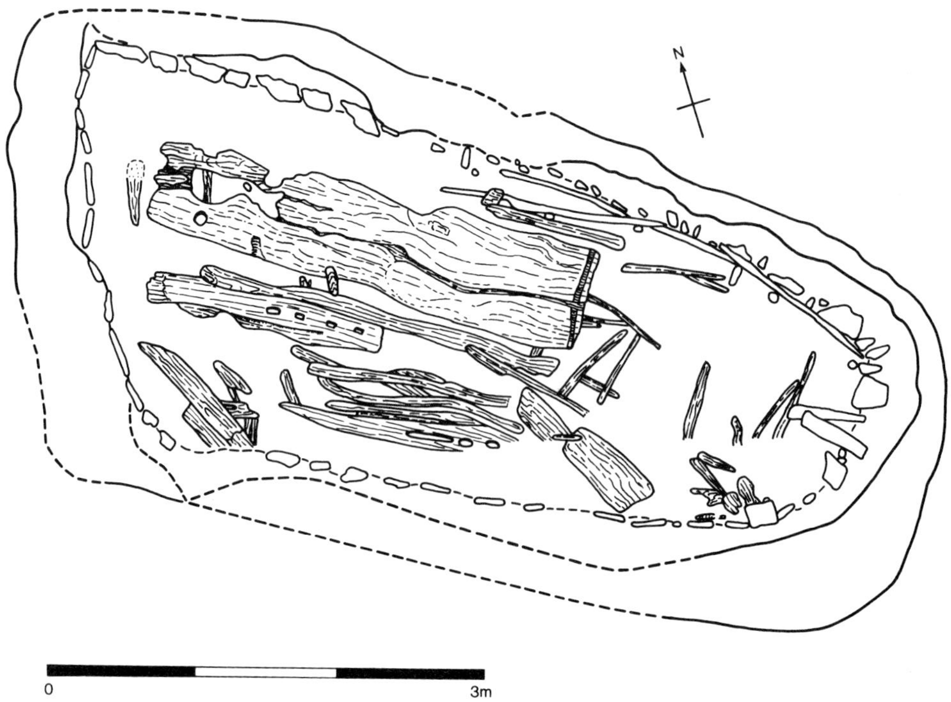

Figure 13 Wood deposited in structure 8

Layers (S97, S106, S138, S147 and S175)

Of the fifteen layers of phase 5, eight contained briquetage as a minor component. The number of layers in this and succeeding phases was not great enough to determine the residuality (or otherwise) of briquetage in the same way as for phases 2 and 4. It is however useful to compare the 77% of layers containing briquetage in phases 1 to 4 (77 layers) with the 25% in phases 5 to 14 (83 layers). It was in phase 4 that the disuse of briquetage and its incorporation as a residual element of layers occurred, contemporary with the disuse of the brine tanks.

Other features (S50, S51, S146, S177 and S209)

A number of postholes were recorded in the southern part of the excavated area, perhaps indicating the presence of either boundaries or buildings. Three posts remained *in situ* (S50 and S146) and measured from 0.20 to 0.40m in diameter. Later activity however, obscured much of the form of these structures. Thus, for example, any eastward extension of structure 50 was removed by the construction of structure 31 in phase 8.

Discussion

Severn Valley ware dominated the pottery assemblage, with briquetage much reduced in quantity (Chapter 3). The range of coarse wares increased only slightly and there was rather more pottery from distant sources (Chapter 3). Occupation of a rather more domestic than industrial character may have occurred close to the site.

Phase 7: Mid 3rd to late 4th century AD (Fig 15)

The foundations of a building dominated this phase, though little else appeared to be associated. A few layers were deposited, one containing sandstone fragments.

Building and associated layers (S26 and S210)

The remains of a building was marked by a trench defining a roughly trapezoidal area with reconstructed sides of 6.96 x 8.04 x 9.40 x 8.88m, and with an average width of *c* 0.95m. The depth was variable (0.66 to 0.91m) but the trench had a generally flat base, the difference in depth largely being accounted for by a step of *c* 0.20m on the southern side. The trench was filled by sandstone rubble with a loam matrix. The size of the stone generally diminished towards the top. One piece of worked stone was retrieved from the rubble foundation (Chapter 9). Two coins dated AD 260–8 and 64–75 were also retrieved from the fill (2:D1).

If the fill of the trench is interpreted as a foundation, this was fairly substantial, though there was little evidence for the method of construction for the superstructure. A doorway may be postulated in the south-west corner, where the sandstone foundation did not appear to exist. This area however had been disturbed by later activity which casts doubt on this reason for the absence of the foundation. Alternatively the fill of the trench may be interpreted as the result of robbing a more carefully founded building. Of the seven layers, one (part of S210) contained sandstone fragments and lay to the south of the building (S26). This may represent waste from the building's construction of which the foundations at least were made of sandstone. There were no deposits that could conceivably be related to the building's use. It is useful here to mention the possible construction of a track (S28), in this or phase 8 (see below), over the layers of structure 210. The track could have provided access for wheeled vehicles to the building.

Discussion

Unlike previous phases there were no structures that could be associated with the production of salt, with any degree of certainty. This would appear to have been a short period, the coin in the building's trench fill providing a *terminus post quem* of AD 260, and pottery a *terminus post quem* of *c* AD 300 for the phase as a whole (Chapter 3). The function of this building must remain obscure largely due to the paucity of associated deposits, though this itself may be explained by a short life for the building.

The artefactual evidence adds little to the structure or function of the building (S26), though there is some possibility that tile was used in its construction. Relatively large fragments were recovered from its constructional elements and from the layers around it. Smaller fragments, indicating increased residuality, were recovered from contexts placed in phases 8 and 9 which themselves did not include structures likely to have incorporated tile (Chapter 4).

Phase 8: Mid 3rd to late 4th century AD (Fig 16)

Two spatial groups of structures may be identified, a northern alignment of reused barrels in pits and a group to the south which included two reused barrels within a trench, 25.80m long. Two similarly sized pits, without barrels, were also recorded. The track fell into disuse during this phase.

Barrels (S15, S16, S17, S18, S19, S20, S21 and S31)

The remains of five barrels recovered *in situ* were recorded. Two consisted only of hoops (S15 and S20), whereas only three staves of structure 21 remained. Structures 16 (Fig 17) and 19 (Fig 18)

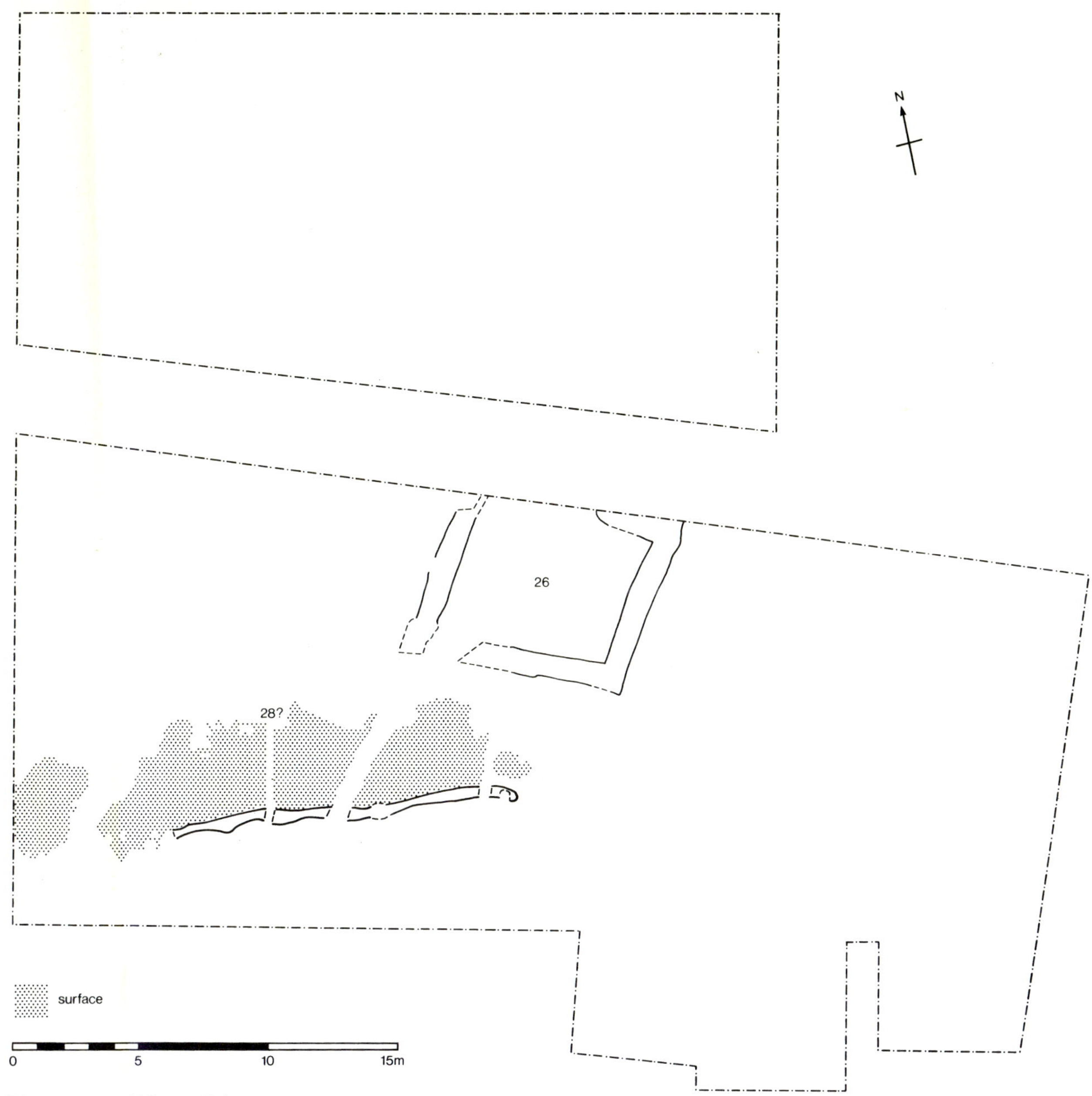

Figure 15 Phase 7 features

were the best preserved (Pl 3), each consisting of *c* 34% of a complete barrel. The barrels were laid within pits or a trench and constructional fills of clay were recorded for three (S15, S16 and S19). Radiocarbon dates of 18 Cal BC–Cal AD 135 (S15), Cal AD 139–343 (S16), 36 Cal BC–Cal AD 129 (S19),Cal AD 139–343 (S20) and 2 Cal BC–Cal AD 214 (S21) were obtained for the barrels (Table M17, 1:E10). Though fragmentary, the fill that lay directly on top of the barrel of structure 21 was of sandy silt with yellow/green flecks and may have been deposited as part of the final use of the structure.

The pits of structures 17 and 18 were of similar dimensions (2.25 x 1.40 x 0.70m and 1.80 x 0.94 x 0.24m respectively). Structure 17 was aligned with structures 16 and 19 and structure 18 lay parallel and just to the north-west.

A straight trench (S31), with near vertical sides and a flat base (25.80 x 1.00 x 0.80m), sloped from either end down towards the middle. It contained two barrels (S20 and S21) and ran parallel to the track (S28). A radiocarbon date of 187 Cal BC–Cal AD 51 was obtained from a disuse fill (Table M17, 1:E10).

Three of the barrels were laid in pits (S15, S16 and S19), the other two (S20 and S21) within the trench (S31). The trench, with its vertical sides, absence of a consistent gradient in one direction and butt ends, would not have had a drainage function. It is probable that structures 20 and 21, as well as others previously removed, were laid

Figure 16 Phase 8 features

within the trench excavated specifically for this purpose. Indeed a similar arrangement may be postulated for the group of barrels to the north. However, in order to extract the most complete barrel (S16), the surrounding area was lowered, which was not conducive to properly recording the evidence of construction. The barrels must have been cut longitudinally *in situ*, as the action of breaking the binding hoops would have required support to retain the shape of the container. Clay in the constructional fills may have served the purpose of inhibiting drainage of the barrels' contents. The dimensions and form of structures 17 and 18 were sufficient to contain similar barrels to structures 16 and 19.

The sewer trench that ran through the excavation (see Chapter 1) exposed the remains of a further barrel, similarly deposited *in situ* (HWCM 3682). It lay *c* 5.00m to the west of the excavation and its long axis was orientated roughly north-west to south-east.

The most obvious function of these barrels was for the storage of liquid and, in this location, the storage and settling of brine. The function of these barrels may therefore be analogous to the brine tanks of phases 2 to 4. Other functions however, such as the processing of animal products, must also be considered. The early radiocarbon dates obtained from the barrels indicate their reuse in this later Roman context. Indeed the radiocarbon

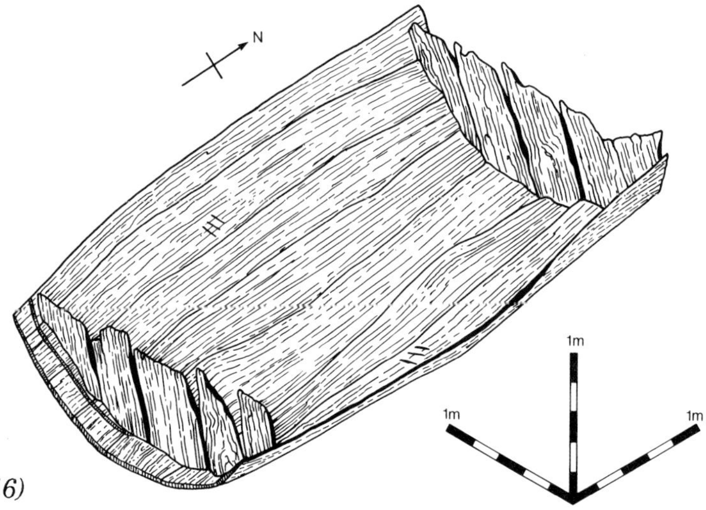

Figure 17 Barrel (S16)

dates suggest that at least some of the barrels had probably been in use for up to three centuries, or even more. The consistency of the dates supports an early date of manufacture and it would seem unlikely that barrels would have been made from reused timber. Even allowing that the barrels themselves would have been reused, this period of use seems surprisingly long.

Plate 3 Barrels (S16 and S19, P8)

A track and an alignment of posts (S28 and S55)

Six postholes and possible postholes formed a single alignment (S55). The postholes varied in size between 0.48 x 0.44 x 0.40m and 0.70 x 0.56 x 0.32m. In addition a stakehole, with a pointed base, formed part of this structure. One of the postholes contained packing stones and another a padstone. The postpipes were all subrectangular, varying in size from 0.24 x 0.18 x 0.20m to 0.16 x 0.14 x 0.40m. This structure ran parallel to the trench (S31) and just to the north of it for approximately 4.50m of its length. Some doubt must remain over the structure's dating, as the mid 3rd to late 4th century *terminus post quem* is based on a small assemblage. It has been placed in this phase largely on the basis of its spatial relationship with structures 28 and 31.

The track (S28), possibly constructed in the previous phase, went out of use in phase 8, indicated by the filling of the drainage ditch to the south. Though the track petered out to the east there was no sign of this having been the result of removal by later activity. The eastern end coincided with the beginning of the line of postholes extending eastwards (S55). The most obvious interpretation was that these formed a barrier continuing from where the small ditch, to the south of the track, ended.

Discussion

No conclusive case can be made for the resumption of salt production in this area, after the period of interruption represented by phases 5 and 7. Prior to phase 7, activity was concentrated in the east of the excavated area, whereas the track (S28) indicates reorientation towards the west, away from the presumed area of brine extraction (see above). This may be seen to strengthen the possibility that the barrels were not used for the production of salt. However structure 57 (P4–8, see below), a possible hearth, may be the contemporary equivalent of the hearths of phase 3. All of these structures fell into disuse during this phase.

As neither the European larch nor silver fir (used for the staves of the barrels) is native to this country, they must have been imports. Their original use will have been in the transport of wet or dry goods, perhaps from the species' native area of Central Europe (Chapter 19). The use of oak for the hoops, as opposed to the more suitable hazel, is an unusual characteristic (Chapter 19). An alternative use for the barrels would be as part of an animal products industry, itself related to the salt industry. Use of the barrels for leather processing or salting of meat must be considered a possibility. However, though the animal bone assemblage was examined with this in mind, the results were not conclusive (Chapter 15). A large fragment of lead sheet (possibly an offcut, Chapter 14) and some inscriptions on pottery vessels (Chapter 10) may not with certainty be associated with the salt or a related industry but this is a tantalising speculation. Tree pollen of alder, hazel and oak from the trench (S31) was found in substantial quantities (Chapter 18).

Phase 9: Mid 3rd to late 4th century AD (Fig 19)

This phase was dominated by the construction of a large rectangular building. Much activity occurred to the south of this building, mostly represented by alignments of postholes and ditches. In the west were two small clay-lined pits, a large ditch reconstructed several times and several other cut features. Five layer groups were also deposited.

Rectangular building and associated structures (S22, S149, S151 and S172)

The padstones that characterised structure 22 formed two parallel rows, of which the three stones to the south were directly perpendicular to those in the north. The westward extent of the southern line was obscured by the limit of the excavation. A possible posthole, but without a stone, was located in the southern row, perpendicular to a stone in the northern alignment. No stone or pit was located perpendicular to the eastern end of the southern side. The stones were all large and of similar dimensions (eg 0.50 x 0.25 x 0.30m). Most had holes cut into one or more faces, probably to facilitate lifting. Only one of the postholes containing stones was largely undisturbed. It was subrectangular in plan and had dimensions of 1.20 x 0.90 x 0.94m. It also contained the poorly preserved remains of a post (*c* 0.30m diameter), *in situ* on top of the stone (Pl 4). The post gave a radiocarbon date of 383–124 Cal BC (Table M17, 1:E10). Two postpipes were also recorded and of these one had a diameter of *c* 0.35m. In two cases, two stones in the same posthole were recorded. The constructional fills were composed of silty clay, silt and loam. Only a few disuse elements were assigned to this phase, and these were composed of silt and ash. Stratigraphically and spatially associated with the building (S22) were a number of cuts (S149 and S172). These were close to the postholes of structure 22 and structure 172 was filled with charcoal. A number of layers (eg S151) were also associated with structure 22 and were composed of briquetage and silt.

The difference between measurements from the centres of corresponding pairs of stones was 0.12m, giving a symmetrical layout to the structure of the building. No internal divisions were identified and there was no obvious location for an entrance. The substantial foundations would have been necessary for any large building in the unstable ground conditions of this area. Again like the barrels of phase 8, an early radiocarbon date was obtained. However this may be more convincingly explained by the reuse of structural timbers, as these may be expected to have had a longer useful life than the

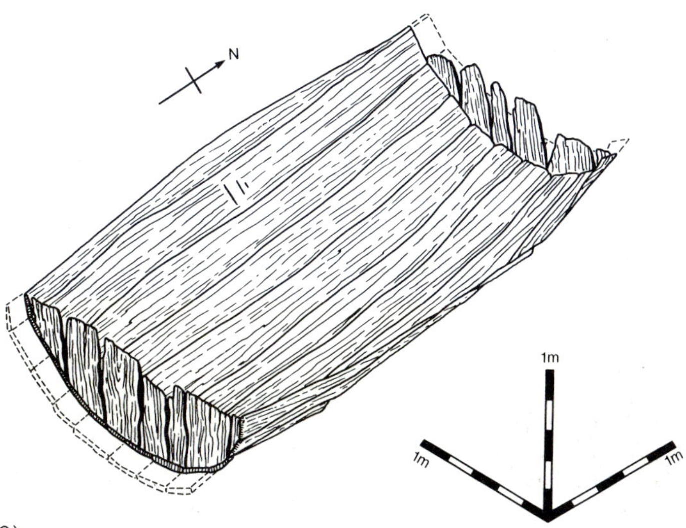

Figure 18 Barrel (S19)

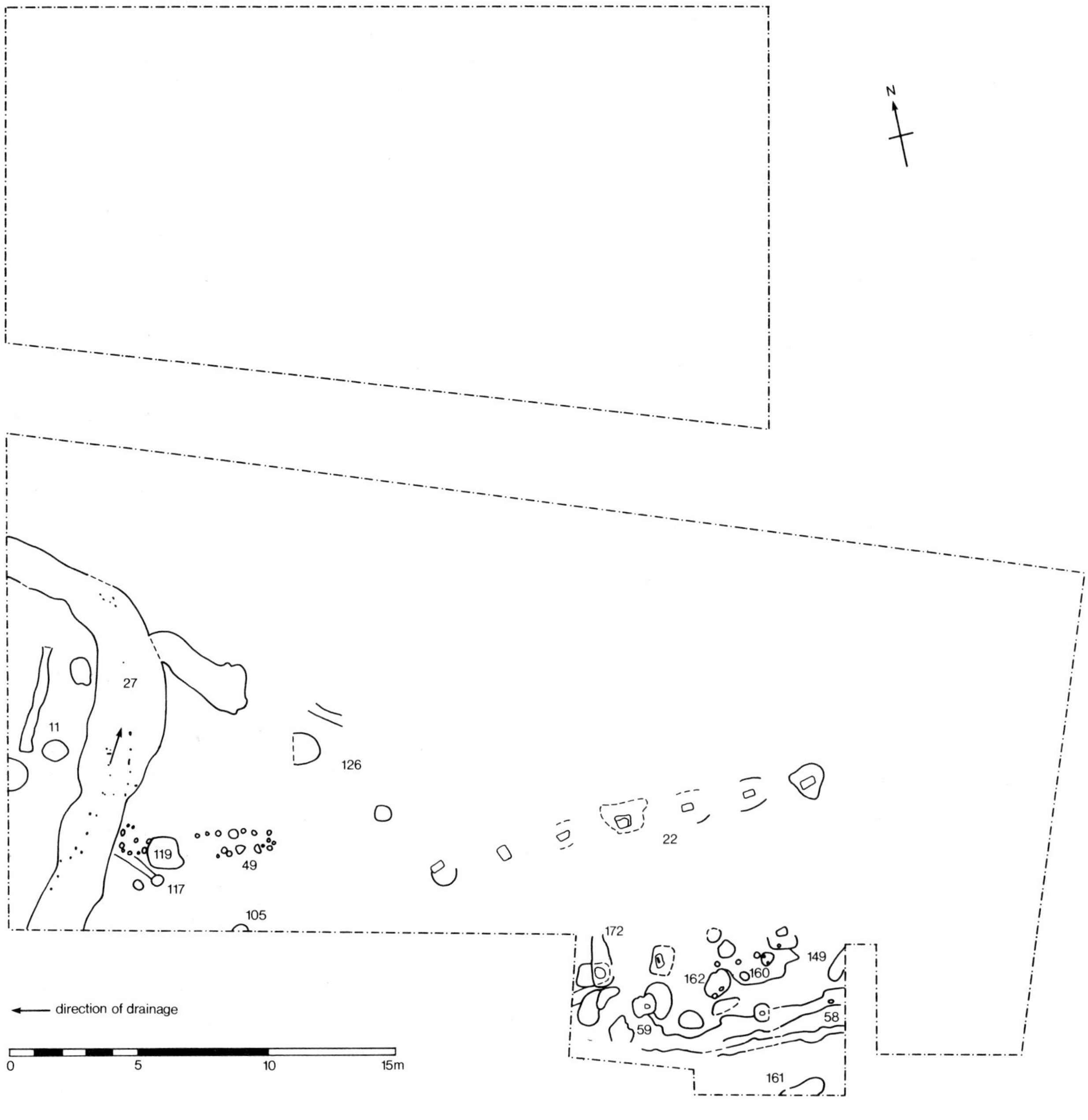

Figure 19 Phase 9 features

less substantial material of the barrels. The building appeared to have undergone reconstruction in this and later phases. Evidence for the replacement of vertical posts may be seen in the cutting of a posthole (S149) and one other feature (S172). The structures stratigraphically between the initial construction and final reconstruction of the building, must have been contemporary with its use (eg S151), but unfortunately they contained little artefactual material that could indicate its function.

Structures to the south of the rectangular building (S58, S59, S152, S160, S161 and S162)

Sandstone and limestone rubble formed the major component of three of a group of five layers (S152) post-dating the rectangular building (S22). The other two were composed of silt but had mortar as a minor component. The earliest layer was composed of sandy silt. The area to the south of structure 22 also contained a number of structures, forming

Plate 4 Post on padstone showing decay of centre (S22, P9)

posthole alignments (S59 and S160), pits (S161) and ditches (S58), all roughly parallel to structure 22. Most of the postholes had post pipes and were large, the largest measuring 1.00 x 0.90 x 0.28m.

The group of layers (S152) to the south were characterised by rubble and mortar which presumably represented the demolition of walls. Whether this was derived from structure 22, from stone filling within a timber framework, or was derived from walls to the south of the excavated area, is unknown. Due to the proximity of the edge of excavation little can be suggested for structures 58, 59, 160 and 161 in terms of their function, beyond their representing boundaries. In the case of structure 59, however, it is also possible that this formed the northern side of a post- built structure, its southern side being beyond the excavated area.

A large ditch and a clay lined pit (S11 and S27)

Reconstructed twice, structure 27 was a ditch enclosing an area largely off the south-western limit of the excavation. The earliest cut had a downward gradient to the north and an even base. The sides and the base were lined with intermittent rows of stakes. A small pit was also located in the base of the cut. The first reconstruction was mostly removed by the second reconstruction, and its dimensions and gradient could not be estimated. It also however, had steep sides and an even base. The final reconstruction had a maximum width of 2.00m and depth of *c* 1.00m, with steep sides and a flat base and a gradient of 1 in 33 downward to the north. Nine stakes and six stakeholes lined the base. Associated with the large ditch were a linear cut, a ditch and two postholes. The linear cut ran north to south, parallel to the large ditch and *c* 2.00m to the west. It had gently sloping sides and a flat base. The ditch continued the east to west alignment of the northern section of the large ditch, and had gently sloping but irregular sides. The postholes had constructional fills of silty clay and postpipes. A clay- lined pit (S11) lay along the western limit of the excavation. Its disuse elements were composed of clay with a charcoal lens.

The steady gradient down to the north indicates a drainage function for structure 27, presumably to the Salwarpe. The change in direction, however, would also imply that it served as a boundary, enclosing an area to the west. The presence of stakes and stakeholes suggest attempts to minimise erosion caused by running water.

Several other features were associated with this structure. A short ditch continued the boundary of the north-west to south-west section of the major ditch. However, it does not appear to have served a drainage function. Other features within the enclosure, two postholes and a shallow linear cut (possibly a beam-slot), could have formed part of a building. Close to this was a small clay-lined pit (S11), indicating the storage of liquid.

Activity between the rectangular building and the large ditch (S49, S105, S117, S119 and S126)

A number of structures lay between the rectangular building (S22) and the large ditch (S27), of which the most complex was structure 49. It consisted of five possible postholes and eight possible stakeholes, forming a tight cluster. To this may be added three possible postholes and thirteen stakeholes shown on a field plan but not otherwise recorded. Though alignments cannot easily be identified, there exist two possible rows running east to west and roughly parallel to each other. It was not possible to assign a function to these structures.

Discussion

Most of the structures were constructed and fell into disuse within this phase, though the two largest (S22 and S27) possibly continued in use into later phases.

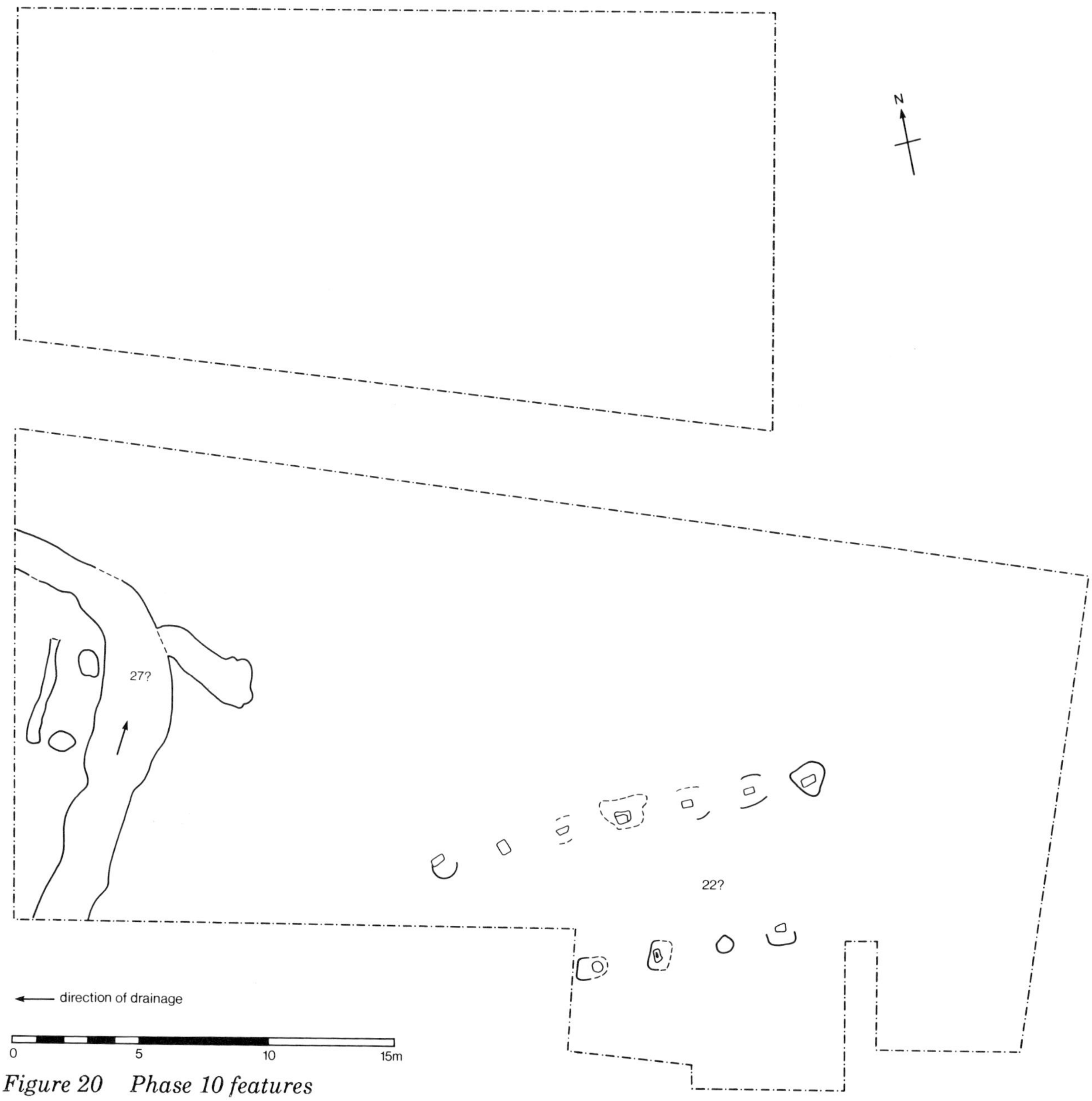

Figure 20 Phase 10 features

The pottery of phases 7 to 9 was very similar (Chapter 3), although the form of the structures indicate very different activities taking place. The function of the Malvernian ceramic objects must remain uncertain, although a slight bias in its distribution has been noted in the fill of structure 27 (Chapter 5). Evidence of an animal products industry may also be seen in the bone assemblage from the large ditch (S27), the fills of which were assigned to phases 9 and 11. The later fills have been assumed to be made up of residual material, as the assemblages were similar for both phases. The animal bone assemblage indicates a proportionally greater disposal of skulls and extremities than of parts of the carcass (Chapter 15). The fill of the large ditch (S27) had a wetland flora and seed assemblage, indicating that the surrounding area was overgrown with weeds and scrub (Chapter 18).

Phase 10: 5th to 11th century (Fig 20)

Several of the major structures of phase 9 possibly continued to be used, though the only certain component of this phase was a single layer.

A layer and possible continued use of other features (S22, S27 and S110)

There was no evidence of activity in the post-Roman period until the 10th to 11th centuries. Some of the structures of phase 9, however, may

Figure 21 Phase 11 features

have dated to this early period if the pottery-derived *termini post quem* were from residual assemblages. One structure, a layer composed of clay with charcoal as a minor component, had a *terminus post quem* of the 10th to 11th centuries, determined by the presence of Stafford-type ware (fabric 48). There is evidence that the rectangular building (S22), constructed in the previous phase, continued in use into phase 11. It was similarly possible that the large ditch (S27) did not fall into disuse until phase 11.

Phase 11: 12th to 14th century (Fig 21)

Concentrated in the eastern half of the excavated area, the structures included one associated with burning, a pebble surface, layers and various cut features.

Disuse of the rectangular building and large ditch (S22 and S27)

The stratigraphic analysis indicated that some of the structures of this phase were contemporary with the use of structure 22. Of these, two were composed of layers (S157 and S158) and one appears to represent the reconstruction or final removal of one of the vertical posts (S159) which contained packing stones. These were stratigraphically earlier than the removal of another post, represented by a postpipe and its fill, which probably marks the end of the building's use. Though this is the most obvious explanation of the

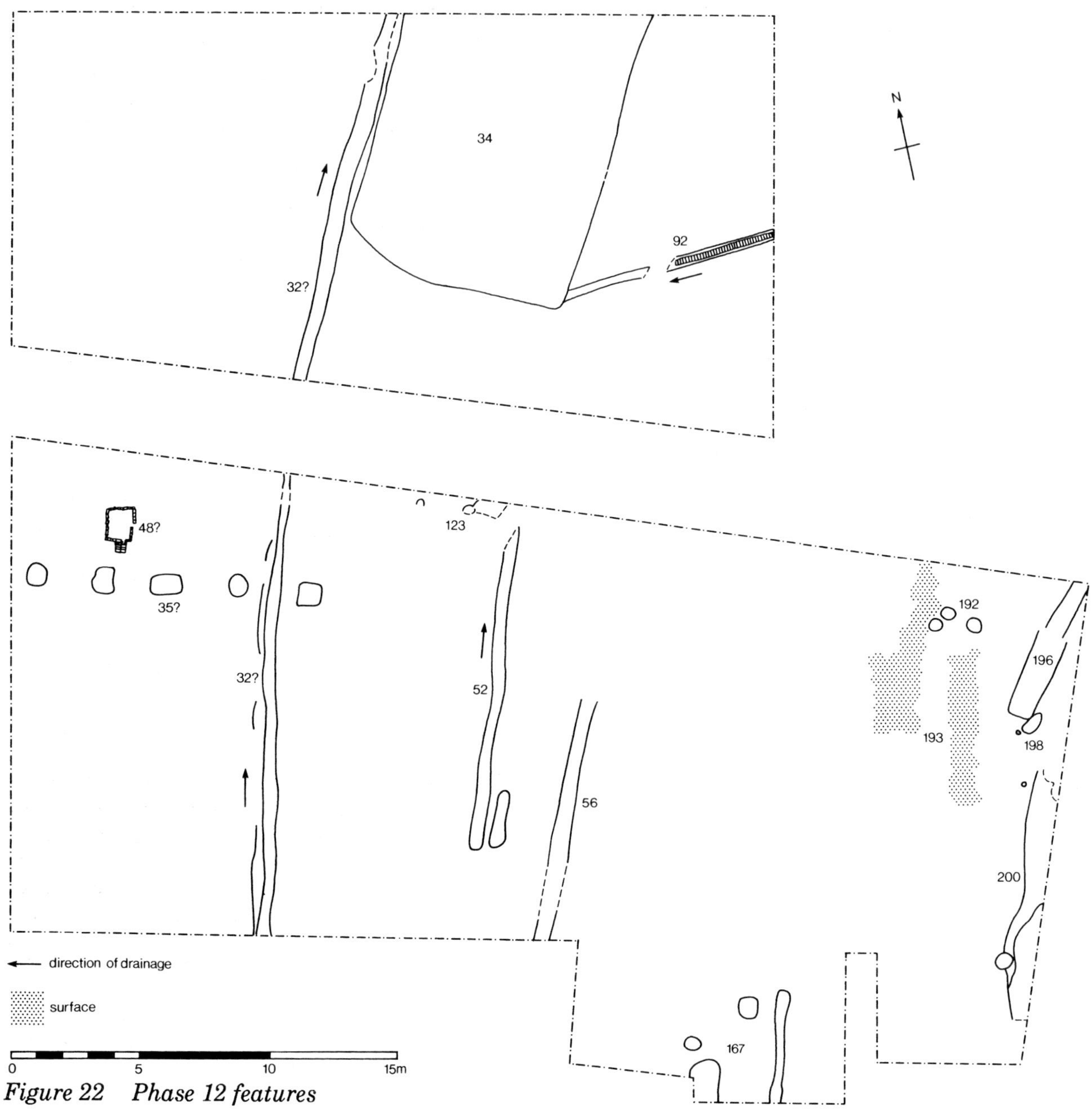

Figure 22 Phase 12 features

building's development, it implies that the building was extant for around 700 years or more and had a post which was *c* 500 years old before it was used in the building (P9). This seems unlikely but is not inconceivable. Moreover, there are two ways in which this apparently long period may be differently interpreted. Firstly, the period of use may have been shorter: it is possible that the building was constructed in phases 10 or 11. However, against this are the substantial number of structures post-dating the building which have consistently late Roman *termini post quem*. Alternatively, the apparent signs of disuse, the postpipe post-dating contexts with 12th to 14th century *termini post quem*, may not be as they seem. As we have only the foundation from which to judge, the building may have been demolished to ground level only, or even left in a ruined state for a long time, while the post rotted. The large ditch (S27) also fell into disuse during phase 11.

Other structures (S13, S38, S76, S141, S186 and S194)

The construction elements of structure 38 consisted of a near-circular pit with vertical sides and a flat base (*c* 1.20m diameter), with two shallow pits or depressions. The fills of the latter included loam with burnt clay and charcoal. The fills of the former were composed of clay and stones. These signs of burning indicate its possible function as a hearth. A pebble surface (S186) existed to the north and it is possible that industrial activity occurred in the area.

The remaining pits (S13, S76, S141 and S194) of this phase were also located in the eastern part of the excavated area, and of these structure 13 was possibly lined with clay.

Final fills of Iron Age features (S4, S9 and S11)

The final fills of structures 4, 9 and 11 were not deposited until phase 11. These features must have survived as slight depressions from the Roman period onwards. Structure 6 was not completely filled until phase 12.

Discussion

The low average weight of the tile from this phase indicates that it was derived from buildings located at some distance from the excavation (Chapter 4). Butchery practices do not appear to have changed from the Roman period (Chapter 15).

Phase 12: 15th to 18th century (Fig 22)

The structures of this phase were characteristically aligned north to south. A very large cut appeared to be the terminus of a number of drains. Another group of structures included a pebble surface and ditches with postholes along the eastern edge of the excavation.

Specialist comment on this and later phases was accorded a low priority, and discussion of the stratigraphic analysis has been kept to a minimum.

Boundary and yard (S193, S196, S198 and S200)

Three structures (S196, S198 and S200) appeared to form a boundary with an entrance, two postholes at the ends of two ditches marking the former position of gate posts. An 18th century plan (reproduced in Crickmore 1984, fig 5) shows that the present Rickett's Lane to the east was established in or before this phase. A cobbled surface covered an area extending from the entrance to the north (S193), presumably forming a yard.

Possible canal wharf (S34 and S92)

This very large cut (S34) had a brick-lined drain running into it (S92). There was no sign of any revetment for the sloping sides, and it had a flat base. One possibility is that it formed a wharf of Brindley's canal, opened in 1771 (Squires 1984). This is supported by the presence of a plant species that grows in mud, and aquatic snails which indicate standing water (Chapter 18). Large amounts of elm pollen from the same context may have been derived from the importation of flowering boughs (Chapter 18).

Possible drains and other features (S32, S35, S48, S52, S56 and S167)

The other linear cuts (S52, S56 and part of S167) may also be interpreted as drains. Despite the absence of a brick lining, these had a gradient down to the north. Mention should also be made of structure 32, a ditch with a gradient down to the north and of similar dimensions. Its construction, however, may have occurred in either this or phase 13. Two other structures (S35 and S48) can be similarly phased, though their function is best discussed later.

Phase 13: 19th to early 20th century (Fig 23)

The main structural components consisted of several pits with brick linings, all in the western part of the excavated area. A linear cut ran north-east to south-west, with two others joining it from the west. All were lined with brick. A number of other features were constructed to the south-east.

Drains, possible hearths and a boundary (S33, S35, S41, S42, S43 and S48)

The most prominent structure of this phase was a brick-lined drain (S33) carrying water away to the north, presumably into the Droitwich Junction Canal (constructed in the mid 19th century, Squires 1984) or into the river. Two further drains joined the main one from the west, where a number of structures were located. Structures 42, 48 and 43 were lined with brick and the latter also had a brick-lined drain associated with it. The existence of a brick superstructures was indicated by the presence of brick rubble as a component of the fills (S41 and S48). Evidence of burning was also noted in the fill of structure 48. Structure 35 probably formed a post-built boundary demarcating the southward extent of the structures described above. The construction of structures 35 and 48 could not be certainly placed in this phase, though their interpretation is best discussed here with structures of similar form. The First Edition Ordnance Survey plan of 1884 shows the site was largely clear of buildings.

Phase 14: Later 20th century (Fig 24)

This phase represents the final function of the site prior to the excavation.

Drains and foundations for a bowling green (S69, S165 and S206)

The ceramic drainage pipes (S165 and S206) would, as the excavator suggested, have drained the bowling green, the latest feature on the site. The excavator also suggested that the laying of

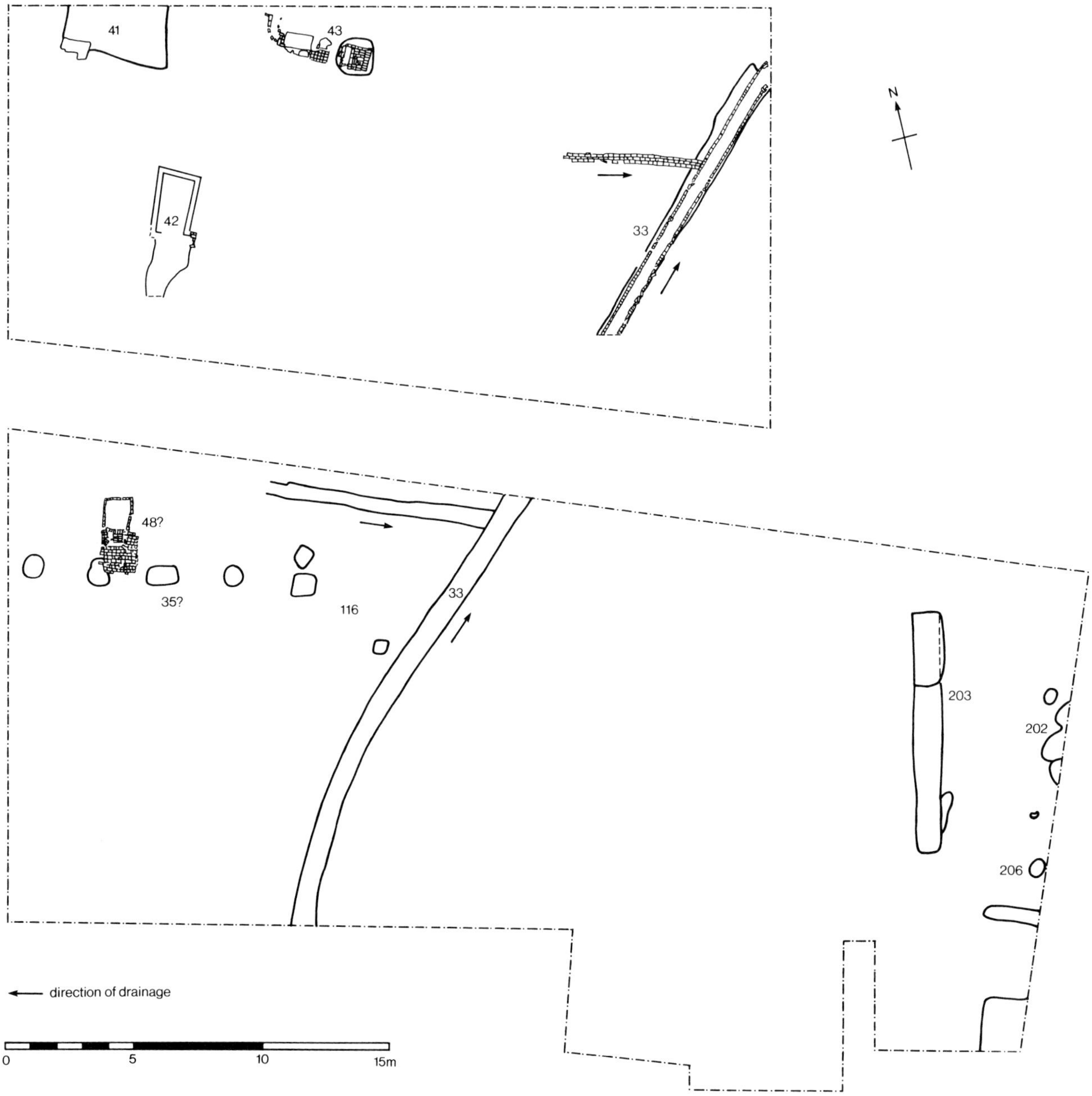

Figure 23 Phase 13 features

foundations (S69) was preceded by soil stripping, truncating many of the archaeological deposits.

Structures of uncertain phase

The structures described and discussed below are of Roman or earlier date, and are considered to be important to the understanding of the site. Some of these structures have only a small range of possible phases and have already been discussed in general terms.

Structure 1 (P2 to P4)

The construction elements consisted of a rectangular pit with steep sides and a curved base (2.00 x 1.55 x 0.75m). The pit contained a clay and wattle lining which did not survive to the top of the cut. The wattle produced a radiocarbon date of 101–72 Cal BC (Table M17, 1:E10). Stratigraphically the structure post-dated contexts with an Iron Age *terminus post quem* but can be placed in either phase 2 or 3. Though similar in construction to structure 2, it was not demonstrably earlier than the clay- and stake-lined pits characteristic of phase 3 and may well have been contemporary with them. The fills also indicated a longer period of use, up to phase 5. The function of this structure may be assumed to be the same as that of structure 2 and the clay- and stake-lined pits (see above): for the settling and storage of brine.

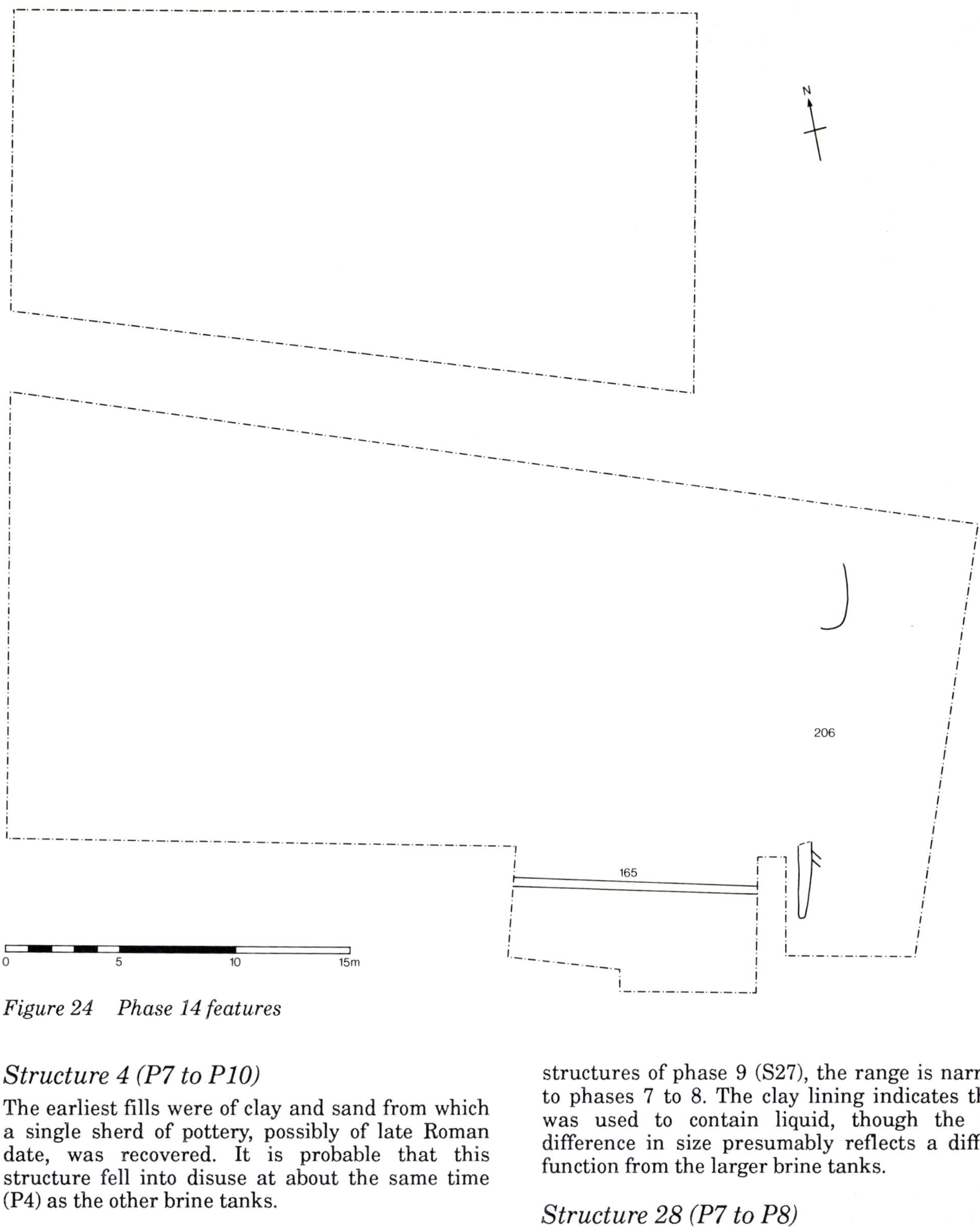

Figure 24 Phase 14 features

Structure 4 (P7 to P10)

The earliest fills were of clay and sand from which a single sherd of pottery, possibly of late Roman date, was recovered. It is probable that this structure fell into disuse at about the same time (P4) as the other brine tanks.

Structure 12 (P4 to P8)

The full extent of this pit was obscured by a temporary baulk. It had a clay lining *c* 0.20m thick and a poorly-recorded disuse fill. Its construction could not be more certainly placed than to somewhere between phases 4 and 8. The disuse fill had a pottery derived *terminus post quem* of mid 3rd to late 4th century and, with this pre-dating structures of phase 9 (S27), the range is narrowed to phases 7 to 8. The clay lining indicates that it was used to contain liquid, though the great difference in size presumably reflects a different function from the larger brine tanks.

Structure 28 (P7 to P8)

This consisted of a surface with a gravel foundation. Two parallel depressions crossed the surface and a ditch ran to the south. The ditch was *c* 0.40m wide and *c* 0.20m deep, with a number of stakeholes in its base. The structure was orientated east to west, fading out to the east and with its full extent obscured by the western limit of excavation. Within the make-up of the surface, a pit and its fill of clay were recorded.

Despite its obvious importance to the site the construction and use of structure 28 was impossible to assign to a specific phase. Stratigraphic and dating evidence indicate construction and use in either, or both, phases 7 and 8. The structure formed a road or track running from the centre of the excavated area to the west. In both phases the structures were so arranged as to be served by the road (S26 and S31) and seems likely that it was in use during both phases. The track appears to have been little used. Only one set of wheel ruts was recorded, where with long term use, many would be expected. Moreover only one level of make-up was recorded and few repairs seem to have been carried out.

Structure 29 (P4 to P10)

This substantial ditch (*c* 2.20m wide and *c* 1.20m deep) ran north-east to south-west and at its southern end turned a right-angle to the south-east. The eastern slope was less steep than that on the western side. The highest part of the base was where its course altered, with a gradient of 1 in 100 to the north-east and 1 in 67 to the south-east. There was also a vertical step down of *c* 0.60m to the top of the latter gradient at this point. A radiocarbon date of 379 Cal BC–Cal AD 116 (Table M17, 1:E10) was obtained from one of the primary fills of clay and silt.

The ditch's construction can be placed as early as phase 4. However it possibly post-dates a context with a mid 3rd to late 4th century *terminus post quem*, which would refine the phasing to between 7 and 11. The radiocarbon date must have been derived from residual material. The change in orientation and profile indicates the structure having enclosed an area to the east of the excavation. The structure may also have acted as a drain, eventually to the Salwarpe, and the fills contained evidence of a wetland flora (Chapter 18).

Structure 36 (P4 to P5)

An alignment of four postholes ran north to south and may be associated with a further possible posthole. Other constructional elements included posts and fills of silt and loam.

Stratigraphic evidence and a radiocarbon date from one of the posts of 2 Cal BC–Cal AD 129 (Table M17, 1:E10) places the construction of the structure in phase 4, though two of the constructional fills have a possible pottery-derived *terminus post quem* of early 2nd to mid 3rd century. However, the date suggested by the pottery is suspect as it was based on very small assemblages or uncertainly-identified fabrics. No parallel row was recorded, which suggests an interpretation as a post-built boundary rather than a building, though the structure ran parallel to, and not far from, the western limit of excavation.

Structure 37 (P4 to P9)

These four stakeholes formed a rectangle with sides measuring 2.50 x 1.50m. Another stakehole lay just to the north-east of one corner. Stratigraphically, construction and use may have occurred at any time between phases 4 and 9. The small volume of the fills made the presence of a datable pottery assemblage unlikely. A single sherd from a fill gave a possible *terminus post quem* of mid 1st to early 2nd century.

Structure 45 (P4 to P5)

Twenty possible postholes were clustered around a linear cut running north-east to south-west. Although it could be later, this structure is best placed in phases 4 to 5. Not much of the structure could be excavated, as it extended beyond the western limit of excavation, though it may have represented a post-and-trench type of construction for a wall.

Structure 46 (P7 to P9)

A rough alignment of six possible postholes ran north-west to south-east. Though their sizes varied considerably (from 0.38 x 0.34 x 0.18m to 0.72 x 0.54 x 0.26m) they were all at the same stratigraphic level. Lack of stratigraphic relationships with the structures of phases 7 to 9 makes this structure difficult to phase with any certainly. However, it was roughly aligned at right-angles to the line of barrels (S17, S16 and S19). A post-built boundary is a likely interpretation.

Structure 47 (P7 to P9)

This single pit was smaller (1.26 x *c* 0.85 x 0.57m) than those containing barrels though it too contained wood, much decayed, over the second fill. If this was *in situ* then the fills of silty clay and sandy clay may be regarded as constructional.

Structure 54 (P2 to P4)

This structure consisted of 31 stakeholes and two possible postholes. The layout of the structure defies a brief written description. The six stakeholes to the west were somewhat similar to those of structure 60, though on a different orientation. The structure may be placed anywhere between phases 2 and 4. Its function is impossible to ascertain though a light structure such as a windbreak or rack may be suggested.

Structure 57 (P4 to P8)

This single pit was of unique form, being subcircular in plan with a central 'pedestal' of undisturbed material. The overall dimensions were 1.90 x 1.50 x 0.27m, the central pedestal being

Figure 25 Reconstruction of salt production in the Iron Age

0.30m in diameter. A fill of charcoal and ash may have been related to the structure's use.

Stakeholes, parts of structure 163 (P4 to 5), may represent a superstructure. The deposit of charcoal and ash indicated use as a hearth and possibly represented a later type, equivalent to structures 23, 24 and 25. A few of the clay- and stake-lined pits, with which structures 23, 24 and 25 were associated, were still possibly in use during phase 4. Alternatively this may be associated with the barrels characteristic of phase 8.

Structure 60 (P2 to P12)

These seven stakeholes formed two diverging rows with a single outlier. Though it may be assigned to a greater range of phases, the similarity with part of structure 54 has already been noted. Again a light structure such as a rack may be suggested.

Conclusion

Activity earlier than the late Iron Age is indicated only by the presence of very small numbers of residual flints and pottery sherds, together with a radiocarbon date. The absence of structures associated with earlier salt production does not rule out salt production elsewhere in Droitwich at this time. Evidence of earlier salt production comes from sites outside Droitwich with briquetage dated to the 6th and 5th centuries BC (Morris 1985, 346).

In the late Iron Age the construction of a number of brine tanks with large capacities, as well as the vast amount of waste briquetage, indicates large-scale production (see Fig 25). The layout too indicates some degree of organisation, though how far this was determined by the method of production as opposed to other organisational factors, is uncertain. Production continued into the early Roman period, though for how long this continued is again uncertain. One of the brine tanks was probably reconstructed during this period.

The precise method of salt production in the late Iron Age and Roman periods is still uncertain. However, there is no evidence to refute the general model, drawn largely from later evidence, outlined in the introduction to this volume (Chapter 1). The brine tanks and hearths were used for storage and or settling, and boiling respectively. Some of the wooden artefacts deposited in the fills of the brine tanks may have been used in the production of salt.

The specific use (if one existed) of briquetage is also still uncertain and further experimental work is needed to test its use for boiling or draining.

It appears that salt production declined (at least within the area of the excavation) in the 2nd or 3rd centuries, as the structures of phases 5 and 7 cannot be associated with salt production with any degree of certainty. Briquetage is probably a residual element of the assemblage from phase 5 onwards. In the absence of evidence for alternative functions, the barrels of phase 8 may most readily be compared in function to the brine tanks of phase 3. There is however, little evidence of the hearths which were complementary to the brine tanks of the earlier phase. The layout of the features of phase 8 also suggests a significant reorientation of the focus of activity in the area from the east to the west. Both of these points prompt consideration of an alternative function to that of salt production. For instance, the processing of animal products, such as tanning, itself an activity requiring brine or salt. Indeed, such secondary industries may be expected to have been located close to the primary salt industry. Again, however from phase 9 there is no evidence of salt production occurring on the site, though the best evidence for animal products processing comes from this phase.

Except in some details of topography, the poor survival of later deposits provided little information on aspects of the town's development. The excavations at Friar Street were much more productive in this respect.

3 Pottery *Helen Rees*

Nearly 56,000 sherds were recorded from the excavation, the majority of which (*c* 51%) were Roman in date. A further *c* 38% were from vessels of briquetage and *c* 10% were of medieval or later date. The remainder comprised *c* 100 Iron Age sherds, *c* 400 sherds which, due to the continuation of certain Iron Age fabrics into earlier Roman times, were not securely ascribable to either period, and four sherds thought to be of earlier prehistoric date.

The assemblage was recorded with three interrelated aims in mind:

1) To characterize the pottery by fabric and, as far as possible, to identify sources for the fabrics so defined.
2) To characterize the pottery by form, and to produce a vessel type series.
3) To note any other characteristics which were significant for the chronology, function or origin of the pottery.

Fabrics were defined macroscopically according to characteristics of technique of manufacture, firing, texture, surface treatment and mineral inclusions present in fracture. They were then checked and ascribed to source wherever possible, through petrological analysis (1:F2–G7) and consultation with other researchers (see acknowledgements). Samian ware and *amphorae* were the subject of specialist reports by Brenda Dickinson (see below and 1:G13–2:C12) and David Williams (1:B8–10) respectively. Amongst the forms, ten major prehistoric and Roman classes were recognised: flagon, jug, beaker, saucepan/tubby cooking pot (Peacock 1965–7; 1968), jar, tankard, bowl, *mortarium*, dish and lid. More detailed typological study allowed the definition of specific forms within these classes. Post-Roman forms were largely paralleled in the already existing type series for the City of Worcester (Morris 1980).

Using the fabric and form classifications thus defined, the material was quantified by number and weight of sherds and recorded by context, noting type of decoration and other characteristics such as the presence of rivet holes, the presence of slip on a normally unslipped fabric and distortion caused by lack of control over firing conditions.

The aims of analysis were threefold:

1) In conjunction with analyses of other types of artefact, to supply absolute dating for the sequence defined by the stratigraphy of the site.
2) Through application of a generally known ceramic sequence, to supplement the stratigraphic evidence concerning the chronological development of the site.
3) To assess site-specific evidence for the chronology, function or origin of the pottery fabrics and forms.

To this end a *terminus post quem* within the broad date ranges of a ceramic phase was assigned to each assemblage. The ceramic phases were defined through reference to external parallels and a *terminus post quem* was assigned to each assemblage on the basis of the presence of certain types and the absence of others (see 1:E11–F1 for the details of the criteria used to define ceramic phases). When the 'spot dating' was considered in conjunction with the stratigraphy of the site and with alternative dating evidence from radiocarbon and dendrochronological analysis, it became clear that there had been a significant intrusion of sherds and that the contribution of pottery dating evidence to the stratigraphic analysis needed careful consideration (1:B7). Discrepancies in the chronological cohesiveness of the pottery assemblages were therefore rechecked against the site data and areas of uncertainty in the stratigraphic record noted. These contexts were excluded from subsequent analyses, together with contexts whose ceramic assemblages were uncertainly phased.

Illustrated pottery

Iron Age and Roman pottery (Fig 26)

Handmade Malvernian ware Group A (fabric 3)

1 Saucepan pot decorated with linear tooling (184)
2 Saucepan pot (2192, S180, P1)
3 Tubby cooking pot decorated with acute lattice (1181, S20, P8)
4 Tubby cooking pot decorated with vertical and horizontal burnishing (1070, S16, P8)
5 Tubby cooking pot decorated with horizontal burnishing (1185, S106, P5(7))
6 Tubby cooking pot decorated with vertical and horizontal burnishing (731, S94, P7–8)
7 Tubby cooking pot (2436, S8, P4)
8 Tubby cooking pot decorated with vertical burnishing (448, S83, P2)
9 Tubby cooking pot decorated with vertical burnishing (1618)
10 Cooking pot with sinuous profile (477, S30, P4)
11 Jar (1689, S8, P4–5(7)?)
12 Handmade jar with everted rim (622, S117, P9)
13 Handmade jar with everted rim (1716, S8, P4)

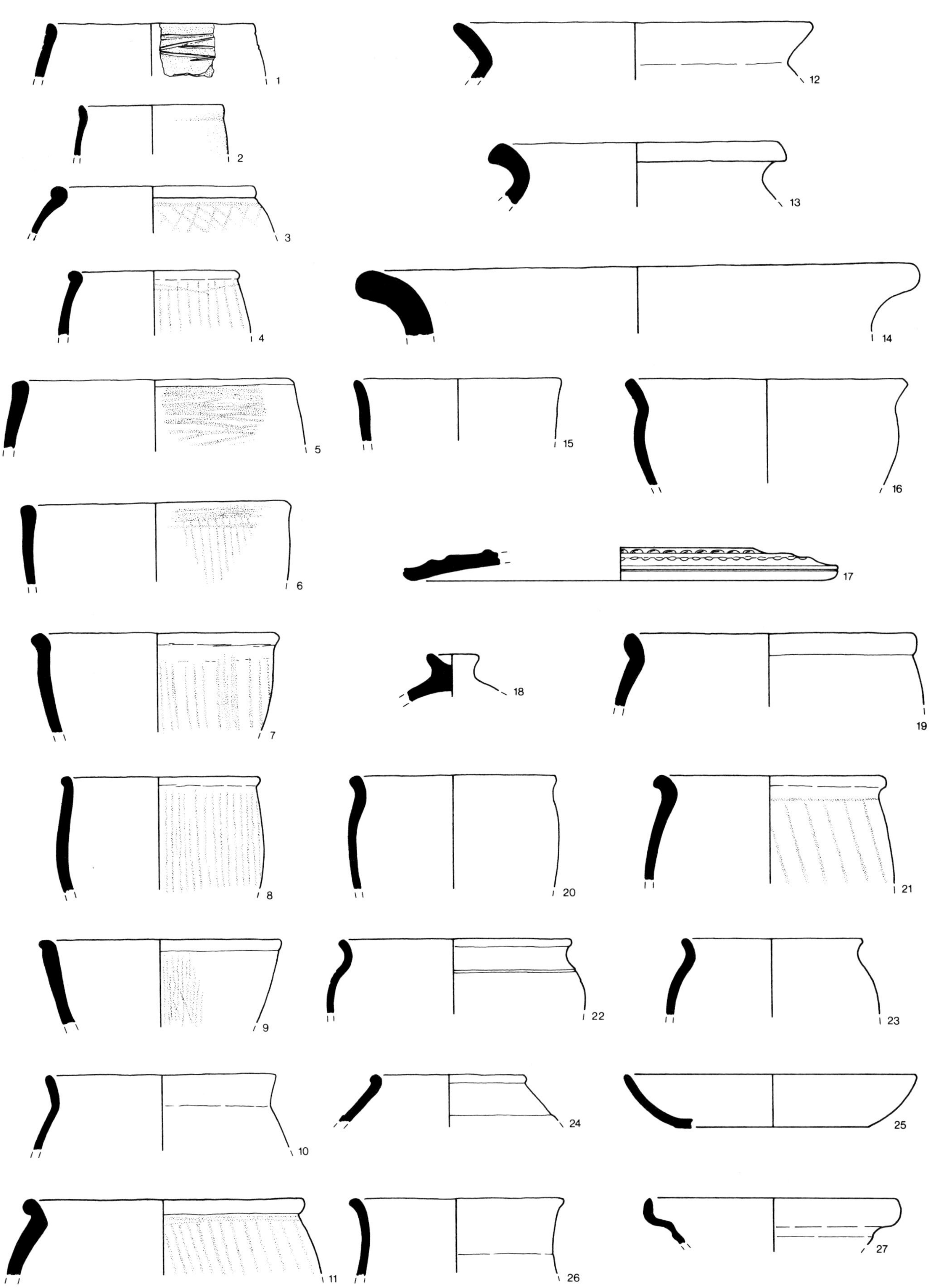

Figure 26 Iron Age and Roman pottery vessels in fabrics 3 (1–18), 4 (19–22), 5 (23), 7 (24, 25) and 8 (26, 27). Scale 1:4

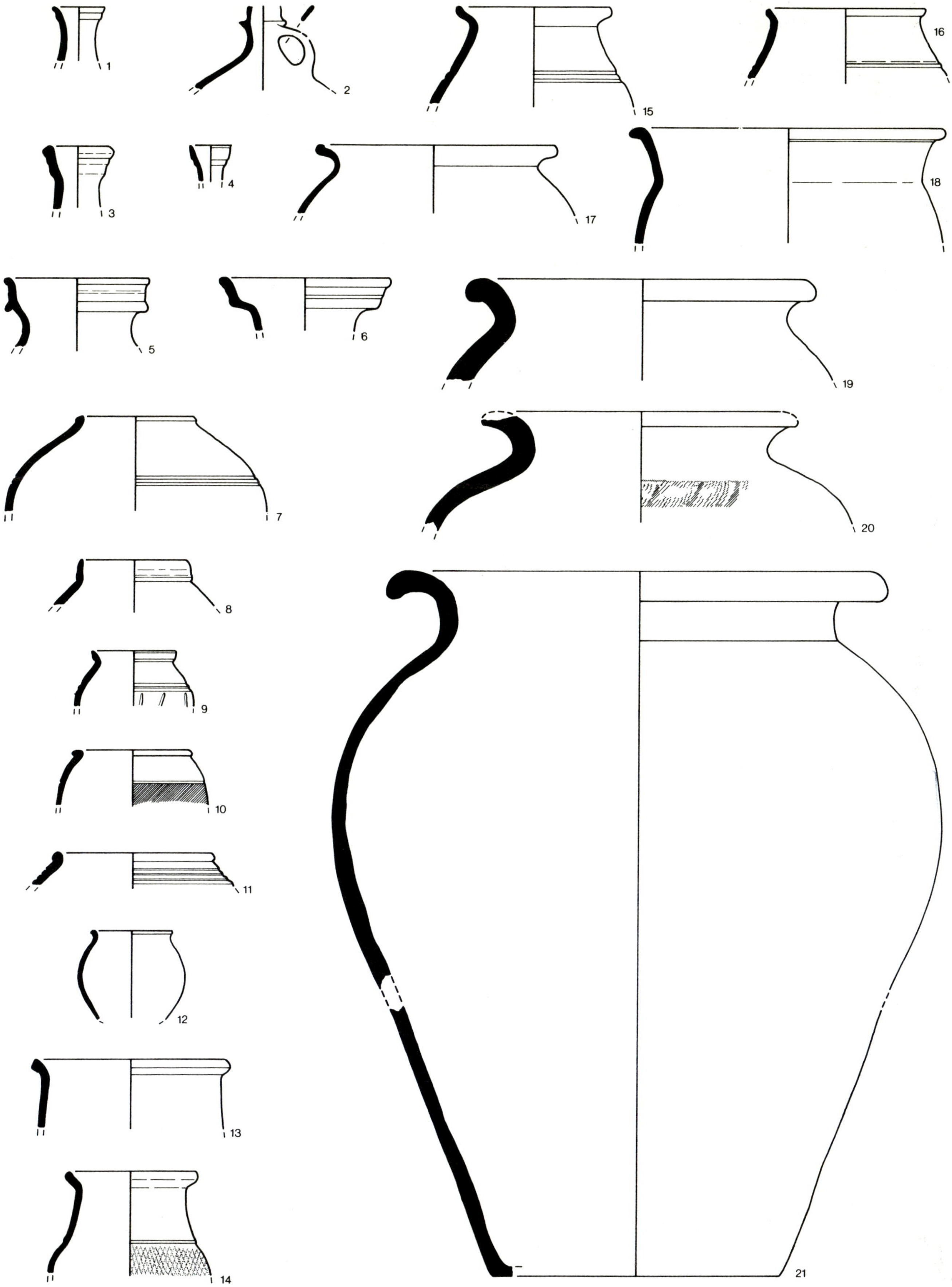

Figure 27 Roman pottery vessels in fabric 12. Scale 1:4

14 Handmade jar with everted rim (1205, S110, P10)
15 Saucepan pot (2193, S181, P2)
16 Cooking pot with sinuous profile (412, S81, P2)
17 Lid decorated with impressed ovals and circles (582, S99, P9–11)
18 Lid (582, S99, P9–11)

Handmade limestone tempered ware (fabric 4)

19 Jar with bead rim (1258, S106, P4–5)
20 Jar with everted rim (2437, S8, P4)
21 Cooking pot (457, S69?, P14?)
22 Cooking pot decorated with two tooled lines on shoulder (490, S90, P3–4)

Handmade sandy ware (fabric 5)

23 Jar with everted rim (2192, S180, P1)

'Belgic'-type ware (fabric 7)

24 Beaker (2327, S152, P9)
25 Bowl with plain rim (564, S27, P9(11))

'Belgic'-type ware (fabric 8)

26 Bowl (2184, S181, P2)
27 Segmental dish (1942/2048, S29, P11)

Roman pottery (Fig 27)

Severn Valley ware (fabric 12)

1 Flagon (2281, S158, P11–12)
2 Flagon (1553, S155?, P12)
3 Flagon (589, S27, P9(11))
4 Flagon with a ring neck (559, S27, P9–11)
5 Jar with double lip (579, S27?, P9?)
6 Flagon (589, S27, P9(11))
7 Beaker (2436, S8, P4)
8 Beaker (531, S27, P11)
9 Beaker decorated with applied vertical strips (581, S27?, P9?)
10 Roughcast beaker (711, S97, P5)
11 Beaker decorated with cordons below rim (2610, S208, P12)
12 Beaker with narrow mouth (294, S63, P4)
13 Beaker with high neck (35, S66, P7–9)
14 Beaker decorated with acute lattice (1938, S192, P12)
15 Jar or beaker (1689, S8, P4–5(7)?)
16 Jar decorated with a cordon (1213, S107, P5–7)
17 Jar (1918, S29, P11)
18 Jar (40, S1, P4)
19 Jar with close mouth (477, S30, P4)
20 Storage jar (2450, S150, P5–9)
21 Jar (50)

Roman pottery (Fig 28)

Severn Valley ware (fabric 12)

1 Jar (1583, S21, P8)
2 Jar with narrow mouth and decorated with a cordon at the base the of neck and three incised rings (3016, S30, P11)
3 Jar with narrow mouth and decorated with a cordon on the base of the neck and the girth (564, S27, P9(11))
4 Jar decorated with grooves and burnished lattice on the girth (2436, S8, P4)
5 Jar with two handles (574, S27, P9(11))
6 Jar decorated with a cordon (49, S66, P7–9)
7 Jar (1175, S114?, P5(11)?)
8 Jar (1002, S28, P7–8)
9 Jar with pulley rim (1040, S127, P12)
10 Jar with pulley rim (582, S99, P9–11)
11 Jar (523, S27, P11)
12 Jar with wide mouth decorated with grooves on the girth (531, S27, P11)
13 Jar with wide mouth (69, S112, P7)
14 Jar (1005)
15 Jar decorated with cordons below the rim (2606, S208, P13)

Roman pottery (Fig 29)

Severn Valley ware (fabric 12)

1 Jar (1070/1156, S16, P8)
2 Jar (1689, S8, P4–5(7)?)
3 Jar (2363/2292, S163, P5)
4 Small jar with wide mouth (559, S27, P9–11)
5 Small jar or bowl (1018, S52, P12)
6 Jar or bowl (1018, S52, P12)
7 Jar with wide mouth (531, S27, P11)
8 Jar or bowl (577, S31, P8)
9 Jar with wide mouth and small lower lip (559, S27, P9–11)
10 Bowl (1689, S8, P4–5(7)?)
11 Tankard with straight sides and decorated with a cordon (1686, S7, P5)
12 Tankard decorated with acute lattice (1178, S109?, P5(7))
13 Tankard with a bead rim and decorated with an acute lattice and cordons (1173, S109, P5–7)
14 Tankard with a bead rim (1023, S114, P7–9)
15 Tankard decorated with acute lattice (4, S68?, P13)
16 Tankard decorated with acute lattice (3012, S1, P5)
17 Tankard (1567, S155, P7–11)
18 Tankard with everted sides and bead rim (3012, S1, P5)
19 Tankard with everted sides and handle (1205, S110, P10)
20 Tankard decorated with white painted 'sun' motifs under rim (1163)
21 Bowl with carination and decorated with cordons on carination (127/128, S3, P4)
22 Bowl with carination (1689, S8, P4–5(7)?)
23 Bowl with carination (1689, S8, P4–5(7)?)
24 Bowl with carination and decorated with lattice (2267, S153, P11(12)?)
25 Bowl with carination (731, S94, P7–8)

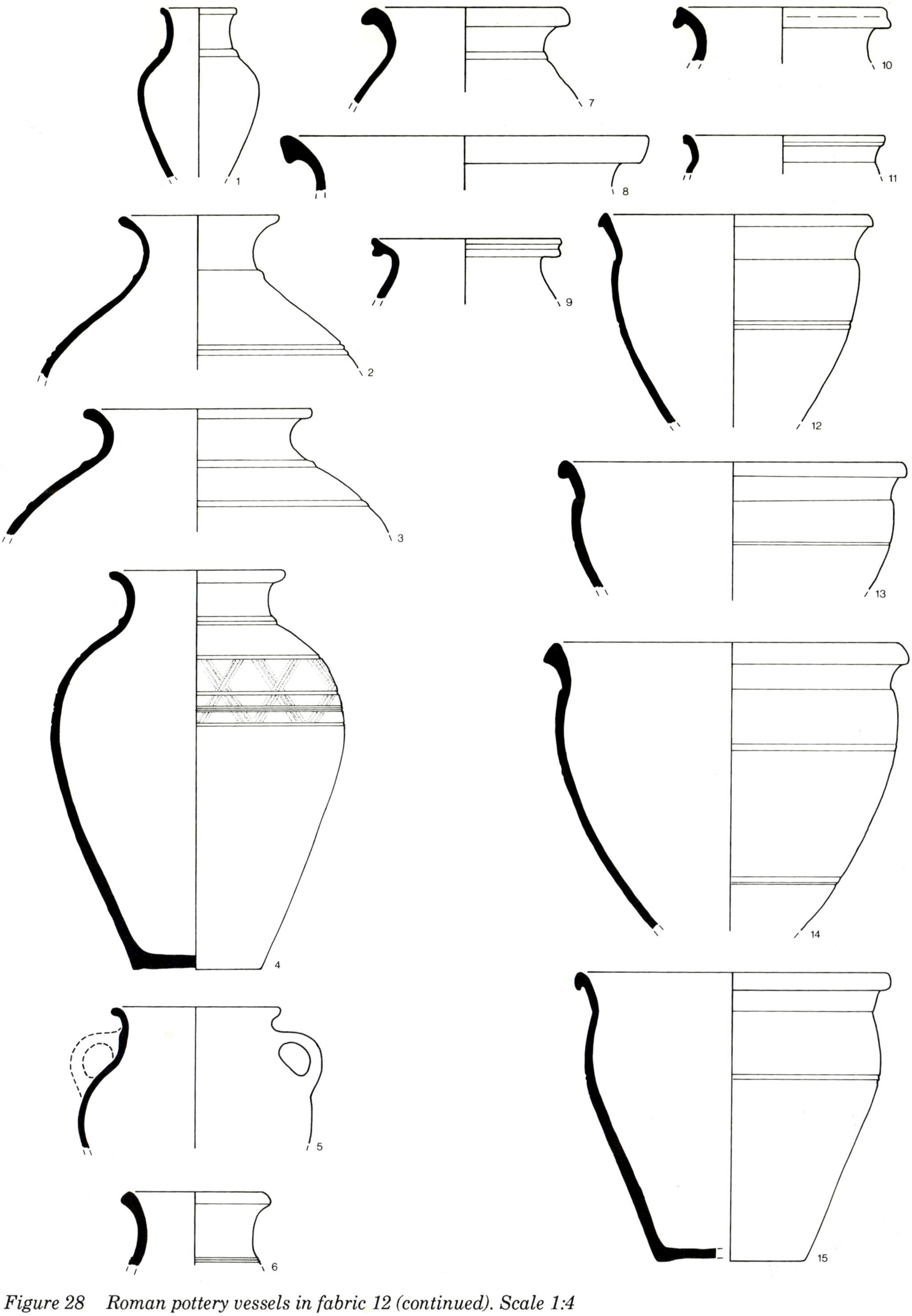

Figure 28 Roman pottery vessels in fabric 12 (continued). Scale 1:4

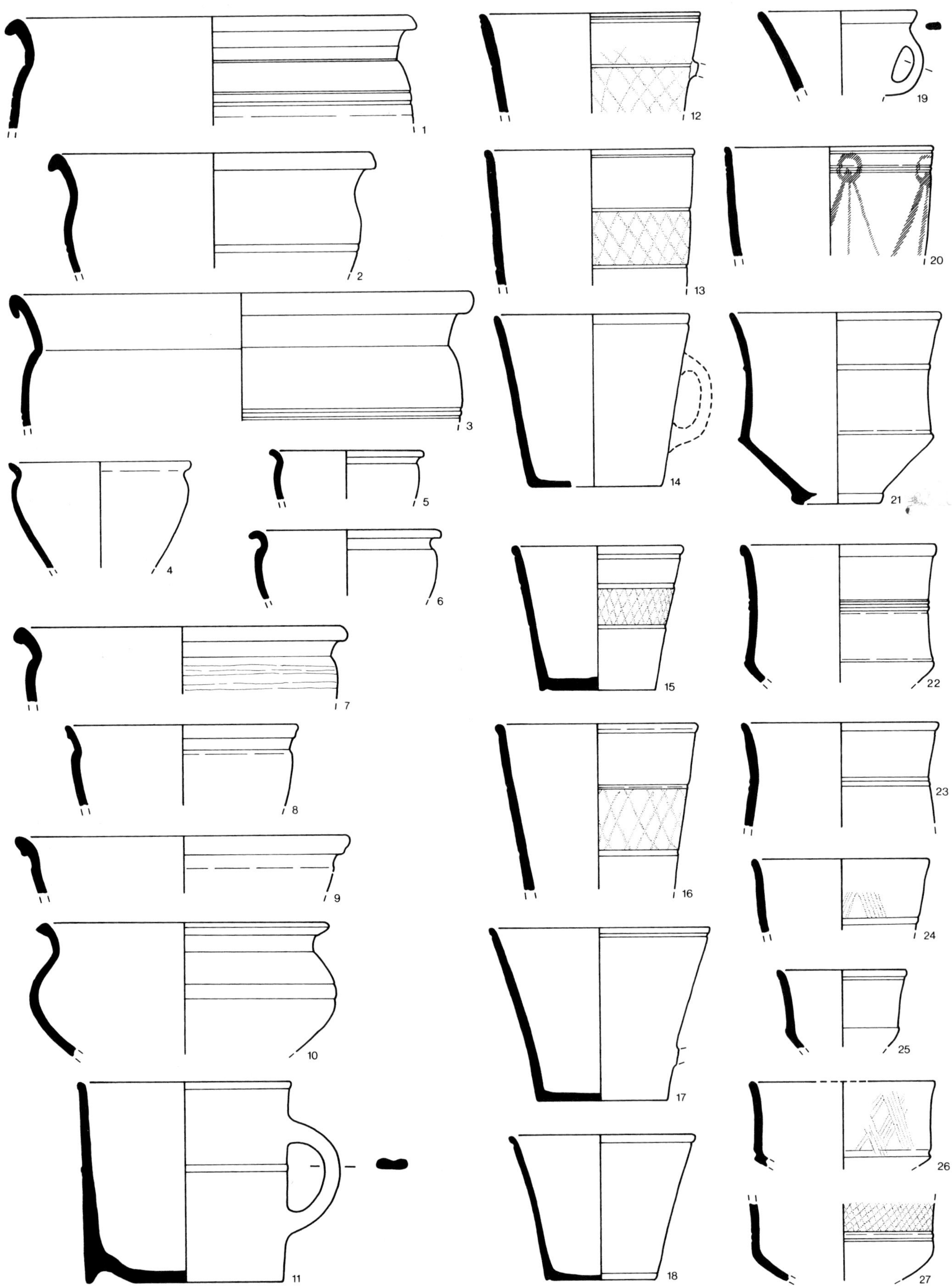

Figure 29 Roman pottery vessels in fabric 12 (continued). Scale 1:4

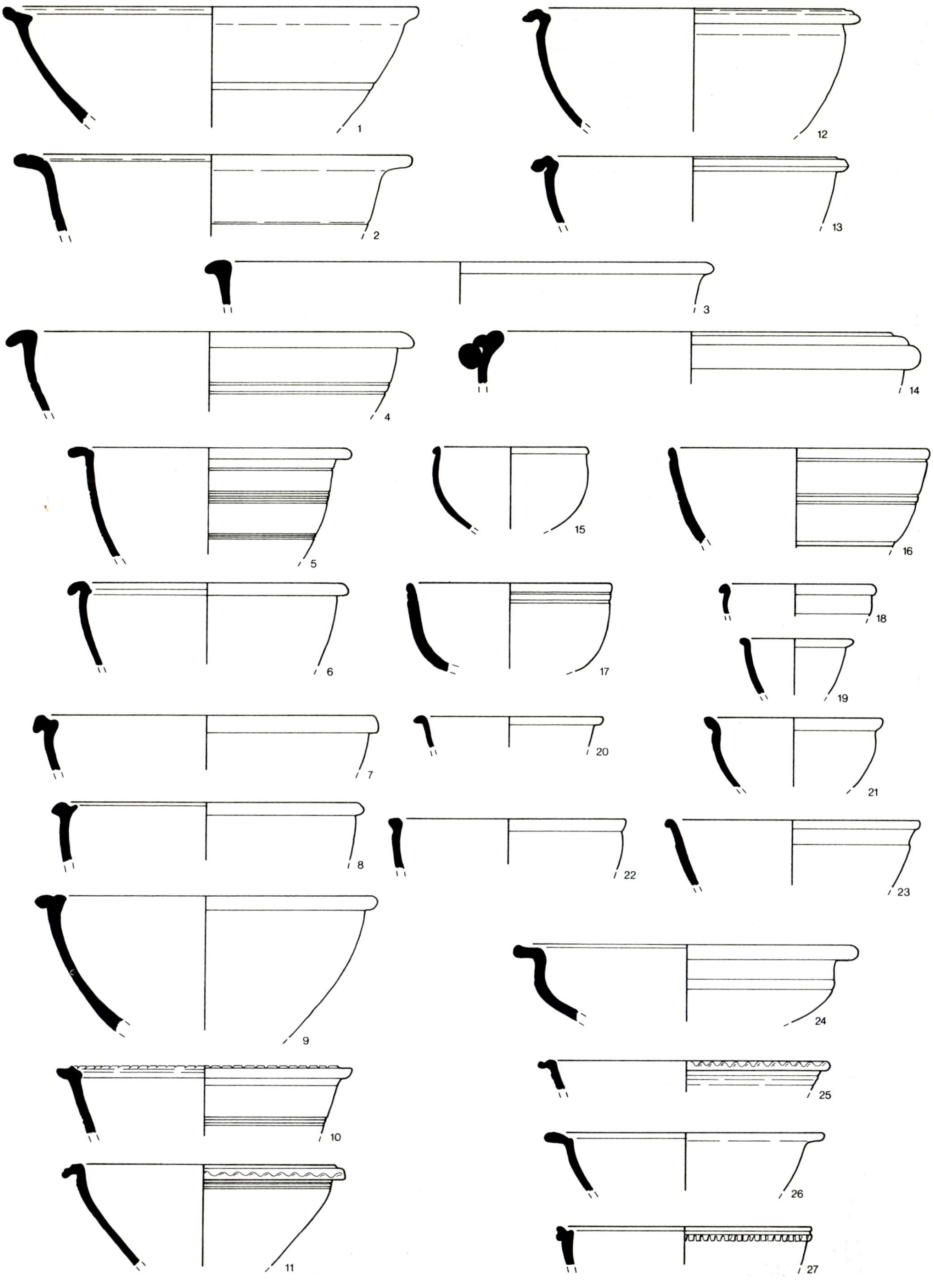

Figure 30 Roman pottery vessels in fabric 12 (continued). Scale 1:4

26 Bowl with carination (2401, S22, P9)
27 Bowl with carination decorated with lattice (4, S68?, P13)

Roman pottery (Fig 30)

Severn Valley ware (fabric 12)

1 Bowl with flat rim and decorated with grooves on rim and body (2382 2351)
2 Bowl with reeded rim and decorated with grooves on the girth (531, S27, P11)
3 Bowl with flat rim (717, S96, P7–9)
4 Bowl with flat rim and decorated with grooves (531, S27, P11)
5 Bowl with flat rim and decorated with grooves on rim and body (2255, S159, P11)
6 Bowl with flanged rim and internal bead (559, S27, P9(11))
7 Bowl with flanged rim (560, S27, P9(11))
8 Bowl with flanged rim (731, S94, P7–8)
9 Bowl with flanged rim (564, S27, P9(11))
10 Bowl with reeded rim and stabbed decoration on reed (2600, S69, P14)
11 Bowl with reeded piecrust rim (585, S31, P8)
12 Bowl with reeded rim (563, S102, P9)
13 Bowl with reeded rim (619, S112, P7)
14 Bowl with reeded rim (630, S27, P9(11))
15 Bowl with bead rim (1070, S16, P8)
16 Bowl (585, S31, P8)
17 Bowl with bead rim (1085, S127, P12)
18 Bowl (1175, S114?, P5(11)?)
19 Bowl (501, S69, P14)
20 Bowl (582, S99, P9–11)
21 Bowl with small everted rim (1143, S17, P8)
22 Bowl with flanged rim (745, S94, P7–8)
23 Bowl with shallow bead rim (1039, S15, P8)
24 Bowl with flat rim (2267, S153, P11(12)?)
25 Bowl with flat piecrust rim and decorated with grooves below rim (711, S97, P5)
26 Bowl with flanged rim (1192, S113, P5(7))
27 Bowl with double lip and decorated with stabbed lines on bottom of lip (2226, S167, P12)

Roman pottery (Fig 31)

Severn Valley ware (fabric 12)

1 Dish with plain rim (629, S27, P9(11))
2 Shallow bowl with carination (582, S99, P9–11)
3 Dish with bead rim (629, S27, P9(11))
4 Dish (1139, S109, P5?)
5 Lid (1023, S114, P7–9)
6 Lid (501, S69, P14)
7 Lid decorated with grooves (547, S27, P9)
8 Pedestal base decorated with a cordon above pedestal and perforated base (39)

Sandy oxidised ware (fabric 13)

9 Flagon (552, S27, P11)
10 Flagon (631, S96, P7–9)
11 Flagon (564, S27, P9(11))
12 Jar (1689, S8, P4–5(7)?)
13 Jar (3, S68, P13)

Fine grey ware (fabric 14)

14 Beaker with rusticated decoration (1576, S31, P8)
15 Small jar or beaker (1151, S114, P5–9(11))
16 Beaker decorated with acute cross-hatched burnishing (28?, S1, P5?)
17 Jar with rusticated decoration (1553, S155?, P12)
18 Jar (1918, S29, P11)
19 Jar with narrow mouth and decorated with incised crosses above a stabbed dotted line on the neck, and burnished all over except for a band on the girth (560–531–564, S27, P9(11))
20 Jar (630, S27, P9(11))
21 Jar with narrow mouth and decorated with burnished bands (564, S27, P9(11))
22 Jar with double lip (531, S27, P11)
23 Bowl with 'waist' (28?, S1, P5?)
24 Shallow bowl with carination and decorated with three wavy lines (28?, S1, P5?)
25 Bowl with reeded rim (630, S27, P9(11))
26 Bowl with reeded rim (1029, S129, P12–13)
27 Bowl with flanged rim (3, S68, P13)
28 Bowl decorated with dimples (501, S69, P14)
29 Dish with plain rim (501, S69, P14)
30 Dish with a bead rim (531, S27, P11)
31 Lid (2330, S57, P4–9)

Roman pottery (Fig 32)

Fine grey ware (fabric 14)

1 Lid (531, S27, P11)

Coarse grey ware (fabric 15)

2 Jar (745, S94, P7–8)
3 Beaker with rusticated decoration (630, S27, P9(11))
4 Bowl with flat rim and decorated with grooves on the rim (731, S94, P7–8)
5 Bowl with flat rim (745, S94, P7–8)
6 Bowl with bead rim (531, S27, P11)
7 Dish with bead rim (559, S27, P9(11))

Mudstone tempered ware (fabric 17)

8 Storage jar (1023, S114, P7–9)

Malvernian derived ware (fabric 18)

9 Jar with close mouth (631, S96, P7–9)
10 Jar (1515, S156, P12)
11 Lid (1023, S114, P7–9)

Roman wheelmade Malvernian ware (fabric 19)

12 Jar with everted rim (2601)
13 Cooking pot (699, S112, P7)
14 Cooking pot (699, S112, P7)

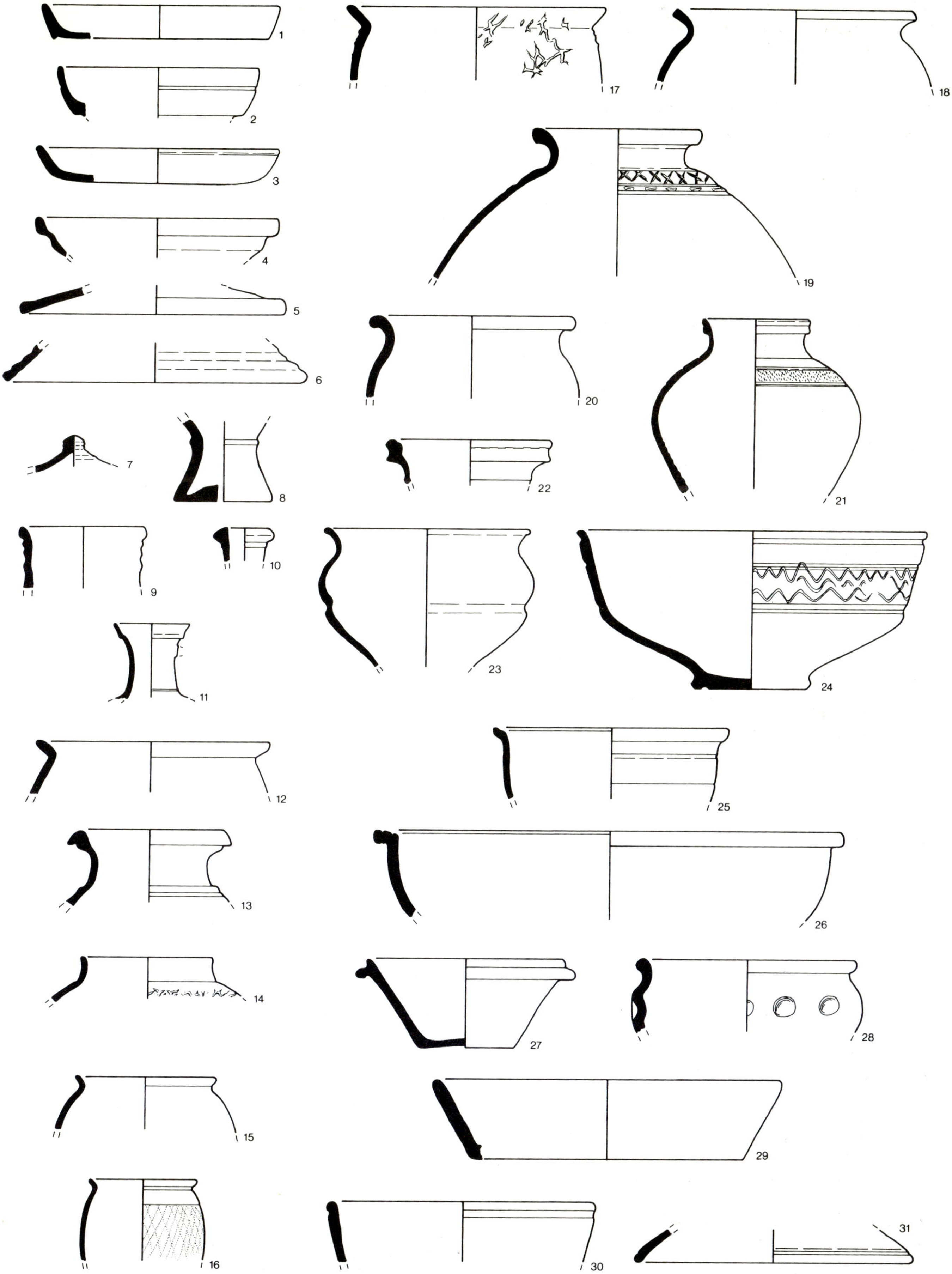

Figure 31   Roman pottery vessels in fabrics 12 (continued, 1–8), 13 (9–13) and 14 (14–31). Scale 1:4

Figure 32 Roman pottery vessels in fabrics 14 (continued, 1), 15 (2–7), 17 (8), 18 (9–11), 19 (12–20), 20 (21), 21.1 (22), 21.2 (23) and 22 (24–8). Scale 1:4

15 Jar decorated with a groove at the neck (1126, S26, P7)
16 Bowl with flanged rim (1190, S31, P8)
17 Bowl with flanged rim (579, S27?, P9?)
18 Dish with plain rim and decorated with a groove halfway between the rim and base (501, S69, P14)
19 Dish or bowl with flat rim (564, S27, P9(11))
20 Dish with flat rim and decorated with lattice burnishing on the inside (1205, S110, P10)

White slipped ware (fabric 20)

21 Flagon (564, S27, P9(11))

Handmade micaceous ware (fabric 21.1)

22 Dish with plain rim and decorated with acute lattice on the inside base (564, S27, P9(11))

Wheelthrown micaceous ware (fabric 21.2)

23 Bowl with flanged rim and decorated with horizontal burnishing (531, S27, P11)

Black-burnished ware, category 1 (fabric 22)

24 Jug (28?, S1, P5?)
25 Beaker (28?, S1, P5?)
26 Cooking pot (2168, S195?, P5–12)
27 Cooking pot (39)
28 Cooking pot (28?, S1, P5?)

Roman pottery (Fig 33)

Black-burnished ware, category 1 (fabric 22)

1 Cooking pot decorated with obtuse lattice and a scored line (1205, S110, P10)
2 Cooking pot decorated with obtuse lattice (574, S27, P9(11))
3 Cooking pot decorated with obtuse lattice and a scored line (1164, S19, P8)
4 Bowl with flat rim and chamfer (28?, S1, P5?)
5 Bowl with flat rim and decorated with acute lattice (751, S94, P7–8)
6 Bowl with flat rim and chamfer and decorated with lattice (722)
7 Bowl with flanged rim and decorated with burnished lines (1536, S155, P11–12)
8 Bowl decorated with burnished lines (559, S27, P9(11))
9 Bowl with flanged rim (564, S27, P9(11))
10 Bowl with flanged rim and decorated with intersecting arcs (559, S27, P9(11))
11 Bowl with bead rim and decorated with inverted chevrons (1192, S113, P5(7))
12 Bowl with bead rim and decorated with acute lattice (1673, S22, P9)
13 Bowl with bead rim and decorated with acute lattice (1076, S120, P8–9)
14 Dish or bowl with flat rim and decorated with acute lattice (2610, S208, P12)
15 Dish with plain rim (560, S27, P9(11))
16 Dish with plain rim and decorated with acute lattice and burnished line on base (2267, S153, P11(12)?)
17 Dish decorated with intersecting arcs on side and loops on underside of the base (559, S27, P9(11))
18 Oval dish decorated with a squiggle on the base (2600, S69, P14)
19 Lid decorated with wavy lines (314, S42, P13)

Shell gritted ware (fabric 23)

20 Jar (523, S27, P11)
21 Jar (559, S27, P9(11))
22 Jar (531, S27, P11)
23 Jar (559, S27, P9(11))
24 Jar (531, S27, P11)
25 Bowl with flanged rim and rilling inside (2601)

Roman pottery (Fig 34)

Indixivixus (fabric 27)

1 Possibly a copy of Samian form Dr 36, with barbotine decoration on top of rim (574, S27, P9(11)

Nene Valley ware (fabric 28)

2 Beaker with high neck (698, S7–9)
3 Castor box decorated with rouletting (1681, S22, P9)
4 Bowl with flanged rim (531, S27, P11)
5 Bowl with flanged rim (531, S27, P11)
a6 Bowl decorated with white paint (501, S69, P14)
7 Bowl with flat rim (531, S27, P11)
8 Dish (564, S27, P9(11))

Oxfordshire red/brown colour coated ware (fabric 29, Young 1977)

9 Jug or flagon (559, S27, P9(11))
10 Beaker with high neck and decorated with traces of rouletting, C22 (577, S31, P8)
11 Bowl with carination, C81 (531, S27, P11)
12 Bowl with flanged rim, C93 (531, S27, P11)
13 Bowl decorated with rouletting, C68 (3, S68, P13)
14 Bowl decorated with semi-rosette stamps, C70 (1025, S125, P12)
15 Bowl,C71 (583, S31, P8)
16 Bowl decorated with rosette stamps, C73 (556, S102, P9)
17 Jar, C75 (1161, S126, P11)
18 Bowl with drop flange, C51 (560, S27, P9(11)
19 Bowl, C44 (1072, S120, P8–9)
20 Bowl, C45 (559, S27, P9(11))
21 Bowl, C47 (1005)
22 Bowl decorated with white paint on flange, C48 (552, S27, P11)
23 *Mortarium*, C97 (564, S27, P9(11))

Brown slipped ware (fabric 31)

24 Flagon (564, S27, P9(11))
25 Beaker (531, S27, P11)
26 Beaker decorated with rouletting (523, S27, P11)

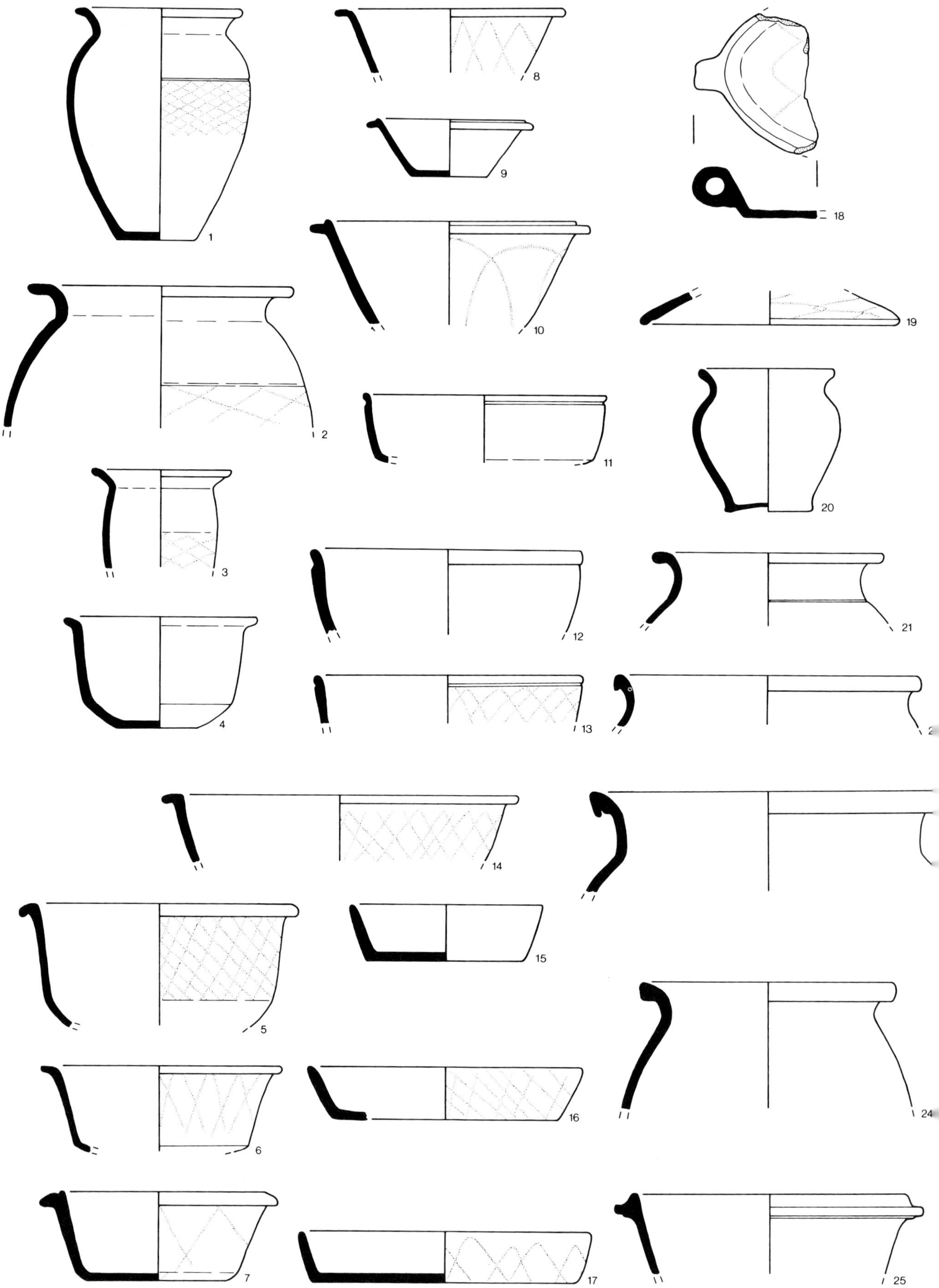

Figure 33 Roman pottery vessels in fabrics 22 (1–19) and 23 (20–5). Scale 1:4

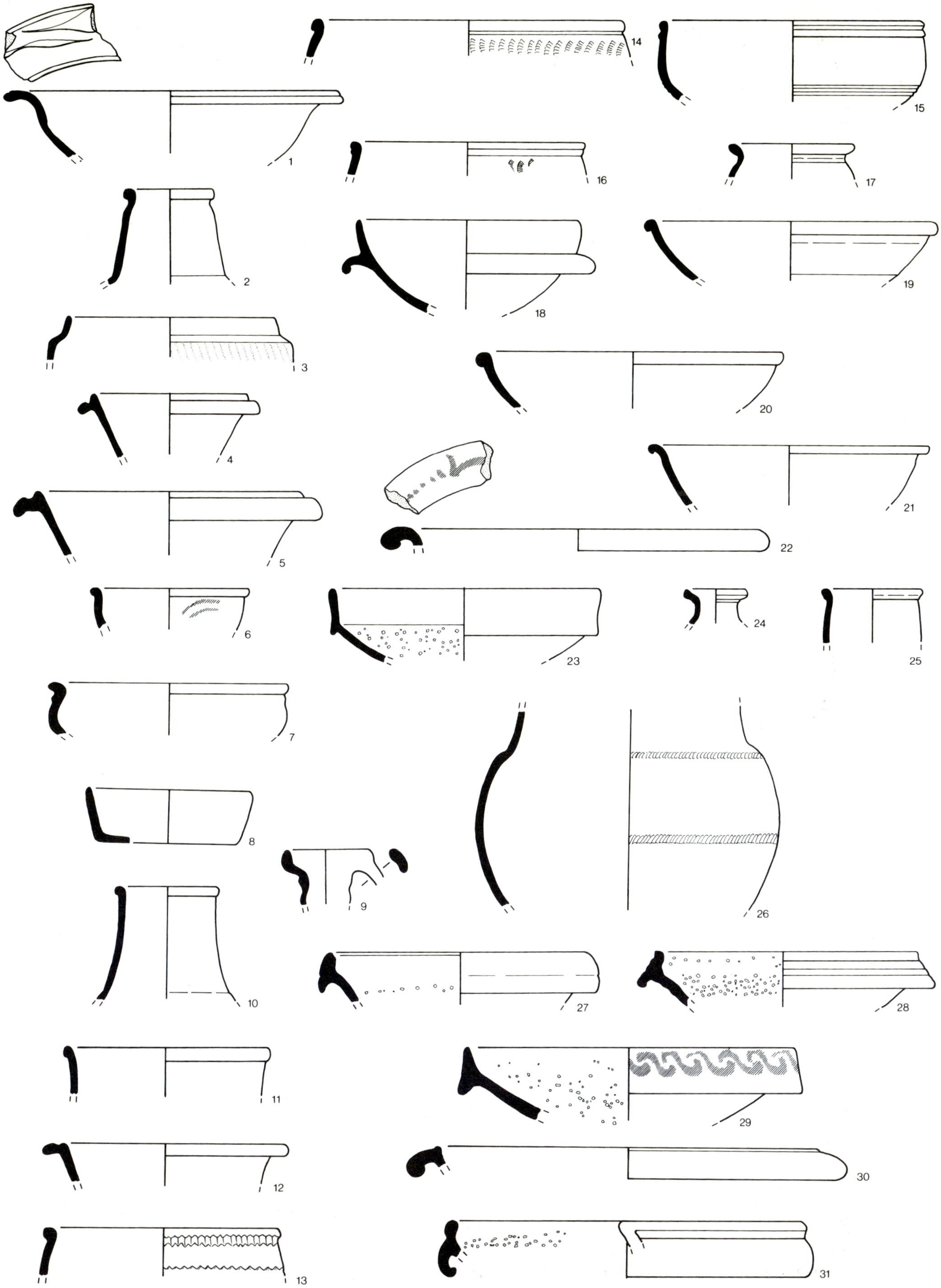

Figure 34 Roman pottery vessels in fabrics 27 (1), 28 (2–8), 29 (9–23), 31 (24–6), 32 (27–9) and 33 (30, 31). Scale 1:4

Mancetter/Hartshill white ware (fabric 32)

27 *Mortarium* with wall sides (2600, S69, P14)

28 *Mortarium* with hammerhead rim and decorated with grooves on the rim and body (559, S27, P9(11))

29 *Mortarium* with hammerhead rim and decorated with painted scrolls (531, S27, P11)

Oxfordshire white *mortaria* (fabric 33, Young 1977)

30 *Mortarium*, M6 (677, S96, P7–9)

31 *Mortarium*, M10 (589, S27, P9(11))

Roman pottery (Fig 35)

Oxfordshire white *mortaria* (fabric 33, Young 1977)

1 *Mortarium*, M11 (523, S27, P11)

2 *Mortarium*, M18 (2600, S69, P14)

3 *Mortarium*, M21 (582, S99, P9–11)

4 *Mortarium*, M22 (564, S27, P9(11))

5 *Mortarium*, M23 (1576, S31, P8)

West Midlands *mortaria* (fabric 34)

6 *Mortarium* with a flange (1190, S31, P8)

Kent/Continental *mortaria* (fabric 36)

7 *Mortarium* with a flange (1167, S109, P5(7))

8 *Mortarium* with a flange (1030, S122, P11(12))

Oxfordshire white ware (fabric 38, Young 1977)

9 Jug, W26 (563, S102, P9)

Oxfordshire parchment ware (fabric 40)

10 Bowl decorated with traces of red paint (559, S27, P9(11))

11 Bowl decorated with red paint (501, S69, P14)

Unprovenanced white ware (fabric 41)

12 Possible bowl decorated grooves (500)

The prehistoric and Roman pottery

The general range is illustrated and tabulated in Figures 26–35, which also serves as a general indication of how common, or otherwise, specific fabrics and forms were. However, it is appropriate in this section to offer a brief discussion of the fabric type series as it applies to this excavation and its implications for the supply of pottery to the site, and to Droitwich in general.

Iron Age fabrics and wares in the Iron Age tradition

It has been established through petrological work that the Malvern region of Hereford and Worcester was the location of specialist pottery production during the Iron Age period (Peacock 1968). Of the three Malvern-region fabrics identified during the course of this study, subsequently expanded to five through further research (Morris 1983), two, fabrics 3 (Peacocks's fabric A) and 4 (Peacock's fabric B1), were present at the Old Bowling Green.

Through the identification of wares containing inclusions available within the immediate vicinity of particular sites, it has also been possible to isolate a more 'local', probably less specialised, component of Iron Age assemblages in the region (eg Morris 1983, 338; Ford and Rees pers comm). The handmade sandy ware (fabric 5) may be an example of this, although the presence of inclusions potentially ascribable to a local source does not necessarily indicate a local origin for the material, as other regional wares may remain unidentified due to their geologically undiagnostic nature.

Previous research has also established the continuation of some handmade Iron Age fabrics into the Roman period, usually in more Romanised forms, as in the case of Durotrigian Black Burnished ware (Williams 1977). This general pattern of continuity in technology accompanied by slight changes in form has been observed amongst the Malvernian wares (Peacock 1965–7), but similar evidence of change within a general framework of continuity for the Palaeozoic limestone-tempered wares at Kenchester (Tomber 1985, 119) and Beckford (Ford and Rees pers comm) is perhaps less well-known. The metamorphic fabric (fabric 3, Peacock's fabric A) is relatively common from the excavation in both Iron Age and conquest period to early Roman forms, whilst the limestone fabric (fabric 4) is present in conquest period forms. The emphasis on fabric 3 may be due to contrasts in distribution between central and northern Worcestershire, and southern Worcestershire and Herefordshire, during the later Iron Age period.

Similarly, the fact that 'Belgic-style' fabrics 7 and 8 are present but poorly represented from the excavation may be a result of limited distribution of these wares in central and northern Worcestershire. However, these fabrics are never common in the Hereford and Worcester area, and their scarcity here may be due to the relatively small sample of material datable to this period.

Regional Roman coarse wares

The mass of the material in this group comprised Severn Valley and related wares (fabrics 12, 13, 16, 17 and 20) with a small quantity of grey wares (fabrics 14, 15 and 18) and late Roman cookwares generally imitating Black Burnished ware category 1 in their forms (fabrics 19 and 21). Since Severn Valley wares are known to have been produced at various locations (Webster 1976), it was hoped that petrological analysis would serve to identify more closely the sources of the material reaching the site. In addition it was observed that certain grey

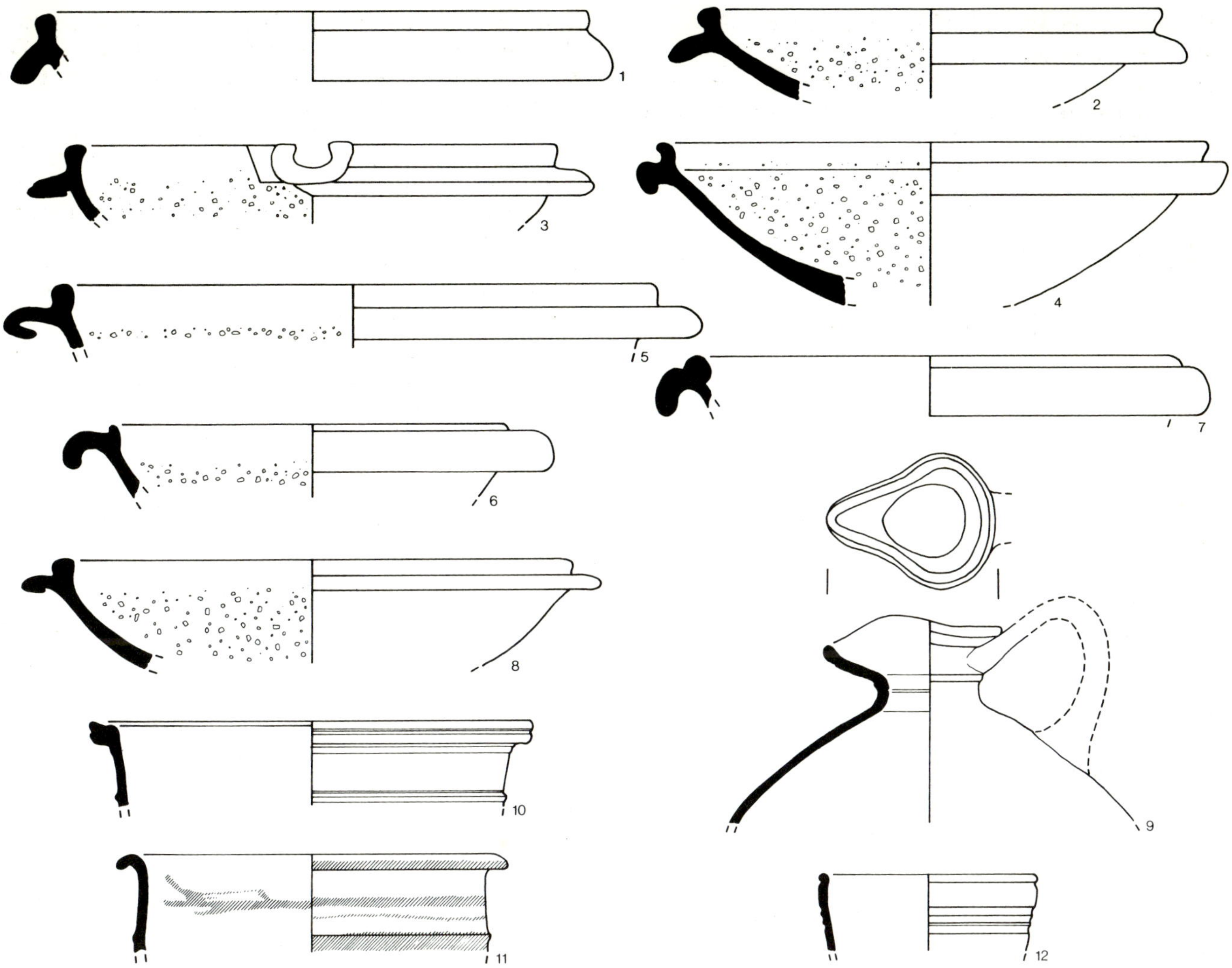

Figure 35 Roman pottery vessels in fabrics 33 (continued, 1–5), 34 (6), 36 (7, 8), 38 (9), 40 (10, 11) and 41 (12). Scale 1:4

fabrics were very similar in fracture, if not always in form, to Severn Valley wares and samples were submitted with the aim of clarifying the relationship between the two.

Initial results were equivocal, showing a possible partial correlation between grey and oxidized wares, but emphasising the smallness of the sample size, the geologically undiagnostic nature of the inclusions and the lack of a general framework of research for the petrological investigation of the Roman coarse wares of the region (1:F2–6).

Subsequent analysis, as part of a regional study of Severn Valley fabrics, went some way to remedying this situation, identifying at least two sources, one of which was probably Malvernian. However, the range of fabrics revealed in this study was far greater at consumer sites than amongst samples from kilns, suggesting that the number of known production centres are at present too few to allow more than the most basic conclusions concerning the marketing of Severn Valley wares to be drawn (1:F13–G2). Indeed, some of the material from the Old Bowling Green itself was sufficiently distorted as to be classified as kiln waste, although no contemporary kiln structure has so far been recognised in the town.

The effect of these constraints on the results presented in this report, together with occasional difficulties in the consistent identification in hand specimens of petrologically-defined fabrics (particularly fabric 18) should be borne in mind.

With hindsight, it may be suggested that the material from the excavation was too diverse in nature to be representative of all periods of Roman pottery in Droitwich. Pottery from excavations with a more limited date, such as Dodderhill (HWCM 603) and Hanbury Street (HWCM 681), which are to be the subject of a further volume in this series, will provide a valuable addition to the fabric type series.

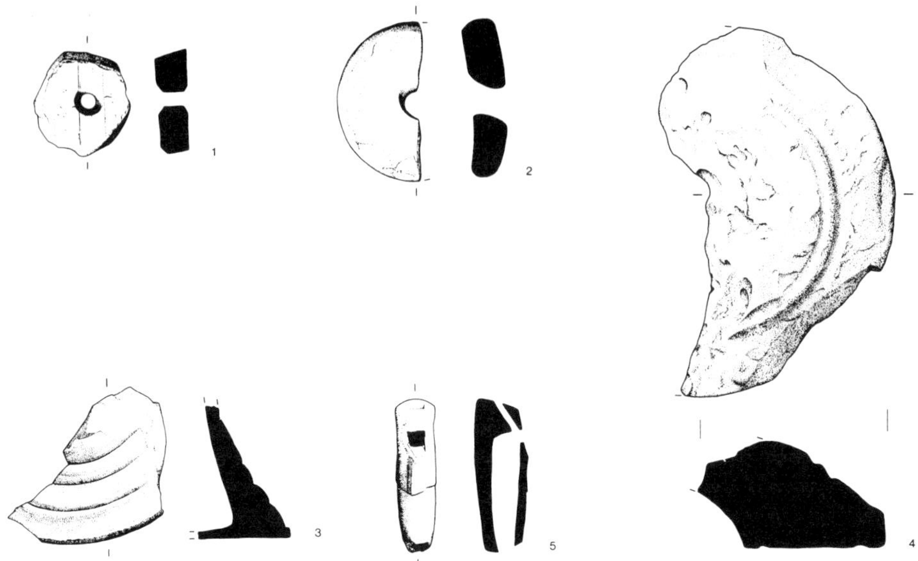

Figure 36 Non-vessel pottery forms: spindle whorls (1, 2), candlesticks (3, 4) and a whistle (5). Scale 1:2

Traded Roman coarse wares

The vast majority of the material in this group consisted of Black Burnished ware category 1 (fabric 22). Late Roman shell tempered wares (fabric 23), which were relatively common, may also be included here as, although it is possible that these were produced and marketed on a fairly localised basis, similarities in form over a wide area suggest a common source (Rob Perrin pers comm). At a late stage in the post-excavation programme, the presence of grey wares from the Alice Holt/Farnham kilns (Lyne and Jeffries 1979) was recognised. They have not been characterized or quantified in detail and have probably, for the most part, been assigned general grey ware classifications (fabric 14 and 15). These wares also form a relatively small proportion of the late Roman assemblage from Beckford (Ford and Rees pers comm). Their presence, so far from source, emphasises the success of this industry during the late Roman period.

Mortaria (kindly identified by Kay Hartley) were relatively rare, especially the earlier Roman types (fabrics 34–37). The presence of Oxfordshire white *mortaria* (fabric 33) during the later Roman period is consistent with the general pattern of supply to the site and the area from this source (Young 1977, and see below). The predominance of Mancetter/ Hartshill products in the West Midlands is reflected in the presence of these wares in increasing quantities from the mid second century onwards.

Roman fine wares

Only the most generally common of the pre-Flavian fine wares, *Terra Nigra* (fabric 25) and Lyons ware (fabric 26) were present, and these in very small quantities. The later 'Rhenish' wares of both Central and Eastern Gaulish origin were slightly more numerous, and the presence of a single vessel from the workshop of the Nene Valley potter 'Indixivixus' (fabric 27) is noteworthy. Only in the later Roman period were fine wares other than the earlier samian at all common, being largely supplied by the Oxfordshire potteries (fabrics 29, 30, 38, 39 and 40), with smaller quantities of Nene Valley (fabric 28) and brown-slipped wares (fabric 31).

Later pottery and other ceramic types

Post-Roman pottery

Almost all of the fabrics contained in the type series (Chapter 35) were present at the Old Bowling Green. The chronological emphasis was, however, on the later medieval to early post-medieval periods, and on vessels post-dating *c* 1800.

Non-vessel forms in pottery fabrics

Five ceramic artefacts not of vessel form were recovered. Two of these, the spindle whorls, had been reworked from vessels, whilst the remainder

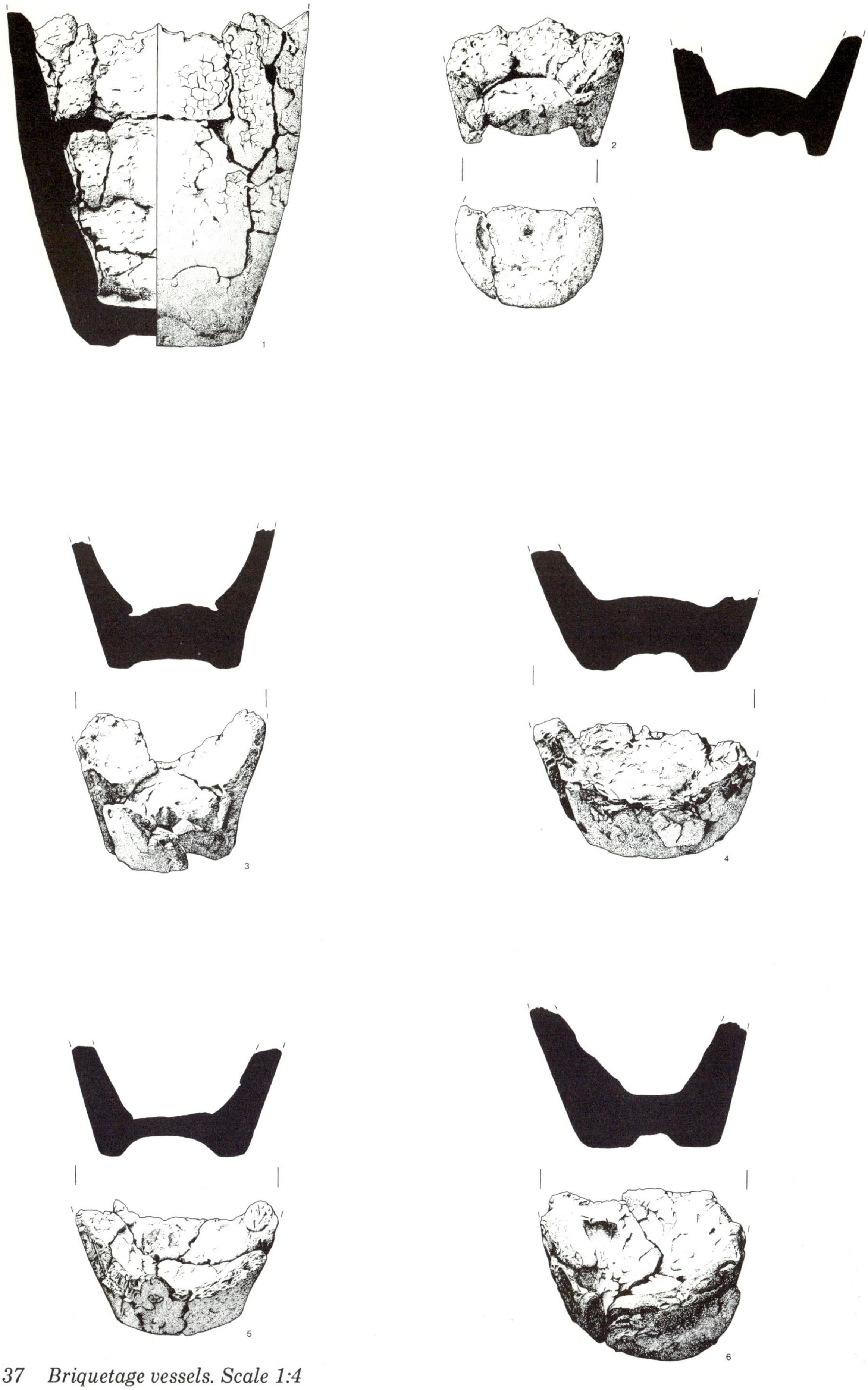

Figure 37 Briquetage vessels. Scale 1:4

had been especially manufactured. All occurred in fabrics represented amongst the vessel assemblage.

The spindle whorls were reworked from sherds in fabrics 12 (Severn Valley ware, Fig 36 no 1) and 43 (samian, Fig 36 no 2), and were derived from contexts of phases 12 (2095, S192) and 10–11 (1230, S111) respectively.

Two wheel- or lathe-turned objects were probably candlesticks. One (Fig 36 no 3) is possibly of Severn Valley ware (fabric 12) but is rather finer. It was retrieved from a layer (1023, S114) assigned to phases 7–9. The other (Fig 36 no 4) was again possibly of Severn Valley ware or Oxfordshire red slipped tableware (fabric 29). If it is the latter then the form is not represented by Young (1977). It was residual, being retrieved from the upper fill (531) of a ditch (S27) of phase 11.

The mouthpiece of a whistle (Fig 36 no 5) was retrieved from a ditch fill (2609, S208, P13). The fabric is Stamford ware (fabric 46) and is glazed to just over half way along its surviving length. However, the remainder may also have been glazed originally as the glaze has the appearance of having been flaked away. The complete instrument would probably have been similar to the modern 'flageolet', but is fractured at one end and no fingerholes survive.

Briquetage

A discussion of the briquetage from this excavation has already been published (Rees 1986) and a more comprehensive survey of its character and distribution has been undertaken by Morris (1985). Further comment is offered on two points only.

There appears to be a slightly greater diversity in both the fabric and form of the collections examined by Morris (1985, 338–44) than amongst the material from the Old Bowling Green (Fig 37). It is possible that these variations reflect the greater chronological span of the former material. Secondly, the pottery dating evidence from the excavation initially suggested that the use of briquetage on the site was wholly of Roman date. This was subsequently refuted by dendrochronological and radiocarbon analysis, and it now seems likely that the main use of briquetage in the process of salt manufacture is more in keeping with the Iron Age to early Roman dating of the material found on the consumer sites (Morris 1985). For further discussion of briquetage and its relationship to the structures on the site, see chapter 2.

Objects in 'briquetage' fabrics

These were too few in number to enable more than very tentative suggestions as to their function to be made. Since the large perforated triangular objects (Fig 38 nos 1–2) were recovered from the fill of one of the clay- and stake-lined pits (1689, S8, P4–5(7)), they may have served to weigh down the roof of a superstructure over this feature, although none of the wood debris from the fill of this feature was unequivocally structural in nature. Alternatively, and more likely, they may have functioned as loomweights. The shallow vessel (Fig 38 no 3) may represent the one surviving example of a pan in which brine was boiled (1752, S8, P4). In neither case is there strict necessity to postulate a direct link with the salt industry, as the local clays may have been used in the manufacture of all kinds of domestic objects, although the likelihood of a connection remains. Various other pieces of non-vessel form were identified in briquetage fabrics, but these were too fragmentary to be reconstructed.

Discussion of the stratified material

Phases 1 to 3: Iron Age (Fig 39)

Clear differences between the range of fabrics and forms present in phase 1 and those of phase 2 were observable. This was probably partially related to differences in sample size, as phase 1 contexts contained only 42 sherds, whilst phase 2 contexts produced 3,219. The factor of sample size may also have been responsible for the apparent lack of ceramic development between phases 2 and 3, since only 799 sherds were recovered from phase 3 contexts. However, this could be a genuine indication that these two latter phases were relatively close to one another in date.

Domestic pottery, in any case, was relatively rare throughout, briquetage accounting for the vast proportion of the assemblage. In order to compare the relative proportion of briquetage in each phase, it has been included with the other pottery in the presentation of quantification by phase. With these constraints in mind, it may nevertheless be of significance that, apart from a sherd of probably residual earlier prehistoric pottery, and apart from briquetage from Droitwich itself (fabrics 1 and 2), only two sources of domestic pottery, one Malvernian (fabric 3) and one unknown (fabric 5.1) were represented amongst material of phase 1. Contexts of phase 2, by contrast, contained four fabrics, with Palaeozoic limestone (fabric 4.1) and 'Belgic'-style pottery (fabric 8) making their stratigraphically-earliest appearance, in addition to the two earlier fabrics. This development was matched in the forms. The range in phase 1 was confined to plain saucepan pots (Fig 26 no 2) in fabric 3 and a single everted jar in fabric 5 (Fig 26 no 23), and the range in phase 2 comprised both plain and linear tooled saucepan pots (Fig 26 nos 1 and 15) and pattern burnished tubby cooking pots (Fig 26 nos 3–4 and 7–10), a single everted jar in fabric 4.1 (Fig 26 nos 20–21) and a wheelturned vessel, probably a carinated bowl in fabric 8 (Fig 26 no 26). This extension in the range of forms and sources may have reflected a general trend towards greater communication between regions and

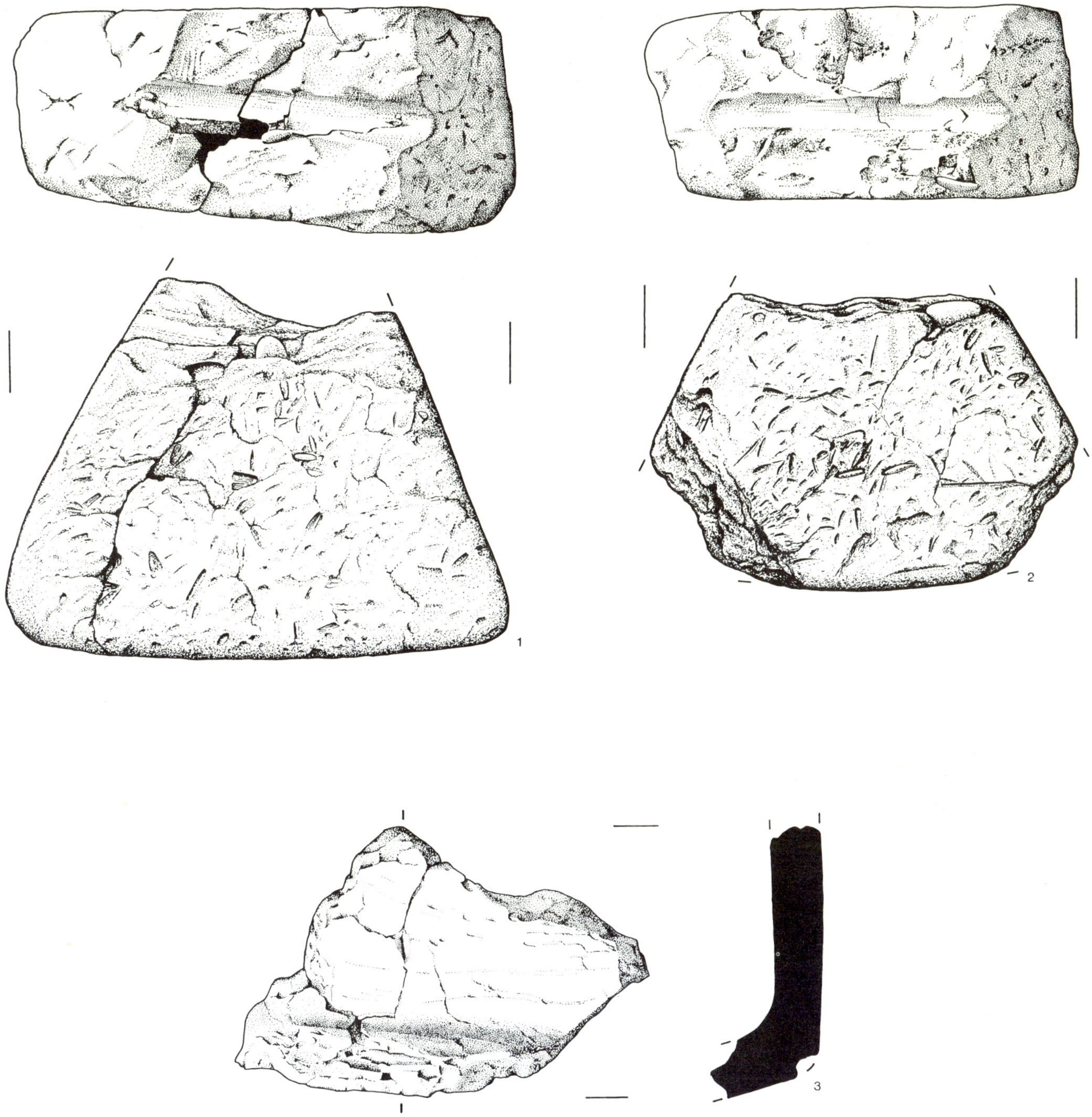

Figure 38 Objects in 'briquetage' fabrics: thatch or loomweights (1, 2) and shallow vessel (3). Scale 1:2

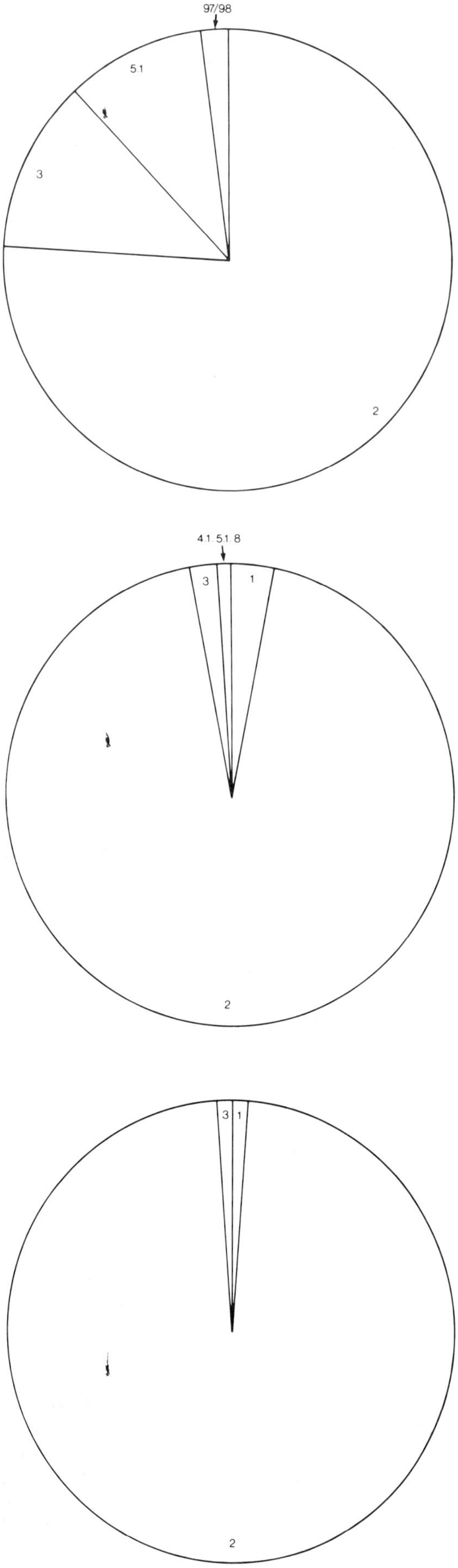

Figure 39 Proportion (by number of sherds) of fabrics present in phases 1–3

increasing functional specialization in pottery forms at the end of the Iron Age. It may also be related to the importance of Droitwich as a centre for the manufacture and export of salt during the later Iron Age period (see Chapter 34).

Phases 4 to 5: Mid 1st to mid 3rd century (Fig 40)

Since assemblages from contexts of phases 4 and 5 were large (3,838 and 5,643 sherds respectively), they not only represented a reliable sample size in themselves but they can also be compared with reasonable confidence.

The pottery of phase 4 consisted of fabrics in the Iron Age tradition (fabrics 3, 4 and 7) together with regional Roman types, predominantly Severn Valley and related wares (fabrics 12, 13, 14 and 15). Continental sources were represented by Rhodian-style *amphorae* and samian ware from Southern Gaul and Les Martres-de-Veyre in Central Gaul.

Since the range of forms in fabric 3 increased in phase 4 to include narrow necked jars and that a necked bowl and a lug, in addition to the everted jars of phase 3, occurred in fabric 4, the evidence from the site corroborates to some extent other indications that these wares continued to be manufactured after the Roman conquest.

The range of Severn Valley ware forms was fairly limited, consisting largely of carinated bowls (Fig 29 nos 21–27), with a smaller number of storage and other jars (Fig 27 nos 7, 16–17, and 19–21) and single examples in flagon (Fig 27 nos 3–4), beaker (Fig 27 no 12), necked bowl (Fig 29 no 10) and bead rimmed bowl (Fig 30 no 21) forms.

In phase 5, the range of coarse wares increased slightly, most significantly to include Black Burnished ware category 1 of Dorset origin (fabric 22), and there was a slightly higher proportion of samian ware (fabric 43), mostly from central Gaul, which was expected for this date range. *Amphorae* (fabric 42) comprised Dressel 20 and *Camulodunum* 186A forms and *mortaria* in fabric 36 (Kent/Continental) were also present for the first time. In addition, it may be significant that briquetage was almost completely absent from the largest assemblage of phase 5 (that from the fills of S1), suggesting that its use had ceased by this time.

Severn Valley ware forms making their stratigraphically-earliest appearance in phase 5 included wide-mouthed jars, some with long upright necks (Fig 27 no 18), others with long out-turned necks and hooked or triangular rims (Fig 28 nos 12–15 and 29 nos 1–3) and tankards with out-turned sides (Fig 29 nos 12–18). Everted jars with short necks (see above) were also much more common in phase 5 than in phase 4.

The assemblage from structure 1 appeared, from the samian, to be of Hadrianic to early Antonine date. Its large and well-preserved coarse ware assemblage included many of the above Severn

Figure 40 Proportion (by number of sherds) of fabrics present in phases 4 and 5

Valley ware forms but lacked the hooked or triangular rimmed jars, which were more regularly associated in phase 5 with Antonine samian ware. This suggested that this form was of later date within the range of phase 5. Structure 1 also produced a jug in fabric 22, a form and fabric combination unique for the excavation (Fig 32 no 24), and a small quantity of possible Severn Valley ware waster sherds. Identification of the forms of these sherds was difficult due to extreme distortion, but jars and a single tankard with out-turned sides and lattice decoration were represented.

Phases 7 to 9: Mid 3rd to late 4th century (Fig 41)

As the sample size in phase 7 was rather small (*c* 185 sherds), no ceramic significance could be attached to the absence therein of types represented in phase 8 (1,078 sherds) and phase 9 (2,122 sherds). There was a slight difference in absolute dating between phases 7 and 8, each with a *terminus post quem* of *c* AD 300 from single sherds in fabric 29 (Oxfordshire red-slipped tableware (Young 1977, C100 and C71), and phase 9 for which a single fabric 29 sherd supplied a *terminus post quem* of *c* AD 325. A coin of AD 320 confirmed a late date for this latter phase (2:D1). However, there was no other indication of ceramic development between phases 7/8 and phase 9, and the phases were therefore considered together.

For the purpose of the discussion, the material is divided into five functional groups:

Storage and everyday wares

Severn Valley wares (fabrics 12 and 13) were predominant in this group, occurring in a wide variety of late Roman forms, most notably squat-necked jars (eg Fig 29 nos 4–9) and bowls with grooved rims (Fig 30 nos 1–14). Amongst the range of narrow- necked jars, the double-lipped jar (Fig 28 nos 9–10) was the most distinctive of this period. A small quantity of grey wares (fabrics 14, 15 and 18) occurred in somewhat similar forms to Severn Valley ware. At a late stage in the analysis, the presence of grey wares from the Alice Holt/Farnham kilns was recognised, and it is likely that at least a few of the grey wares occurring in contexts assigned to phases 7 to 9 were of this origin. It may be of significance that fabrics 14 and 15, the groups most likely to have contained unrecognised Alice Holt/Farnham wares, did not occur in cookware forms.

A fabric related to Severn Valley ware containing mudstone (fabric 17) was reserved for distinctive large storage jars with heavy rims (Fig 32 no 8).

Cookwares

The two major cookware fabrics in currency during this period were Black Burnished ware category 1 (fabric 22) and shell tempered wares, perhaps of eastern Midland origin (fabrics 23 and 24, Fig 33 nos 20–25). Another common fabric in this group, which usually occurred in forms imitating Black Burnished ware (eg Fig 32 nos 15–16 and 19–20), was wheelthrown Malvernian metamorphic (fabric

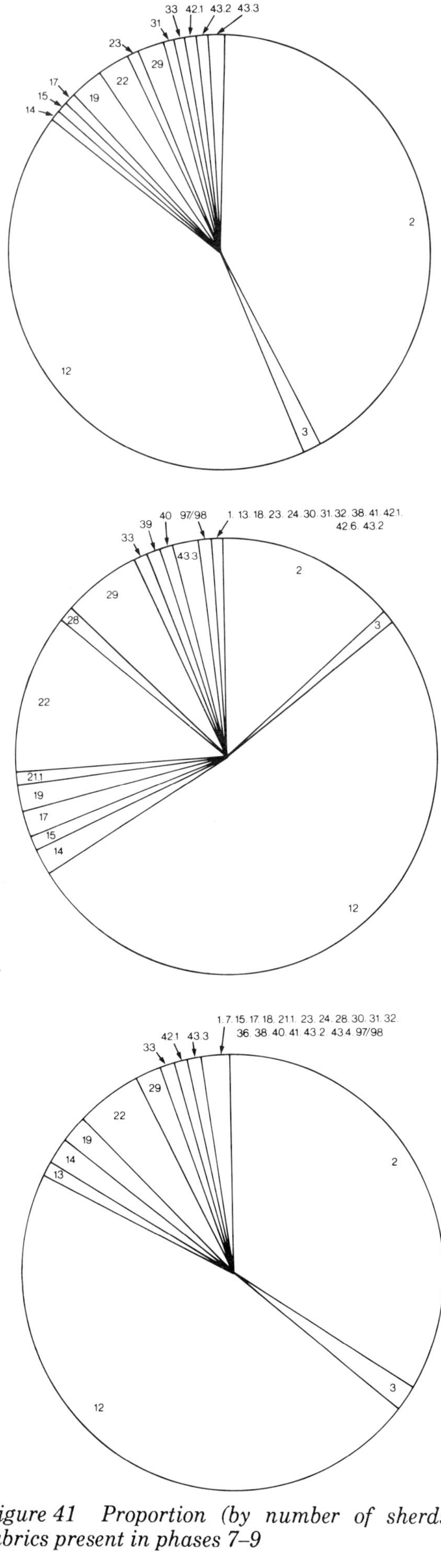

Figure 41 Proportion (by number of sherds) of fabrics present in phases 7–9

19). A sandy micaceous fabric (fabric 21.2), which was present in flanged bowl form (eg Fig 32 no 23), formed a small proportion of the assemblage.

Finewares

Oxfordshire red slipped tableware (fabric 29) was the most common fabric in this group but Nene Valley ware (fabric 28) and south-western brown slipped ware (fabric 31) were present in smaller quantities. White wares were rare, and where the source was identifiable, were also of Oxfordshire origin. One Oxfordshire parchment bowl (fabric 40) occurred in a variation not represented in Young's (1977) type series (Fig 35 no 11).

Mortaria

Mortaria from Oxfordshire (fabric 32) and Mancetter/Hartshill (fabric 33) were present in relatively small quantities and in later Roman forms in phases 7 to 9 (eg Fig 34 nos 27–9 and Fig 35 no 4).

Amphorae

Dressel 20 (1:G8) was the most common form amongst the small quantity of *amphorae* in this period.

Phases 10 to 13: 5th to 20th century (Fig 42)

The post-Roman pottery development of Droitwich was much better represented at Friar Street (Chapter 22). It may be of significance that, despite the presence on the excavation of a small quantity of pottery of the late Saxon and early medieval periods, few assemblages could be assigned with certainty to this date-range, suggesting that the site was not intensively occupied at this time.

Conclusions

With the exception of the Anglo-Saxon period, the material from the excavation presented a continuous ceramic sequence from the prehistoric period to the present day. The absence of early and middle Saxon material is unremarkable for central Worcestershire, and does not necessarily imply that the site was unoccupied at this time. Unusually, however, sherds of this date were recovered from the nearby excavations at Upwich (HWCM 4575, Derek Hurst pers comm) and Friar Street (Chapter 22).

Within this broad span, the quantitative distribution of pottery by period suggests that occupation of the site was more intensive during the Iron Age to early Roman, the late Roman and the late medieval to post-medieval periods than during mid-Roman and early medieval times. That the early medieval period was poorly represented is in contrast to the nearby Friar Street excavation

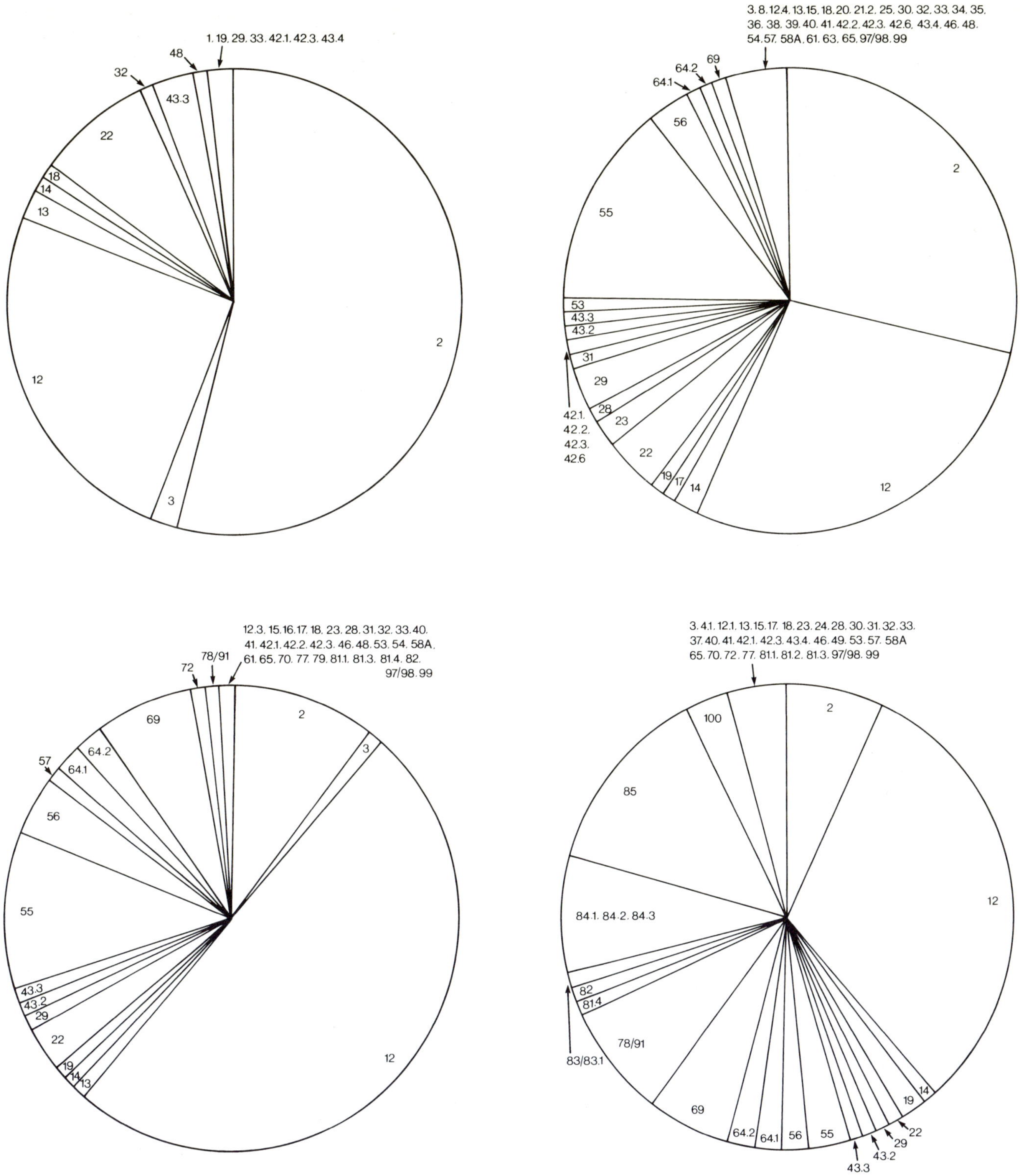

Figure 42 Proportion (by number of sherds) of fabrics present in phases 10–13

(Chapter 22). The increasing paucity of material throughout the 2nd century and the absence of pottery demonstrably of the earlier to middle years of the 3rd century, would seem to be a more general phenomenon (see eg Fulford 1975, 105–11; Ford and Rees pers comm).

To turn to specific aspects of the pottery, two points, both concerning the Iron Age period, may be of wider significance. Firstly, whilst it has been suggested that wheelturned 'Belgic' pottery (fabrics 7 and 8) is exclusively of Roman date in this area (Swan 1975), its presence in phase 2 implies that such material was in currency, albeit in very small quantities, before the Roman conquest. The pottery assemblage of phase 2 also suggests that the Malvernian tubby cooking pot form, although most likely to be Roman in date (Peacock 1965–7), originated in the late pre-Roman Iron Age. This is consistent with the evidence from Beckford (Ford and Rees pers comm), where such material is present in contexts of probable Iron Age date but is much more common in the early Roman period. Secondly, as has already been suggested, briquetage ceased to be used for the production and export of salt during the first two centuries of the Roman period (Rees 1986, 51–3).

In general, the pottery evidence from the Old Bowling Green is consistent with a pattern of regional pottery supply in the later Iron Age and early Roman periods, increasingly supplemented by material from outside the region throughout the Roman period.

In future studies, it will be of interest to compare Roman material from the industrial phases of the Old Bowling Green and Friar Street with pottery from the fort at Dodderhill, the villa at Bays Meadow, the civil settlement at Hanbury Street and the later Roman phases of Friar Street, in order to assess the impact of site function on patterns of pottery supply in Roman Droitwich.

Samian, by Brenda Dickinson

Over eight hundred sherds of samian (fabric 43) were recovered from the excavation, of which approximately 26% were South Gaulish ware, 68% Central Gaulish and 6% East Gaulish.

With only two or three exceptions, all the 1st century material comes from La Graufesenque. Another possible South Gaulish sherd seems to be in Montans fabric. Samian from this factory, while not uncommon in Britain in the 2nd century, is distinctly rare in the 1st century, except perhaps at London. First century Central Gaulish ware, represented here by two decorated bowls from Lezoux, does not occur in great quantities at any site in Britain, but its distribution pattern has widened considerably in the past few years and examples have been noted as far north as Camelon, in Scotland. In the Midlands it is known from sites such as Alcester, Little Chester and Tiddington, and there is another bowl from Friar Street, Droitwich (Chapter 22).

None of the earliest decorated ware from the site suggests that samian was in use here before the 60s. This impression is reinforced by the scarcity of the commoner pre-Flavian plain forms, such as 24 and Ritterling 8, while the contemporary forms Ritterling 9 and 12 are absent altogether.

The quantities of samian discarded on the site increased considerably in the Flavian period but, again on the evidence of the decorated ware, declined sharply towards the end of the 1st century.

Such small quantities of Trajanic South Gaulish ware as there may have been were supplemented by contemporary material from the Central Gaulish factory of Les Martres-de-Veyre, whose products account for approximately 8% of the samian as a whole and 10% of the 2nd century and later material. This means that considerably less samian was being discarded in the early 2nd century than in the Flavian period and this amount does not increase appreciably before *c* AD 150.

The Antonine Central Gaulish ware, of which the bulk comes from Lezoux, amounts to almost twice the quantity of 1st century South Gaulish material. Much of it is not closely datable, but it includes forms not made before *c* AD 160, such as the dish forms 31R and 79, and several of the later 2nd century potters are represented by decorated bowls or plainware stamps. They include Advocisus, Casurius ii, Do(v)eccus i, Iullinus ii and Paternus v and his associates. There are at least eight decorated bowls in a style of Cinnamus ii current in the period *c* AD 150–180.

All the East Gaulish samian, with the exception of one possible sherd of La Madeleine ware, is late 2nd or early 3rd century. In the absence of any sherds with surviving decoration, attribution has been made on the evidence of fabrics and glazes. These suggest that most of the material comes from Rheinzabern, but there is one sherd of Trier ware. Approximately 9% of the 2nd century and later samian is East Gaulish, but scarcely any will be earlier than *c* AD 160.

The fluctuations in the quantities of samian discarded on this site over an unbroken period of more than a century and a half do not differ greatly from the standard pattern for Britain, though it may be noted that there is rather less 1st century material than might be expected in a collection of this size.

The following is an extract of the full catalogue (1:G13–2:C12).

Illustrated sherds (Fig 43)

1 Form 37, burnt. The fabric is very micaceous, with a dull orange glaze, both typical of 1st century Lezoux ware. The motif is very like Rogers G9. The piece is not closely datable, but is likely to belong to the Flavian or Flavian–Trajanic period, since Lezoux samian was not exported regularly to Britain

Figure 43 Decorated samian (fabric 43). Scale 1:2

before then. Context 2267, structure 153, phase 11(12)?.

2 Form 29, South Gaulish. The upper and lower zones are completely filled with decoration, in the manner of the Bassus i-Coelus firm, though with a different range of motifs. Double wreaths in the upper zone are not common and, when they occur, are usually identical. Gallicanus used them on some of his bowls. The motif in the basal wreath occurs mainly in the Claudio-Neronian period, and the use of wavy lines through it, and dividing the two lower wreaths also suggests this date. No parallels have been found for the exclusive use of wreaths on form 29 except on bowls stamped by Gallicanus, but an unprovenanced bowl at Mainz, of which only the lower zone survives, has two wreaths separated by beads. Its basal wreath, very similar to the one on the Droitwich bowl, is divided in the same way, but by beads. *C* AD 50–65. Context 564, structure 27, phase 9(11).

3 Form 29, in 1st century Lezoux ware, with orange fabric and a dull, orange glaze. The moulding is exceptionally good, though the finishing of the rim is not to the same standard. The upper zone occurs on bowls stamped by, or in the style of Atepomarus ii (*Bull Hist et Scient de l'Auvergne*, **LXII** (1942), 200, 15). The motifs in the lower zone appear to be made up of repeated impressions of similar large and small leaves. Neronian. Context 564, structure 27, phase 9(11).

4 Form 30, in the fabric of Les Martres-de-Veyre. A freestyle scene, with ovolo (Rogers B14) and figures used by mould-makers for Donnaucus. The warriors to left (0.210) and right (not in D or O) are on a bowl from London (Walters 1908, M1441). The kilted figure (D. 125, with a detached sword instead of a spear) and amazon (0.251) are on a bowl from Les Martres (Terrisse 1968, pl xxxi, 606) *c* AD 110–120. Context 28C, structure 1, phase 5.

5 Form 37, Central Gaulish, mended in several places with lead rivets. The two repeated panels are: 1) a horseman (0.251). 2) a boar (0.1668) over a lion (0.1497I). On the decoration, this bowl would have almost certainly have been attributed to Drusus ii of Lezoux. However, since it is in Les Martres-de-Veyre fabric, it was either made there in a Lezoux mould or, more probably, Drusus began his career at Les Martres, as his connections with Sacer i and his associates would have suggested. It is in his familiar Lezoux early style, with alternating wide and narrow panels without subdivisions (cf Stanfield and Simpson 1958, pl 89, 12, from Heronbridge). His small bowls all seem to be in this style. The type of footring is more common at Les Martres than Lezoux, and also suggests that this is one of his earlier products. A range *c* AD 120–125 is likely for his activity at Les Martres-de-Veyre. Context 731, structure 74, phase 7–9.

6 Form 37, Central Gaulish, with a freestyle scene. The Mercury (D. 288 variant), siren (D. 499), leaf (Rogers J146) and nine-petalled rosette all occur on a stamped bowl of Maccius ii from High Cross (unpublished). The ovolo (Rogers B109) is probably the one on a stamped mould of his from Lezoux (Coll Oswald-Plicque), which was also used by Butrio. The other figure-types are a Diana in a chariot and a small, naked figure (neither in D or O), another Diana (D. 69, also recorded for Maccius) and a bird (D. 1041). Very few of Maccius's decorated bowls are known, and it is useful to have one which adds some new figure-types. He clearly has connections with Butrio, but one of his decorated bowls and a plainware cup occur in a pit-group of *c* AD 150–160 at Alcester (forthcoming). His plain forms include 18/31, 18/31R and 27. *C* AD 125–150. Context 1023, structure 114, phase 7–9.

7 Form 37, Central Gaulish. The unusual ovolo, with thick tongue, occurs at Castleford in a layer dated after AD 150. The trilobed motif (of the type Rogers G152–185) is not closely identifiable. Antonine. Context 1553, structure 155?, phase 12.

Stamped sherds (not illustrated)

1 Form 31, Central Gaulish, stamped]I or I[. Antonine. Context 24, structure 66, phase 7–9.

2 Form 18/31R, stamped TA[SGILLVSF] by Tasgillus ii (die 2a). This potter worked at both Les Martres-de-Veyre and Lezoux, but all the examples noted of this stamp are in Lea Martres fabric, and there are several examples from the kilns there. One is under the base of a form 29. The stamp has also been noted at Malton. *C* AD 110–125. Context 28A, structure 1, phase 5.

3 Form 37, Central Gaulish, stamped A[DVOCISI] by Advocisus of Lezoux where the die (8a) is known. There are many bowls with this stamp on Hadrians Wall and at its hinterland forts, but only one has been noted in Scotland (at Kelso). The trifid motif occurs on a stamped bowl from the Wroxeter forum destruction (Atkinson 1942, H3). *C* AD 160–190. Context 541, structure 47, phase 7–9(11).

4 Form 31, Central Gaulish, stamped]ILVI ·M or]IIVI ·M. Antonine. Context 552, structure 27, phase 11.

5 Form 33, stamped [AC]VRIO ·F by Acurio of Lezoux, where the die (5a) is known. There is no site dating for Acurio, but his repertoire includes forms 18/31R, 27 and 38. This is probably one of his later stamps, since he used it on forms 31 and 80. *C* AD 150–170. Context 559, structure 27, phase 9(11).

6 Form 31, Central Gaulish, stamped]I· or ·I[. Hadrianic–Antonine. Context 563, structure 27, phase 9(11).

7 Form 18R, South Gaulish, with very faint rouletting. The profile of the wall is nearer to form 18. There is still kiln-grit on the inside of the base and under the footring, so the dish may have been broken before it was ever used. The sherds have been fitted together with a large number of lead rivets, presumably because it was worthwhile to mend a new vessel. Unfortunately, although most of the dish has survived, the complete stamp has not, and only C[or] remains. Neronian or early Flavian. Context 564, structure 27, phase 9(11).

8 Form 31, stamped MATERNI by Maternus iv of Lezoux (die 11). This particular stamp has not been recorded before, but his work is known at the hinterland forts of Hadrians Wall and at Pudding Pan Rock. His repertoire includes forms 31R, 79, 80 and, occasionally, 27. *C* AD 160–190. Context 630, structure 27, phase 9(11).

9 Form 33 or 80 etc, stamped [FΛ]BIΛNIMΛ retrograde, by Fabianus ii of Lezoux, where the die (2a) is known. Fabianus probably only used this one die, though others have been attributed to him. It occurs mainly on form 33, but single examples of forms 31 and 31R have been noted. He worked in the Antonine period, but no closer dating is possible. Context 714, structure 96, phase 7–9.

10 Form 32 etc, East Gaulish, perhaps stamped LI[. Late 2nd or early 3rd century. Context 1098, structure 31, phase 8.

11 Form 29 base, South Gaulish, stamped O[or]O. Pre-Flavian. Context 1194, structure 106, phase 4–5.

12 Dish, stamped [DOVIICC]VS by Do(v)eccus i of Lezoux (die 11e). This stamp has been recorded at sites in northern Britain reoccupied *c* AD 160. The forms include 31R. *C* AD 165–200. Context 1240, structure 106, phase 5.

13 Form 27, stamped S DAT.M by Sedatus IV of Lezoux, where the die (2c) is known. This stamp was used on forms 18/31, 27 and 33. His work appears occasionally in the Rhineland, suggesting activity before *c* AD 150 and one of his signed jar moulds also has a signature of Paullus iv. *C* AD 125–150. Context 1919, structure 29, phase 11.

14 Form 18/31, stamped [GONO]I ·M with die 2a of Gongius of Central Gaul, presumably Lezoux. Forms 18/31, 18/31R and 27 are among his commonest forms, but the presence of this stamp at Camelon and Old Kilpatrick shows that he worked in the Antonine period. One of his stamps occurs in a pit-group of the 150s at Alcester. *C* AD 140–170. Context 2212, structure 157, phase 11.

15 Form 31, Central Gaulish, stamped C[. Mid to late Antonine. Context 2600, structure 69, phase 14.

16 Form 29 base, stamped OFMODES[TI] by Modestus i of La Graufesenque (die 2g). One of his commonest stamps, though not recorded in dated contexts, apart from the Cirencester fort ditch (*c* AD 45–65). Context 2609, structure 208, phase 13.

4 Ceramic building material *Simon Woodiwiss*

An assemblage of 2,141 fragments, weighing 246.21kg, was recovered from the excavation. The fabric reference series used for identification was as defined by Hurst (Chapter 23) for Friar Street. This discussion covers only tile, as brick formed too small a fraction of the assemblage to enable valid conclusions to be drawn. An element of doubt concerns the tile assemblage as this too was of small, though not insignificant, size.

Chronological development

Roman tile was recovered in small quantities from both phases 4 and 5 (7 and 58 fragments respectively), and of these one layer (711) in phase 5 contained 34 fragments

Most of the tile from phase 7 (28 fragments, weighing 3,166g) was recovered from two layer groups (S112 and S210). Another eight fragments came from the foundations of a building (S26). The average weight of the fragments (136g) was the highest of phases 5 to 9 and perhaps indicated that the tile was close to its place of use, possibly tile broken during the construction of a building (ie S26).

Phase 8 contained the most tile of the Roman period, recovered from the disuse elements of structures which did not themselves appear to have been roofed. The average fragment weight (76g) was small, indicating that the assemblage was probably largely residual. This tile may have derived from the demolition of the building (S26) which was constructed in phase 7.

The tile of phase 9 also had a low average weight (57g), indicating a largely residual assemblage. Very few of the fragments were from contexts post-dating the construction of the post-built building (S22) suggesting that tile from this phase may also have been derived from the building constructed in phase 7 (S26).

A single example of a fabric with limestone inclusions, indicating a Cotswold source, came from phase 10.

A sizeable proportion (37% by weight) of the phase 11 assemblage was made up of the sandy fabric (fabric 2a), though residual material was still dominant (see Chapter 23 for discussion of the early date of this fabric). Most of the fragments were unglazed. The small average weight (71g) again indicated a largely residual component. The

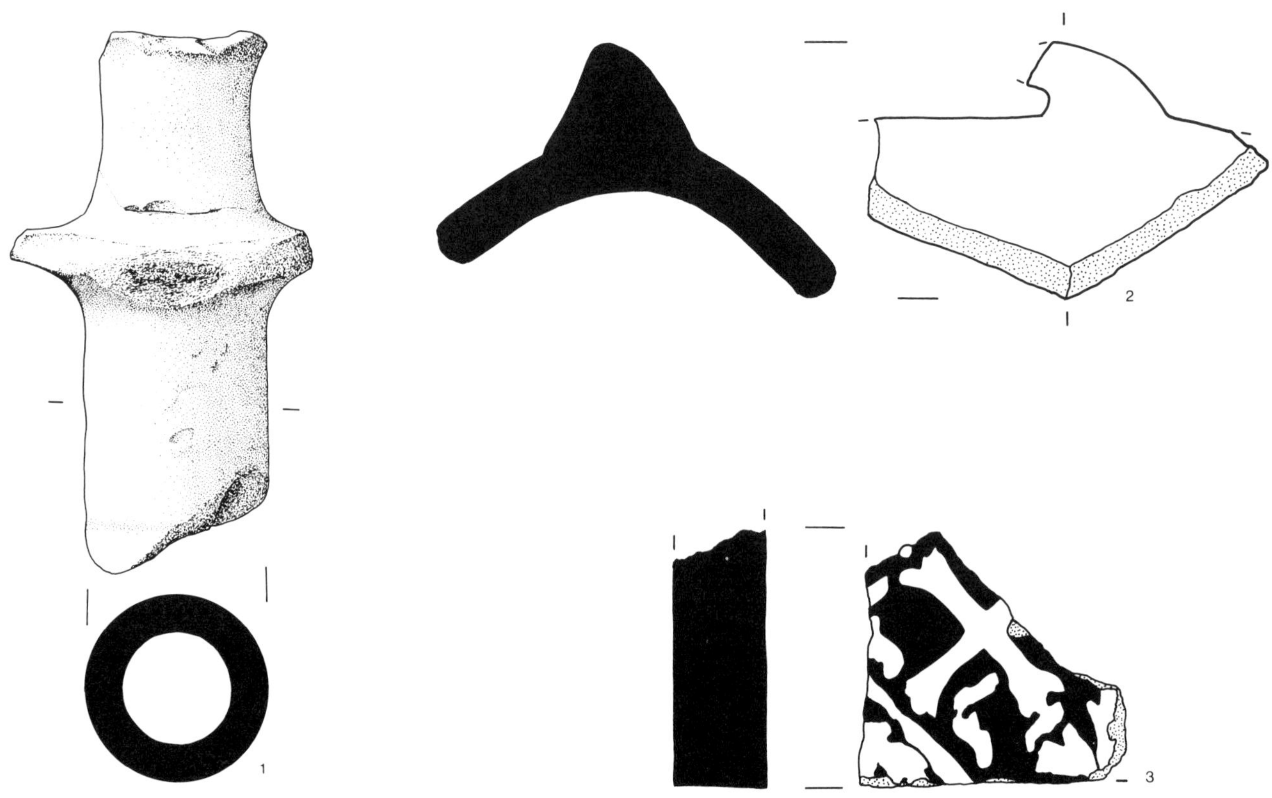

Figure 44 Ceramic building material. Scale 1:2

average weight of fragments of fabric 2a was also small (62g) and suggests the tile was used at some distance from the excavated area.

In phase 12 fabric 2a increased to 77% of the assemblage by weight and had a fairly high average weight at 140g. Glaze, pegholes and nibs were present, as well as a roof finial (Fig 44 no 1) and a ridge tile of characteristic form (Fig 44 no 2). A small proportion of the assemblage (4%) was made up of fabric 3 with Malvernian rock inclusions.

Hard-fired machine-made tile (fabric 1) was added to the assemblage in phases 13 and 14, though fabric 2a remained the largest component.

Tile types, stamps and reuse

The range and date of tile types conformed closely to that observed at Friar Street (Chapter 23). However the assemblage was too small for complete certainty in relating the tile to specific structures. The forms of Roman tile were *tegula*, *imbrex* and box tile, in decreasing order of quantity. Nibs and pegholes were both features of the medieval tile.

The tile stamps, all from phase 12 contexts (1917, S191, 1565 and 2214, both from S170), may, as Hurst suggests (Chapter 23) have been as early as the 15th century. The stamps were of two types, a circle and a partial circle (Fig 45), pressed into the tile before firing. They bore some resemblance to one example from Worcester (Carver 1980, 213). A fragment of inlaid floor tile (Fig 44 no 3) was recovered from the foundations of the bowling green (100, S69, P14). It bears a coat of arms: a

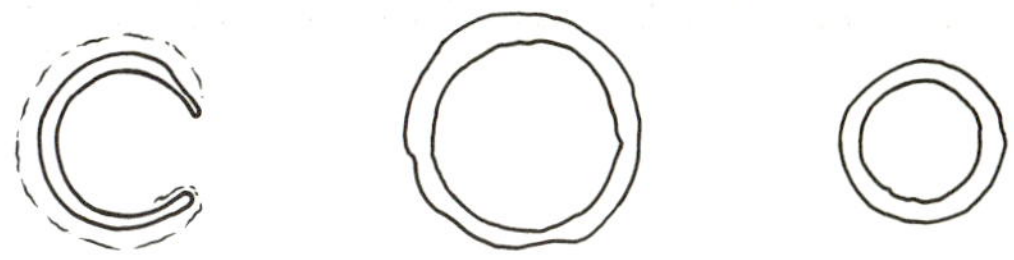

Figure 45 Tile stamps. Scale 1:1

cross surrounded by martlets and is probably of late 13th or 14th century date (Hilary White pers comm). Floor tiles, apparently *in situ*, have been found behind the Hop Pole Inn on Friar Street (Hodgkinson 1931, 70. HWCM 610) and a kiln is known from Witton (Allies 1852, 102. HWCM 660).

Two roughly-shaped disks (65 and 70mm in diameter) were recovered from phases 11 and 12 (531, S27 and 2209, S167), and two from uncertainly-phased contexts (559, S27, P9(11) and 1992, S9, P4–5(11?)). The earliest may have come from Roman contexts (perhaps as early as phase 4) and three examples were made from Roman tile. The one reshaped tile in fabric 2a would indicate a medieval date for the reshaping, though all other examples cannot be assumed to be contemporary. Similar, though smaller, pottery disks from other excavations have in the past usually been interpreted as 'counters' (eg Crummy 1983, 93). The presence of counters used for reckoning and games has been well established from the later prehistoric period onwards. On this particular site it is an attractive speculation to relate the tile disks to reckoning used in the production of salt.

5 Other ceramic objects *Derek Hurst and Simon Woodiwiss*

A group of handmade fragments in the Malvernian pottery fabric (fabric 3) was recognised, and these were characteristically *c* 15–20mm thick (see also Chapter 24). Three possible forms were represented: large vessels with thick heavy rims, straight sides and rounded corners (Fig 46 no 1), straight-sided plates (Fig 46 no 2) and circular plates. One other form was represented by a single example (Fig 46 no 3). There were no complete examples of any of these types and dimensions could not be established.

Evidence from other sites in the region, for example Beckford (Hurst pers comm), has indicated that this type of ceramic is 3rd to 4th century in date. The earliest material from the Old Bowling Green consisted of two fragments from the earliest fill of a brine tank (2438, S8) assigned to phase 4 (mid 1st to early 2nd century). Phases 9 (mid 3rd to late 4th century) and 11 (12th to 14th century) produced the most material (13 and 14 fragments respectively). Twenty-nine fragments, of a total assemblage of 117, came from the fills of a large ditch (S27).

Fragments were also recovered of a possible shallow vessel in a fabric similar to pottery fabric 2, but better mixed and formed. There was no sign of coil building. It was recovered from the fill (1752) of structure 8, one of the brine tanks of phase 4 (Chapter 2).

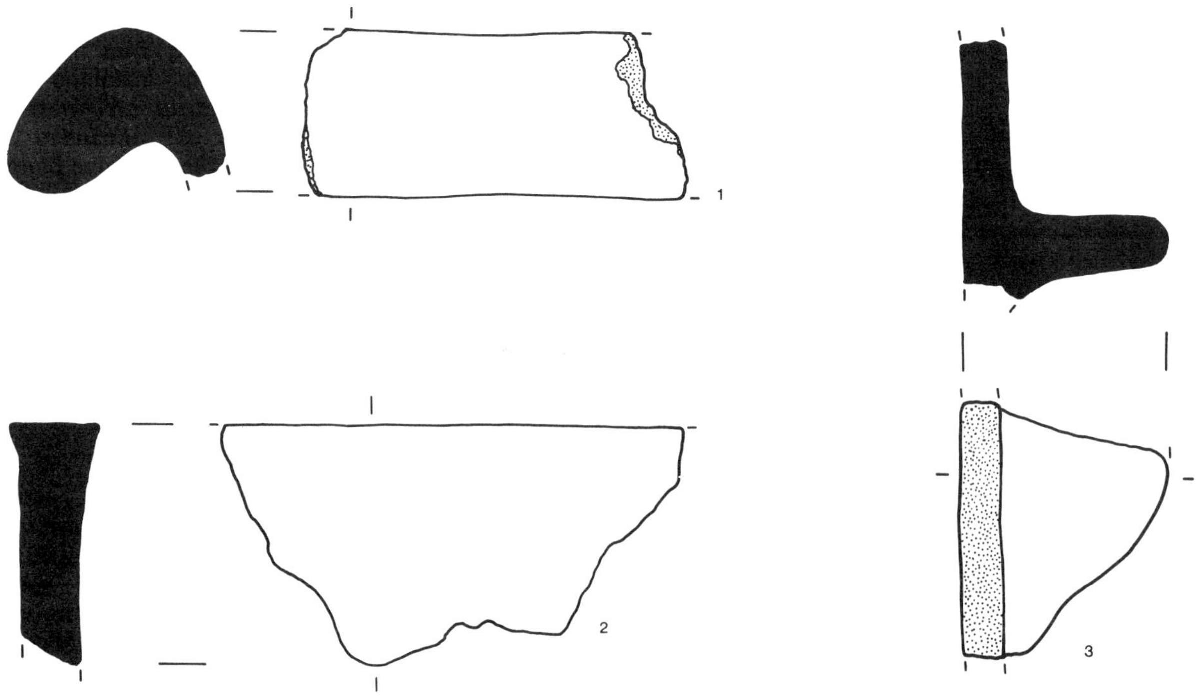

Figure 46 Ceramic objects. Scale 1:2

6 Fired clay *Helen Rees and Simon Woodiwiss*

Twenty-five fragments of fired clay, most with holes for wattle, were recovered from a layer (1065, S122) assigned to phase 11(12). It was not possible to suggest from which structure these were originally derived.

Fragments were also recovered of a possible shallow vessel in a fabric similar to pottery fabric 2, but better mixed and formed. There was no sign of coil building. It was recovered from the fill (1752) of structure 8, one of the brine tanks of phase 4 (Chapter 2).

7 Clay pipe *Derek Hurst*

Clay pipes were common in post-medieval contexts from phase 12 onwards, but especially during phase 13. Several early examples were present with small bulbous bowls comparable with Broseley type 1A and dating to *c* 1630–40 (Atkinson 1975, 24). Until *c* 1800 some of the pipes were from Broseley, South Shropshire, but more locally produced pipes seem to have been extensively available.

In addition there were several examples of pipes with a wheel stamp impressed on the underside of the heel. This stamp type has been dated by Oswald (1975, 66) to *c* 1650–90, and is commonly found on pipes throughout the Welsh Marches region.

Clay pipe marks are summarised in Table 1.

Fig 47

1 Novelty pipe of 19th century (Oswald 1975, 111); context 3119, structure 42, phase 13.

2 Bowl of 17th century date marked with FB in a heart on the heel; context 111/112, structure 92?, phase 12.

3 Broseley type 4A bowl (dated *c* 1720) with small SR mark of Samuel Roden on spur; context 111/112, structure 92?, phase 12.

4 Broseley type 7A pipe (dated 1720–40); context 1825, structure 203, phase 13.

Table 1 Clay pipe marks

Phase	Mark	Pipe type	Date	Maker	Reference
11–12?	IO	Broseley 1A	*c* 1630–40	—	—
12	FB	?	17th C	—	—
12	SR	Broseley 4A	1720–50	Samuel Roden	Atkinson 1975, 78
12	IP	?	17th C	—	—
13	RP	Broseley 2B	*c* 1680	Randle Peak	Atkinson 1975, 72
13	1[—]GFORD WORCESTER	Worcester	—	—	—
13	C.HARD[—] WORCES[—]	Worcester	*c* 1850	C Hardwick	Oswald 1975, 199
13	IAMES/ GETH/ING	Broseley	*c* 1740–70	James Gething	Atkinson 1975, 53
13	RUSS[—] WORCES[—]	Worcester	*c* 1835	—	Oswald 1975, 199
13	EP	Broseley 4B	1690–1720?	—	Atkinson 1975, 90
13	AB	Broseley 4A	1690–1720	Andrew Bradley	Atkinson 1975, 46
13	FB in heart	—	—	—	—

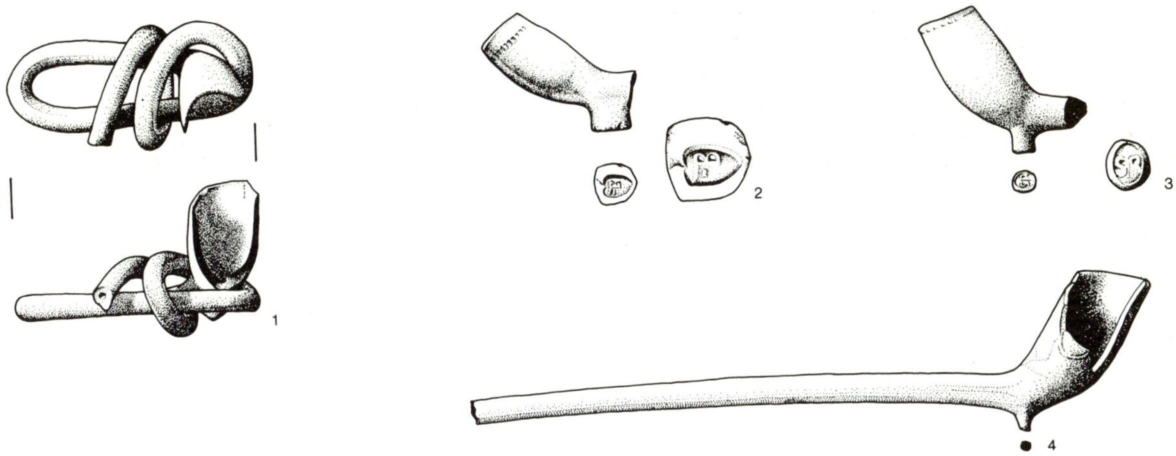

Figure 47 Clay pipe. Scale 1:2

8 Glass *Julian Henderson*

A total of 375 fragments and objects of glass were recorded, ranging in date from Roman to post-medieval. Chemical analysis was limited to qualitative x-ray spectroscopy of the beads. The assemblage is largely residual and too small for either a detailed analysis of development through each phase or to indicate social and economic aspects of the site. Discussion consists of the detailed description and drawing of parallels for selected glass fragments and objects from the excavation. A full catalogue is held as part of the archive.

Roman glass

Vessels

One of the most complete Roman vessels (Fig 48 no 1) is a ribbed bowl in a pale 'natural' green glass which was probably made in the mid 1st century and may be of Syro-Italian origin. It is probably mould-blown with a rim which has been rounded in the flame. There are parallels from *Verulamium* (Charlesworth 1972, 196 and fig 74, 2) and Camulodunum (Harden 1947, 299–300, no 48 and pl lxxxvii) in yellow glass, and others have been found in London (Wheeler 1930, 122, fig 42, 5). The Old Bowling Green example, in 39 pieces, was redeposited in a post-medieval ditch fill (3009, S67, P12). Also of 1st century date, or possibly later, are the remains of blown bottles in a natural greenish glass which were recovered from post-Roman contexts (501, S6, P14, 2212, S157, P11 and 2303, S170?, P11). The bulk of these common green prismatic or square bottles occur in the period AD 70–130, although they are found both before and after this period (Charlesworth 1966; Harden and Price 1971, 361). Only three of the bottle fragments from the Old Bowling Green were securely stratified earlier than the medieval period and, like the other glass from the site, redeposition is common. The bottle fragments are often scratched on the outer surface, probably as a result of being stored in wooden crates.

Another vessel fragment is part of a cobalt blue bowl or jar which has probably been cast and which bears a horizontal band. Other examples have been found in contexts of 1st century and later dates, and of other colours (Charlesworth 1981, 293, no 11). While the glass metal is of good quality, suggesting a 1st century date, this fragment was found in the fill of a late Roman ditch (579, S27?, P9?) and may well date to that period; the fragment is too small to be certain. A further bowl or cup fragment, although residual (1800, S69, P14), can be assigned a 1st to 2nd century date with more confidence. The fragment is of a lightly-weathered colourless metal with a thickened rounded rim. This high quality glass was probably cast and has been ground and polished on both surfaces. A comparable fragment was found at Fishbourne from a Period Two context dated *c* AD 75–100 (Harden and Price 1971, 331, no 23, pl xxv, fig 137). Comparable colourless glasses of Flavian date have also been found in London.

A further rim fragment, of a pale green bowl or cup of Hofheim type, was recovered from a contaminated context (1576, S31, P8). It was probably blown because it exhibits the characteristic gloss of blown glass on its inner surface. The outer, slightly convex, surface is wheel-abraded with a band of wheel incisions around the rim, the latter having been knocked off and polished. Comparable bowls have been recorded from Fishbourne (Harden and Price 1971, 345, fig 139, no 48) and *Camulodunum* (Harden 1947, 303, nos 74 and 76). They have a roughly hemispherical shape and mainly occur in the 1st century (including some examples from Fishbourne and *Camulodunum*), but they also occur in later Roman contexts, like the fragment from the Old Bowling Green. An example from Fishbourne which derived from a context datable to as late as AD 270 was also of pale green glass (Harden and Price 1971, 345, no 480). Two further bowl fragments, of insufficient size to be wholly diagnostic, were tubular ring bases from bowls or beakers in natural green blown glass. They were derived from a late Roman pit fill (1039, S15, P8) and redeposited in a medieval layer (582, S99, P9–11) respectively. Both have the remnant of a slightly domed base and outsplayed vessel walls attached. These bases were common 2nd to 3rd century forms and comparable with examples from Exeter (Charlesworth 1979, 224, fig 70, no 19) and from Housesteads (Charlesworth 1971, 34, nos 4, 5 and 6).

Beads

A brilliant translucent red globular bead from the Old Bowling Green has an opaque white substance lining the perforation. The probable technique of manufacture has been to wrap a filament of red glass around a wire coated in the opaque white substance. The exact nature of the opaque white substance is difficult to ascertain. It is of a granular non-vitreous and friable consistency and will have enabled the bead to have been easily separated from the wire it was made on. The small difference in measurement between the two ends of the perforation (0.5–0.7mm) may indicate that a tapering wire also facilitated removal of the bead. In order to regularise the shape of the bead it was probably reheated until the filaments fully fused.

Figure 48 Glass. Scale 1:2

Another bead of very similar dimensions and the same colour has recently been excavated from a disturbed context on the surface of a Bronze Age site on Harden Moor, West Yorkshire. The bead has precisely the same white lining of the perforation. Other red glass beads are not uncommon from British Roman contexts such as from Trentholme Drive, York (Wenham 1968, 99 and fig 36 no 46).

This bead is of a soda-lime-silica composition and its colourant is probably copper, which is present at relatively high levels. It is however, translucent, and not opaque as would be expected if copper was present in a reduced (cuprous) coordination within the glass. It is conceivable that the cuprous oxide or metallic copper crystals which would normally cause a red colour are extremely small and well dispersed. The bead was derived from a post-medieval layer (1515, S156, P12) but since it is of a Roman composition and type, with no known continuity of production into the medieval period, it is likely to have been redeposited. Soda-lime-silica glass is similarly rare in the high medieval period.

A second translucent glass bead is green and has been produced by winding a strand of glass around the end of a tapering metal wire from which it had been pushed off when completed. It is very roughly made and may originally have been part of a necklace. The type of bead is common in the late Roman period in Britain (see Brodribb *et al* 1971, 106 and fig 52, no 258 for a comparable example of 4th century date). The bead was recovered from a late Roman fill of a foundation trench for a stone building (1015, S26, P7). Analysis indicates that the bead is of a soda-lime-silica composition and that it is coloured by a combination of copper, iron and possibly manganese. Chromium was detected in trace quantities and may indicate that it was introduced in the sand as the mineral chromite.

Ten green vitreous fragments from a third bead appear to be highly weathered since they contain only a minor proportion of silica and any alkalis which may originally have been present have been leached out. Their surface is iridescent and the interior crystalline, with weathered pits visible in section. Analysis indicates that the green colour is due to iron and manganese with some copper. It was impossible to suggest what shape the bead might originally have been. The fragments were recovered from the late Roman fill of a foundation trench for a stone building (1015, S26, P7).

A counter

An opaque white glass counter has striations on the flat surface. These may indicate that it was manufactured by dropping a gob of glass on a wooden surface, since the striations are similar to those produced by the grain of a wooden surface. Many Roman sites have produced counters of this type, including Ospringe (Whiting *et al* 1931, pl lxi), Exeter (Bidwell 1979, fig 71, 59–64 and 231) and Shakenoak Farm (Brodribb *et al* 1971, 106). It was recovered from the Roman fill (1961, S29, P5–9) of a large ditch. This is a manganese-rich glass (*c* 0.1–0.2% oxide), which was typical of the Roman period, and it contains a similar amount of iron oxide. Traces of chromium and cobalt were also detected. This ratio of manganese to iron oxide has also been detected in the glass counters from Castleford, West Yorkshire (Henderson forthcoming). The counter is however, badly weathered, having a surface layer of iridescent glass.

The medieval and post-medieval glass

It is possible to be more certain of the form and date of manufacture of the post-medieval and medieval glass than of the Roman fragments.

Goblets / bowls

One extremely interesting fragment of vessel glass is of an early 16th century millefiori Venetian goblet or bowl (Fig 48 no 2; Pl 5). It is formed from three joining fragments of a vertically-ribbed convex vessel wall, with a rim rounded in the flame, and an applied strip of pale green glass which is double at one end. Since this type of millefiori was manufactured for only a short time there is little doubt about its early 16th century date, although this was not necessarily the date of its deposition in a pit fill (1553, S155?, P12) dated to the 17th or 18th century. Although also found in the early Roman period, this particular form of millefiori decoration is in the Renaissance style, using sections of chevron bead canes which were probably made in a series of dip moulds. Muranese glass beads were made in the colours used here: translucent pale cobalt blue, turquoise and green, and opaque red and white. A characteristic of these chevron rods is the use of alternating layers of

Plate 5 Early 16th century Venetian millefiori goblet or bowl

opaque white glass with other colours, which are observed in the Old Bowling Green example. It is actually possible to see both ends of some rod sections embedded in the colourless matrix. In the production of this vessel, sections of the cane would have been picked up on a gather of colourless glass from a suitable surface, such as a marver, and then blown into a ribbed mould: the sections of cane lie on the outer surface of the vessel, and the interior wall of the vessel is in the shape of the outer form. These characteristics would seem to be best explained by the suggested production technique. The applied strip of glass can be paralleled in a ribbed bowl in the Kunstgewerbemuseum, Berlin, K 428 (Hollister 1981, fig 2, left) which also has a comparable rim diameter (80mm) to the example discussed here (75mm). Unfortunately there is insufficient of the Old Bowling Green bowl surviving to be able to suggest what the complete vessel form might have been, although the Berlin example noted above is similar in shape to examples in the Courtauld Institute and in Dusseldorf (Hollister 1981, 223). Hollister (1981, 225) lists vessel forms with this design as bowl, goblet (chalice), cup, biconical glass, tazza, jar, flask, bottle, sprinkler and handled jug (ewer). One example, a miniature ewer is in the British Museum (Harden *et al*, 1968, no 251) and has a colourless ground colour like the Old Bowling Green vessel. The Old Bowling Green vessel appears to be the first published excavated example from Britain.

A single goblet fragment was recovered from the excavation. It is in a smoky-brown cristallo glass (3, S68, P13) and is a 17th century product, perhaps from Robert Mansell's glasshouse in London (Oswald and Phillips 1949, 32033, pl v). For further examples see Charleston 1977, 285–6 and Hüme 1962, 270–71 and pls 7 and 8). Another drinking vessel fragment from the Old Bowling Green is a pale green domed base of a probable ale glass of late 16th century date with a tubular base rim formed from a double thickness of glass at the same time as the dome of the base was created (10, S68, P13). Such ale glasses were found at Woodchester glasshouse (Daniels 1950), the Denton glasshouse (Hogan 1970, 25–25 and fig 2), Sidney Wood, Surrey (Hüme 1962, 270) and from beneath Honey Lane Market, Cheapside, London (Hüme 1962, 270 and pl 3). As Charleston has noted (1977, 290) this form of foot is a classic type from the forest glasshouses of the Surrey/Sussex Weald and the glasshouses listed above.

Beakers

Another common form of drinking vessel of 16th to 17th century date is the beaker. The pushed-in conical base generally survives because it is made of the thickest glass. Three examples of beaker bases are recorded from the excavation in both pale green (503, S68, P13) and colourless glass (1800, S69, P14 and 2232, S170, P13). They are paralleled for example from the fill of a 17th century pit at the Black Gate, Newcastle-upon-Tyne (Ellison *et al* 1979, 173, fig 7, no 43). The actual filling of the pit took place between *c* 1645 and 1675: the Old Bowling Green examples are liable to have dated from the first half of the 17th century.

Bottles

One fragment in pale green glass formed part of a hexagonal bottle of late 16th to early 17th century date. It was from a bottle shoulder and has parallels at the Woodchester glasshouse (Daniels 1950, pl v, 26–30), Gracechurch Street, London (Oswald and Phillips 1949, 34 and pl xii) and at Winchester (Charleston 1964, 150 and fig 50, no 11). It was recovered from the fill of a pit dating to the 15th to 18th centuries (1518, S155, P12). A small number of pharmacy phials or bottles were recovered from the excavation (517, S48, P13 and 1049, S121, P11(12)) dating to the 16th to 19th centuries, of both octagonal and cylindrical shapes.

Many of the 17th to 18th century so-called 'wine' bottle fragments are in a dark green glass with heavy dark brown surface weathering. They can be dated using their form by referring to other independently dated examples which have dates stamped on them. Such bottles have come from the Oxford taverns for example (Leeds 1941). Work by Hüme (1961 and 1970) has also established that these bottles were in use in colonial America, among other places. The contents of an exhibition held in Truro Museum in 1976 on the history of the English bottle (Anon) has provided further dating evidence for these bottles, and Wills (1977) has published in a popular form. Rim, neck, wall and base shapes are all useful in determining the age of these bottles, although many of the fragments from the Old Bowling Green are too small for close dating. It also seems likely that there may well have been greater variability in the shape of these bottles than has been recognised up to now and, while the generalisations about their shape are valid in most cases, there are liable to be exceptions to the rule. One example of a late 17th century onion bottle in fragments was recovered from the fill of a brick-lined drain (503, S33, P13) and is distinguishable from other later fragments by using its base and wall shape (Anon 1976, no 2; Wills 1977, fig on p 46). The other examples of 18th century wine bottles include one with a low kick in its base which probably dates to *c* 1700 and was also recovered from the fill of the drain. The remaining fragments, from their shape, are liable to be of early to middle 18th century date.

The machine-made wine bottle glass from the excavation, of 19th to 20th century date, is mainly fragmentary, although there is a single complete example (517, S48, P13). It is a bottle which was produced using a technique patented by H Ricketts of Bristol in 1821. 'H.RICKETTS AND CO.,GLASS WORKS, BRISTOL' was stamped onto the base and 'PATENT' on the shoulder. The three-part iron bottle-making mould was perfected in 1802 by Charles Chubsee of Stourbridge, and a patent was granted to Ricketts in 1821. This bottle is likely to date soon after 1821.

Window glass

The medieval and post-medieval window glass is too fragmentary to reconstruct a complete pane. In the catalogue the glass has been grouped according to thickness, colour and weathering characteristics. One fragment (502, S33, P13) was probably of Crown glass manufacture because it has weathered, leaving circumferential iridescent surface striations, and may be of medieval date. This technique of manufacture is described by Harden (1961, 39–43). The remaining glass is probably muff glass.

Conclusion

The archaeological association of the glass from the Old Bowling Green, Droitwich reveals that it has been disturbed in many cases producing, for example, Roman vessels in medieval contexts. The few datable Roman vessels fall closer to the start of the Roman period (1st to 2nd centuries AD) than to the end. The post-medieval glass is of interest because the collection includes an early 16th century Venetian millefiori vessel, probably a goblet. This appears to be the first excavated example from Britain to be published.

9 Worked stone

Flintwork, by Alan Saville (Fig 49)

1 Multi-platformed core, possibly Neolithic or Bronze Age. Context 184, unstratified.
2 Retouched flake, possibly pre-Bronze Age. Context 418, structure 86, phase 2?.
3 Waste flake. Context 170, structure 72, phase 2 (not illustrated).

Shale, by Anne Crone (Fig 49)

4 Fragment of spindle whorl. Context 1015, structure 26, phase 7.

Whetstones and rubbers, by Fiona Roe (Fig 49)

5 Rubber. Context 1143, structure 17, phase 8.
6 Whetstone. Context 1712, structure 177, phase 5.
7 Whetstone. Context 1917, structure 191, phase 12.
8 Whetstone. Context 1923, structure 195, phase 12.
9 Whetstone. Context 2069, structure 196, phase 12.
10 Whetstone. Context 500, unstratified (not illustrated).

The rubber is large, being 180mm in length and irregularly shaped with two worn surfaces. It was retrieved from the fill of one of the barrels of phase 8 (mid 3rd to late 4th century). It is made from a fine-grained, dark grey-green slightly micaceous sandstone. Thin sectioning has shown that the rock is a lithic ararite. A possible source for this would be amongst the Pre-Cambrian rocks of the Longmynd. The whetstone from phase 5 (early 2nd to mid 3rd century) is ovoid in cross-section with all of its surfaces polished. It is made from Kentish Rag, a calcareous sandstone likely to be from the Lower Greensand Hythe Beds of Kent. A second whetstone from an unstratified context (500) is also made from this material. These identifications are based on macroscopic examination only; however both of these whetstones approximate to the cigar shape often seen amongst Roman finds made out of the Kentish Rag. The other three were all from contexts placed in phase 12 and all appear, on macroscopic examination, to be made of Norwegian Rag or Eidsborg schist from the Telemark area of Norway.

These findings are consistent with results from other sites in the area. Two Kentish Rag whetstones were recorded from Roman contexts at Beckford (Roe 1987, 15), confirmed in one case by thin sectioning. Whetstones of the Norwegian Eidsborg schist are additionally known from medieval and post-medieval phases at Friar Street, again confirmed by a thin section (Chapter 27).

Querns, by Fiona Roe (Fig 50)

11 Saddle quern. Context 2193, structure 181, phase 2.
12 Rotary quern. Context 2189, structure 181, phase 2(4).
13 Rotary quern. Context 1023, structure 114, phase 7–9.
14 Rotary quern. Context 1041, structure 15, phase 8.
15 Rotary quern. Context 1059, structure 35, phase 11–12.
16 Quern. Context 403, structure 84, phase 2 (not illustrated).

The saddle quern was retrieved from a context placed in phase 2 (late Iron Age) and appears to be made from a fine-grained variety of Old Red Sandstone or Millstone Grit. Of the rotary quern fragments, one came from an uncertainly phased early context (2189). This is made from a slightly conglomeratic felspathic sandstone likely to derive from the Old Red Sandstone, probably from Wales. Old Red Sandstone is also used for one of two rotary querns derived from Roman contexts (1023). It is now pinkish and appears to have been burnt, perhaps by secondary use as a hearthstone. The other Roman quern (1041) is made from a rather friable conglomerate that bears some resemblance to the Upper Devonian quartz conglomerate found in the Forest of Dean and other areas. A final rotary quern fragment was retrieved from a post-Roman context (1059). This is made from a dark purplish, fairly fine-grained sandstone resembling the Old Red Sandstone brownstones. One other quern fragment (403) may either have been of rotary or saddle type. The material is the same as that used for the saddle quern from context 2193.

The materials used for querns can also be linked with findings from other excavations. Purplish-pink gritstone was extensively used at Beckford in Iron Age contexts and a fragment occurred in the Iron Age phase at Friar Street. The source remains unknown, but it is likely to come either from the Old Red Sandstone or the Millstone Grit, both of which have extensive outcrops. Varieties of Old Red Sandstone from south Wales were also found at Beckford, mainly from late Iron Age and Romano-British contexts. The Upper Devonian quartz conglomerate, which also occurred at Beckford, has been widely noted from excavations, usually from Roman levels.

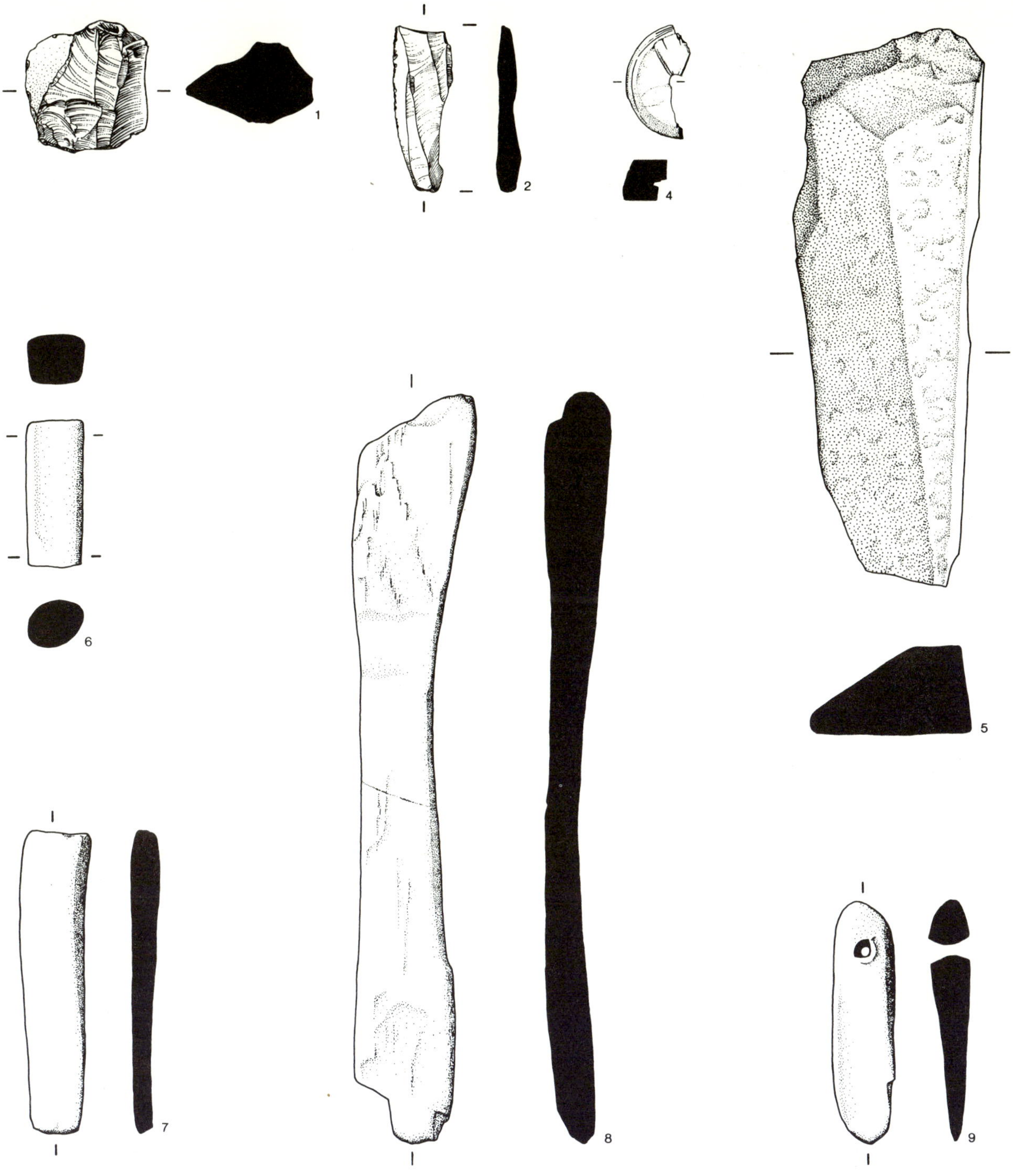

Figure 49 Worked stone: flintwork (1,2), shale (4) and whetstones and rubbers (5-9). Scale 1:2

Shaped building stone, by Anne Crone

Of the fifteen pieces of shaped building stone that were retrieved from Roman contexts, twelve were padstones forming part of the construction of structure 22 (P9). The three remaining pieces came from disuse elements of phases 4–5 and 8, and the constructional fill of a building (S26, P7). The latter did not form part of the wall of the building, as it was from within the foundations and had presumably been reused. One side was covered with toolmarks (Fig 50 no 17). The stone was presumably all local Triassic Lower Keuper sandstone (Mercian Mudstone).

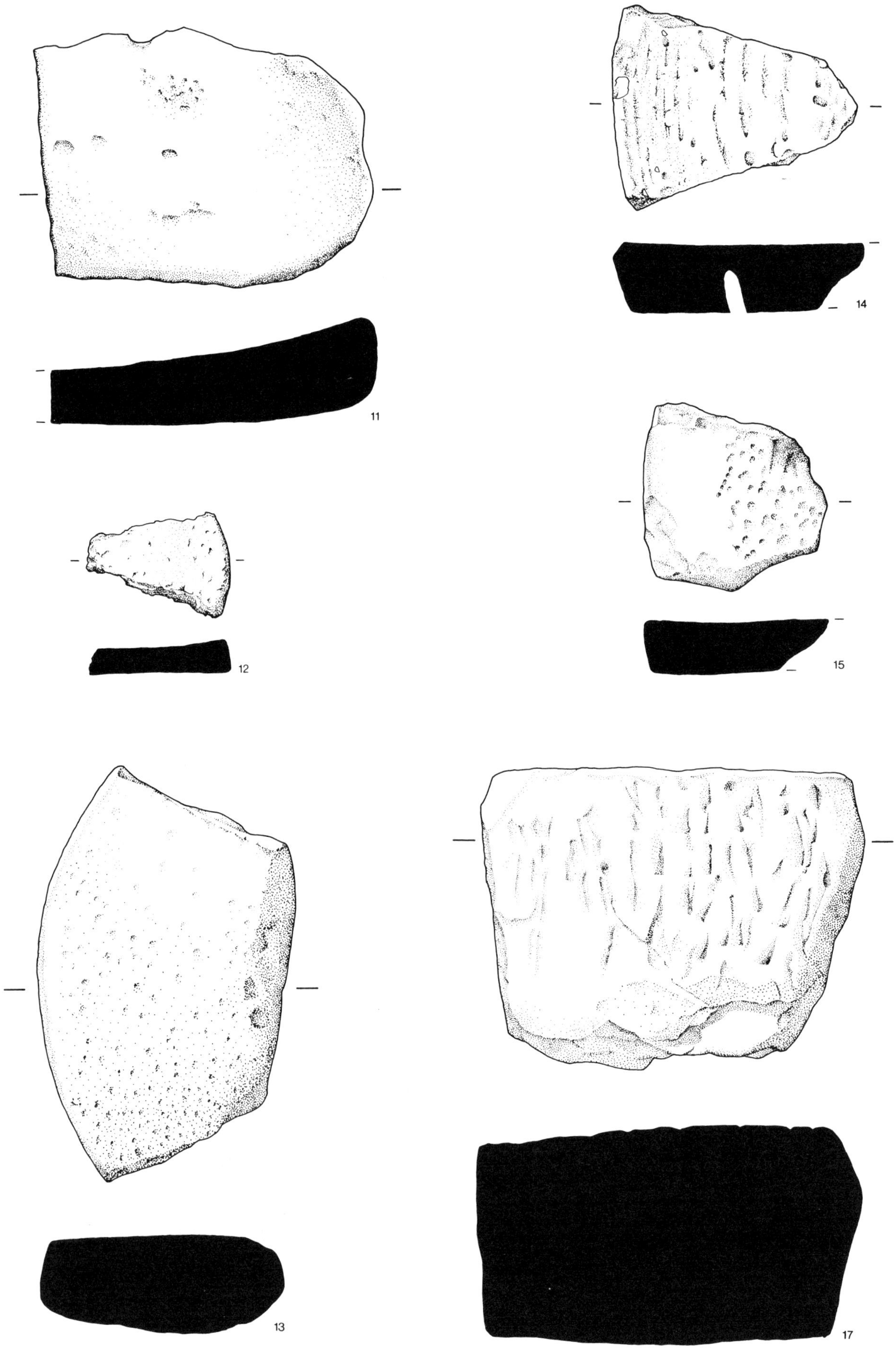

Figure 50 Worked stone: querns (11-15) and shaped building stone (17). Scale 1:4

10 Inscriptions *Mark Hassall*

A sherd from a flanged Severn Valley ware bowl (fabric 12) had an incomplete graffito cut below the rim before firing. It reads: ...]*SC*[..., perhaps for *Va]SC[ella* (Fig 51 no 1). The sherd was retrieved from a layer (1081, S210?) of phase 7–12. A bodysherd of Severn Valley ware (fabric 12) also had a graffito, neatly cut in capitals before firing on the burnished body of the vessel. It reads: *VASCIILLA*, *vascella*, 'vessel' (Fig 51 no 2). The sherd was retrieved from the upper fill of a barrel (1138, S17) of phase 8.

Vascella from *vascellum*, an alternative for the more common *vasculum*, the diminutive of *vas*, the form being attested epigraphically on a funerary inscription from Rome, *CIL* vi 3428 of AD 214, where it refers to an ash chest. The word is

Figure 51 Inscriptions: on pottery (1, 2) and stone (3). Scale 1:2

probably neuter plural (rather than first declension feminine singular), but with singular meaning, cf *vasa*, the neuter plural of *vas* which can be so used (OLD volume 2). Both *vas* and *vas(s)a* (the latter used in the sense of cinerary urn) are attested from Roman Britain (Wright 1954, 107, no 25 and *EE* ix 1351(a)). Note, however, that in medieval Latin *vascella* can be both neuter plural and feminine singular (Niermeyer 1954–76). John Sawle, the director of the excavation, draws our attention to Hooke (1981, 136–7), where it is shown that *vasculum* was used in the medieval period to describe vessels holding brine. If that usage is anticipated here, it suggests that these vessels were in some way associated with the process of salt manufacture. However, since the fabric and form of neither sherd is in any way unusual, it is possible that they should be viewed as secondary domestic or funerary debris, bearing no direct relationship to the salt working function of the site.

Inscribed lines were noticed on a badly eroded sandstone block (200 x 160 x 55mm, Fig 51 no 3). The lines may represent sharpening activity or possibly crudely formed letters. ...]*VI*[.... The orientation of the possible letters or numerals is indicated by deepening of the groove and a more steep cut to the top as would be produced by working the stone towards the body. The block was retrieved from a layer (718, S99) of phase 9.

11 Brooches *Donald Mackreth*

Fig 52

1 The brooch is in relatively poor condition (570, S95, P7–9). The spring is mounted on a loop behind the head of the bow. It had been held in place by means of a rolled sheet-metal tube through the coils in whose ends were once located the ends of a wire loop which once rose above the head of the bow. The waist of the loop would have been caught in a collar and the whole would have been prevented from falling over the head of the brooch by the cast-on nib on the head. The trumpet head is decorated with relief ornament. At the bottom are three bosses, each probably outlined with a groove round its base. These were bordered by a step which ran as a ridge up the centre of the head and then divided to enclose a platform, on each side of which were two more bosses placed on a 'comma'-like element whose lower edges were swept back into the middle of the head. The knop is made up of three elements, the central one being bordered by a thin moulding on each side, the upper and lower elements each has a groove. The lower bow is set off from the knop by a cross-moulding which returns down each side and the centre as ridges. In each of the side panels so formed is a wavy ridge. Beneath these is a cross-moulding divided by a groove. The foot-knob is a cone with a concave profile and with a line of 'petals' round it. The underside has a grooved border with a central recess, from which projects a prominent boss.

The design is the commonest amongst those with relief ornament, and the close similarity between brooches such as this in copper alloy with the highly elaborate silver-gilt example from Carmarthen (Boon and Savory 1975) suggest strongly that all of these emanated from a single workshop, if not craftsman. The distribution lies in the lower Severn Valley running from Wroxeter to the south and spreading out across Wales. None is dated, but the floruit of other relief-moulded brooches should provide a clue as it is likely that all belonged to the same stylistic fashion: Wroxeter, before 130 (Kenyon 1940, 224, fig 15,4), 110–30 (Bushe-Fox 1913, 26, fig 9,7); Holcombe, Devon, 70–180 (Pollard 1974, 138–40, fig 22,2); Derby, 150–75 (Mackreth 1985, 191–3, fig 127, 28); a silver-gilt loop and collar analogous to the Carmarthen brooch from Bourton-on-the-Water was found with pottery dating to *c* 80–160 (Boon 1978, 152–3, pl xlix, a–b); Derby, late 2nd century (Mackreth 1985, 191–3, fig 127, 27), late 2nd-early 3rd century (*ibid* fig 127, 26). The range is basically in the second century and, in common with the bulk of ordinary Trumpet brooches, almost certainly had ceased to be made between 150 and 175, with a few surviving in use to the end of the century, and perhaps a little beyond.

2 The spring lies between two pierced lugs which hold the axis bar through the coils (501, S69, P14). The chord is external and held by a hook behind a cast-on loop mounted on a pedestal on the head. The sides of the bow sweep out to near the ends of the wings, which appear to have intentionally rounded ends, and there are seven grooves on each of the curves. The bow has markedly chamfered sides with a very narrow central face which has a ridge running down it. On each side are four equally spaced divided mouldings. These stop just below the top of the catch-plate. The foot of the bow is pointed, with a sharp lateral projection on each side.

The design is unusual and does not reveal a close affinity with any variety of a standard type. The writer has come across only one parallel, from Eccles, Kent (excavations, A Detsicas, in prep), which is such an exact repeat that it must have come from the same hand. Dating is hardly to be expected and all that can be suggested here, based on the cast-on loop and the general type of the piece, is that it is 2nd century.

3 Now in four pieces (2338, S161, P9), the spring was essentially the same as that on a Colchester example. However, while the hook is part of the body of the brooch, the spring had been made separately and fastened to the bow by inserting its start into a hole in the head. The corrosion hides any real trace of decoration. The wings are short and damaged. The profile of the thin and wide bow is recurved with a cross-moulding at the point of inflection. There may have been some mouldings down the centre of the upper bow. The lower bow is heavily pitted, but seems to have the slightly bowed sides and, possibly, the rounded foot to be expected on the type.

The brooch is simple and is related to the series which includes the Pannonian and the Augenfibel. Never very common in Britain, no very clear picture of either the dating or distribution emerges of the simple form. One from Baldock was dated 90–120 (Stead and Rigby 1986, 112, fig 42, 49). However, the signs of decoration on the upper bow of the present brooch should mean that it had been beaded and the British dating is a little better in this case: Colchester 49–61, two examples (Hawkes and Hull 1947, 321, pl xcvi, 120–1); Towcester, *c* 100 (Lambrick 1980, 60, fig 12, 3), Baldock, two examples, 180–220 (Stead and Rigby 1986, 112, fig 42, 47–8). This is hardly impressive, but one feature appears to mark the distribution of both forms dealt with here. None has been recorded by the writer north of the Humber or Derby, and this could be taken as evidence that the simple form should not really date after *c* AD 75, thus any later should have been residual in their context. The small number recorded from Saalburg and Zugmantel may show that a few lasted beyond *c* 80 and this was the conclusion reached by Böhme (1972, 10–1, Taf 1, 7–14). The few dated ones from Augst tend to support this view and show that the two instanced from Colchester are most probably within their proper floruit (Riha 1979, 68, 69, Taf 7, 207–9, Taf 8, 223–6). The best range in Britain is probably Claudian–early Flavian.

4 This is a collection of fragments amongst which can be seen the head of a brooch with a sprung-pin (559, S27, P9(11)). The spring was probably mounted in the Polden Hill manner, a version of which was used on brooch 2. The corrosion accretions are such that it is difficult to see any diagnostic features which might illumine the type to which the piece belonged. Another fragment appears to be from a bow and the corrosion suggests that there may well have been a series of lozenges down the front. The remaining bits are too small and indeterminate to be of any value. No further comment is advisable (not illustrated).

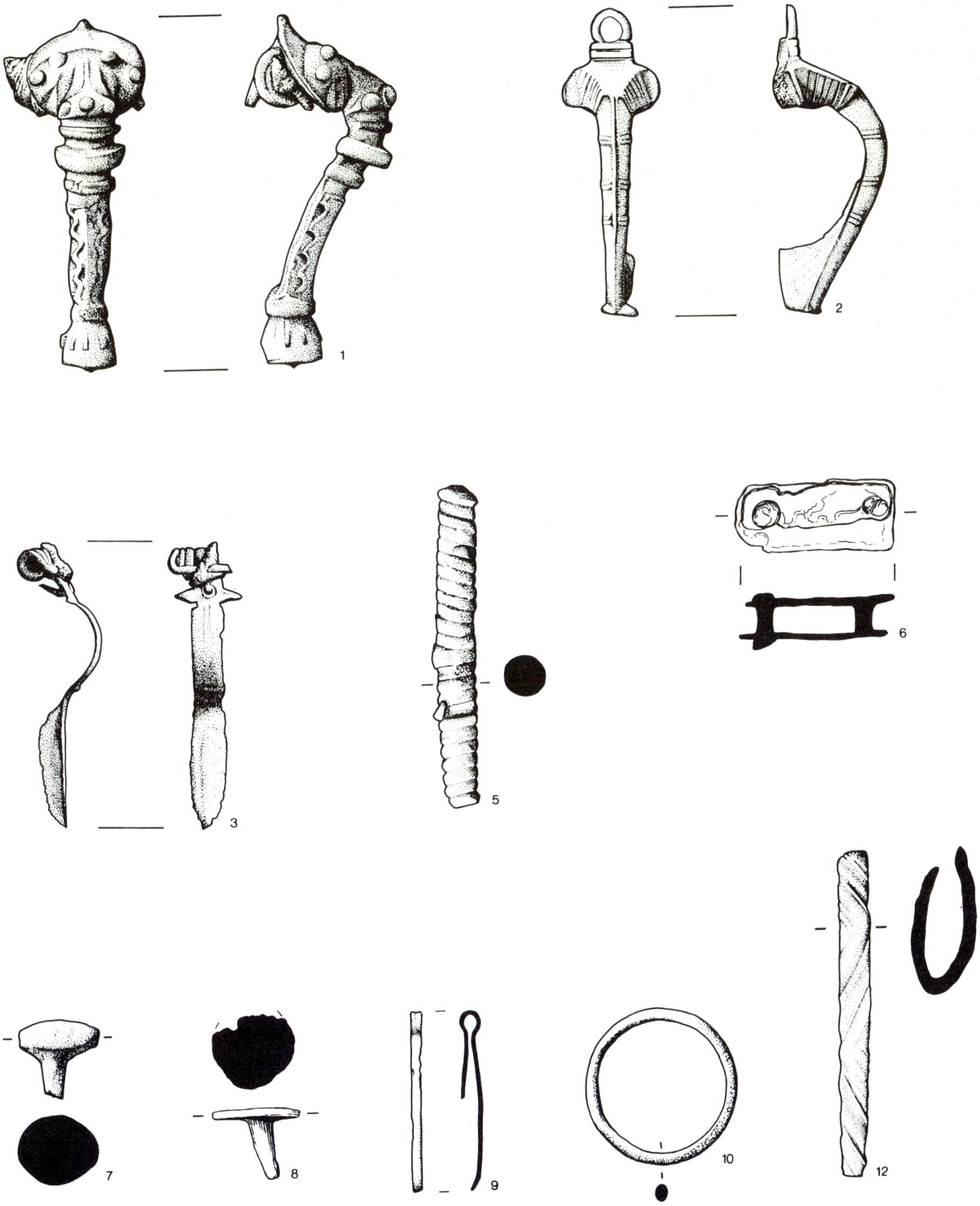

Figure 52 Copper alloy: Roman brooches (1-3), other Roman objects (5-10) and a post-medieval object (12). Scale 1:1

12 Other copper alloy objects *Anne Crone*

Sixteen objects, including the brooches, were recovered from contexts of Roman date (nos 1, 3, 5, 6 and 7), eleven of which were unidentifiable (631, 1023, 1041, 1125, 1171, 1239, 1257, 1508(2), 1688 and 2340). A further nine were from contexts possibly of Roman date (nos 4, 8, 9, 10 and 11), four of which were unidentifiable (559, 564, 574 and 582). One residual Roman object is illustrated (no 2). Objects of later date are recorded in the full archive catalogue. The salt content of the deposits is probably the main reason for the poor condition and limited size of the assemblage. A ring of undetermined date and from a post-medieval context (501, S69, P14. Fig 52 no 12) is worthy of note.

Fig 52

5 Coiled wire. Context 698, structure 95, phase 7–9.

6 Riveted plates. Context 723, structure 112, phase 7.

7 Stud or nailhead. Context 1130, structure 31, phase 8.

8 Stud. Context 1522, structure 22, phase 9(11).

9 Tweezers. Context 589, structure 27, phase 9(11).

10 Ring. Context 1151, structure 114, phase 5–9(11).

11 Twisted wire. Context 1551, structure 155?, phase 7–11 (not illustrated).

13 Ironwork *Anne Crone and Simon Woodiwiss*

A full catalogue appears in archive and only the pre-medieval material is discussed here. Iron is especially adversely affected by the saline environment of Droitwich, and so the condition of the ironwork was poor (no illustrations were prepared for publication). Most small items may be expected to have decayed altogether.

Only one ornament was recovered (2318, S149, P9), although its form and function is obscure. Of the 40 other iron objects from contexts which were assigned to the Roman period (though not certainly phased), 22 were nails and the rest were unidentifiable. Again most of these were found at the western end of the excavation and were retrieved from contexts as early as phase 4. Seven hob nails were recovered together from a layer (677, S96, P7–9) and were presumably from a shoe, the leather having decayed.

14 Lead *Anne Crone and Simon Woodiwiss*

Only objects from contexts which may be definitely placed in pre-medieval phases are described in the following catalogue (the archive contains a full catalogue).

Fig 53

1 Tube formed from two strips. Context 225, structure 63, phase 4.
2 Trimming from sheet, folded, extended length 2.40m. Context 1174, structure 120?, phase 8–9.
3 Handle. Context 2193 (contaminated), structure 181, phase 2.
4 Folded sheet. Context 586, structure 28, phase 7–8 (not illustrated).

In addition unidentifiable scraps were recovered from contexts 566, 1060, 1070, 1023 and 2318.

Though the assemblage was far too small for any definitive statement, there is an apparent absence of containers for brine boiling in other materials and, as lead was used in later periods (eg Rastel 1678, 1062; Thorn and Thorn 1982, nos 1, 1a; 5), its earlier use must be considered a possibility. A lead pan reputedly of Roman date has been recorded from Middlewich, Cheshire (Bestwick 1975, 66). The scraps in later Roman phases, appear to be drops of lead, possibly from the melting of pans which had become too hot. The lead sheet fragments also may be derived from pans. The large trimming (no 2), of late Roman date, may represent wastage from the cutting of a lead sheet, possibly supplied without finished edges. Domesday Book mentions the presence, in Droitwich, of *fabrica plubi* (lead works; Thorn and Thorn 1982, no 2, 50). Though no physical evidence of their existence has been found, it is possible that lead working was carried out in the town during the Roman period. It may be significant that none of the scraps and sheet fragments are earlier than phase 7 (mid 3rd to late 4th century AD) and they may even be contemporary with the barrels that characterise phase 8.

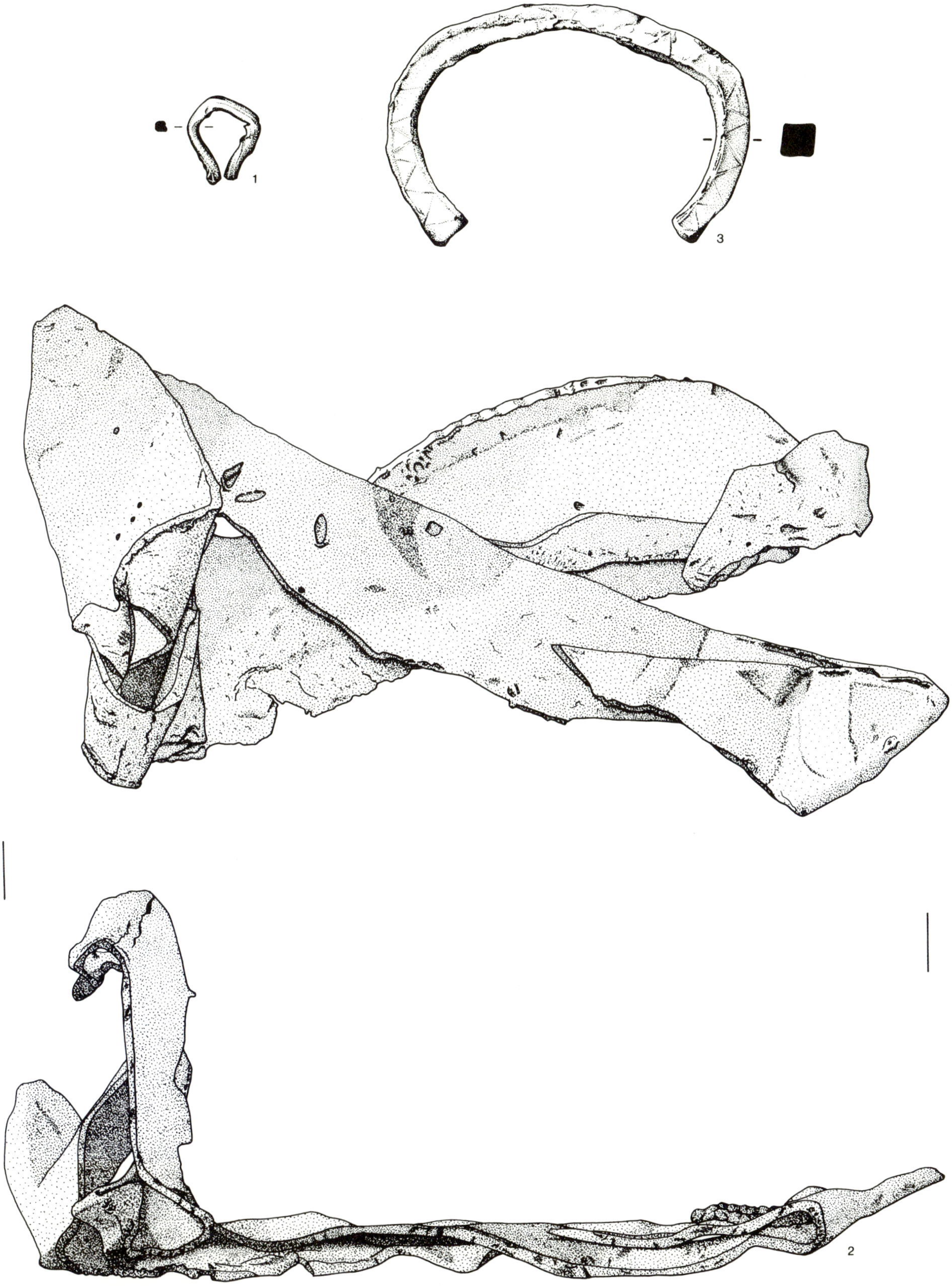

Figure 53 Lead. Scale 1:2 (except no 3)

15 Animal bone *Alison Locker*

Summary

Some 10,000 animal bones were recovered from deposits dating from the late Bronze Age to the modern period. However, bones of later than 18th century date have not been included in this report. Activity on the site from the Iron Age to late Roman period is largely associated with salt extraction. The bone is usually associated with the disuse of structures. Ox and large ungulate fragments are dominant throughout these phases (1 to 9), except phase 3, in which the small sample shows ovicaprid and small ungulates to be dominant, and in some structures of phase 4. The importance of ox and large ungulate continues in the medieval period and in phase 11 there is some evidence for preferential representation of certain parts of the carcase of ox. Butchery practices do not show significant changes through the phases, and pig is poorly represented, between 3% and 6% of all samples. Butchery marks were observed on some horse bones but there is no real evidence that horse was eaten. Dog was present both as isolated bones and skeletons. There were also skeletons of cat and foetal pig, both of medieval date. A few goat horn cores were identified but the majority of the ovicaprid sample is thought to be sheep. Ox and large ungulate, and ovicaprid and small ungulate comprise most of the bones from the excavation, representing butchered bone refuse. The distribution of animal bones by phase is summarised in Table 2.

Introduction

The animal bones discussed in this report were recorded in 1984 by Miranda Armour-Chelu (British Museum, Natural History), the data being subsequently written up in 1987 by the author. Any idiosyncrasies in the interpretation should, therefore, be attributed to the author.

All the bone was recovered by hand picking on site. No sieving was carried out, so there is likely to be a bias in favour of the larger fragments.

The bone was recorded using the method of Jones *et al* (1981). Although some modern material was recorded, bone of later than 18th century date (phase 12) has not been considered in this report. Similarly, bone from contexts that were not certainly phased, or from contaminated contexts, have not been included. Tables of the quantities of bone from each structure are presented in microfiche (2:D3–E12).

The following species were identified; ox (*Bos* sp domestic), sheep (*Ovis* sp domestic), goat (*Capra* sp domestic), pig (*Sus* sp domestic), horse (*Equus* sp domestic), red deer (*Cervus elaphus*), dog (*Canis* sp domestic), cat (*Felis* sp domestic), fowl (*Gallus* sp domestic), goose (*Anser anser*) and buzzard (*Buteo buteo*).

Ovicaprid has been used when sheep and goat were indeterminate, as they were in the majority of cases. Both species were present however, positively identified from horn cores, although application of Boessneck's (1969) index to the metapodials suggests these were largely from sheep. Large and small ungulate refer to bones which, though not specifically identifiable, are most likely to be ox and ovicaprid respectively.

Weathering and canid gnawing occur in all phases and may be evidence of some bone lying around on the surface before being disposed of.

In the following sections the commonly occurring species, their distribution and butchery evidence are discussed followed by the ageing, metrical data and occurrence of other species.

Phase 1: Possible late Bronze Age to Iron Age

The bone from this early phase was found in three layers of dark soil, mostly from layer 2461. From a total of 62 bones for this phase, over 50% were attributed to ox and large ungulate.

The butchered bone included two ox mandibles chopped through the alveoli, three scapulae chopped across the neck and split axially, and a few chopped rib fragments. An ox femur shaft was from an immature animal. An ovicaprid pelvis was chopped, detaching the ilium, and a mandible had been cut towards the diastema. Shaft fragments of radius and metacarpal were also found.

No pig remains were found in this phase. Horse included six skull fragments, part of a scapula and a radius which had been split axially and superficially chopped across the shaft. The red deer antler fragment had been cast and had a cut mark on it.

A few large bones showed evidence of gnawing by dogs, for example some large ungulate rib fragments and an ovicaprid mandible had been gnawed. A horse scapula was more weathered than the rest of the bone.

Phase 2: Late Iron Age

Some of the structures associated with this phase were used for the storage of liquid and possibly for the settling out of silt from brine before boiling. The primary activity during this phase was industrial and the presence of animal bone in disuse contexts incidental.

Table 2 Animal bone summary

Phase	1		2		3		4		5		7		8		9		10		11		12	
Ox	15	24%	92	20%	2	3%	88	18%	68	26%	47	26%	453	33%	1010	44%	–		303	32%	55	13%
Sheep	–		–		–		1	<1%	–		–		–		1	<1%	–		–		1	<1%
Goat	–		–		–		1	<1%	–		–		–		–		–		–		–	
Ovicaprid	14	23%	112	25%	35	47%	195	40%	72	27%	37	20%	158	12%	373	16%	2	67%	160	17%	52	12%
Pig	–		8	2%	2	3%	14	3%	9	3%	7	4%	45	3%	84	4%	–		43	4%	88	21%
Horse	12	19%	34	8%	1	1%	16	3%	5	2%	5	3%	8	<1%	15	<1%	–		8	1%	3	1%
Red deer	1	2%	–		–		1	<1%	–		–		2	<1%	4	<1%	–		3	1%	–	
Dog	–		–		1	1%	3	1%	2	1%	1	<1%	231	17%	10	<1%	–		10	1%	2	<1%
Dog/fox	–		–		–		–		–		–		–		1	<1%	–		–		–	
Cat	–		–		–		–		–		–		–		–		–		–		95	22%
Hare	–		–		–		–		–		–		–		1	<1%	–		–		–	
L Ungulate	19	31%	159	35%	16	21%	111	23%	78	30%	67	37%	369	27%	563	26%	1	33%	351	37%	85	20%
S Ungulate	1	2%	42	9%	18	24%	58	12%	25	10%	16	9%	88	6%	155	7%	–		75	8%	42	10%
L Mammal	–		2	2%	–		–		2	1%	–		–		6	<1%	–		–		–	
Fowl	–		–		–		–		–		3	1%	4	<1%	10	<1%	–		3	<1%	–	
Goose	–		–		–		–		–		–		1	<1%	–		–		2	<1%	–	
Buzzard	–		–		–		–		1	<1%	–		–		–		–		–		–	
Bird	–		–		–		–		–		–		–		1	<1%	–		–		–	
Unident	–		–		–		1	<1%	–		–		3	<1%	17	<1%	–		–		–	
Total	62		449		75		489		262		183		1362		2551		3		958		423	

In Phases 9 and 12 pig and cat respectively are almost complete skeletons

A total of 448 bones were recovered from this phase, over half of which came from the layers which compose structure 131. Ox and large ungulates are dominant (56%), ovicaprid and small ungulates forming 34% of the total sample.

Two fragments of ox skull showed evidence that they had been cleaved, three mandibles were chopped proximally around the condyles and two in the region of the alveoli. Scapulae were chopped around the neck, separating the humerus, for which one specimen had been smashed in the midshaft and cut across the distal articulation. Metacarpals showed evidence of axial splitting as well as knifecuts across the proximal, midshaft and distal surface. On the hind limb three pelves were chopped on the ilium and ischium, femora and tibiae were smashed in the midshaft, as were a few metatarsals, and one was also split axially. A few rib fragments were chopped and a vertebra chopped laterally.

Ovicaprid scapulae were chopped across the neck and glenoid cavity, radii smashed across the midshaft and one was split axially. Metacarpals showed evidence of both chopping and smashing across the shaft and axial splitting. On the hind limb two pelves were chopped across the ilium and acetabulum, eight tibiae were recorded as smashed across the midshaft and one was split. A few metatarsals were chopped and split.

Only three instances of pig butchery were noted in this phase: a mandible chopped across the condyle, and a radius and a humerus smashed across the midshaft.

Three fragments of horse skull may have been chopped: possible evidence of chopping was observed on the proximal end of a jaw, and also on two pelves. This is not definitive evidence that horse was eaten, but might only represent dismemberment of the carcase after skinning.

All the bone from this phase, with the possible exception of horse, is food debris and has no direct association with the activities on site at this period.

Phase 3: Late Iron Age

In this phase the structures were also used for holding brine. The sample of bone was small, totalling 75. Little butchery was recorded in this phase, largely influenced by the small sample size. This was the first phase where ovicaprid and small ungulate fragments outnumbered ox and large ungulate, though this may not be significant because of the small sample size. All the bone appeared to be food debris.

Phase 4: Mid 1st to early 2nd century

Salt production continued into the Roman period, and in phase 4 bone was recovered from a number of structures. In the sample of 489 bones from this phase, 41% were from ox and large ungulate and 52% were ovicaprid and small ungulate: continuing the trend for favouring the latter seen in phase 3. Goat and sheep were specifically identified from single horn cores. However, ox and large ungulate bones were predominant in some structures. For example, in structure 63 they form 53% of the sample of 214 bones and ovicaprid and small ungulate are only 38%, so there are some differences in distribution within the phase. Pig is only present in small numbers, as in the preceding phases, and only accounted for 6% of all bone.

Many bones were butchered. Ox skull fragments had been cleaved axially and horn cores chopped at the base. Mandibles were chopped and cut around the condyles, under the alveoli and occasionally across the diastema. Scapulae were chopped across the neck and glenoid cavity and sometimes split axially. Humeri were chopped across the distal ends and smashed across the shaft, and radii were largely broken across the shaft. Metacarpals tended to be more complete, some had cut marks, and had been broken or chopped. On the hind limb, pelves were chopped across the ilium and acetabulum, femora and tibiae were broken across the midshaft. Metatarsals, like metacarpals, tended to be more complete. A few vertebrae were laterally trimmed and ribs were chopped.

Ovicaprid skulls had also been axially split and horn cores (including one of goat) chopped through at the base. A few mandibles were chopped across the condyles, beneath the alveoli and diastema. Scapulae showed marks and chops across the glenoid cavity, as well as across the neck. Humeri, radii and to a lesser extent metacarpals, were often broken midshaft, with some cutmarks across the distal surfaces. On the hind limb, pelves were cut across the ischium and chopped on the acetabulum and ilium. Femora were broken across the shaft as were tibiae, which were also cut and split. Metatarsals were broken and occasionally cut.

Butchery marks were seen on a small number of pig bones; a mandible was chopped under the alveoli, two scapulae were chopped across the neck. On the hind limb, two pelves were chopped across the acetabulum, two tibiae were split axially and one chopped across the proximal surface.

Some horse bones had also been chopped: a mandible was chopped across the diastema, three pelves were chopped around the acetabulum, and a femur was split.

A radius and a femur of a dog was found in structure 63.

Phase 5: Early 2nd to mid 3rd century

Only one of the brine tanks possibly continued in use into this phase, and did not contain animal bone. No single structure contained more than 50 bones (from a total of 262 for the phase) so the sample is small. Ox and large ungulate are dominant over the whole phase (forming 56%), 37% belonging to ovicaprid and small ungulate. The butchery for ox and ovicaprid is essentially the same in this period as in phase 4. Pig is again poorly represented.

A single buzzard ulna was found in a layer (1561, S147). This bird used to be a common scavenger in Britain.

Phase 7: Mid 3rd to late 4th century

Phases 7, 8 and 9 can be considered largely as contemporary, but show different activities over different areas of the site.

In phase 7 there is no evidence for salt production. The main structure is a stone building (S26) and possibly a track for access (S28). Only 183 bones were associated with this phase. Largely ox and large ungulate were found and, to a lesser degree, ovicaprid and small ungulate. The bone from the building is from the construction trench fills and the other two groups (S112/S210) are composed of layers.

The number of ox and large ungulate fragments is increased in this phase by loose teeth and skull fragments as well as limb extremities, suggesting that the remains are not high quality cuts of meat. Similarly pig is represented by a few mandible fragments, loose teeth, a metacarpal and one radius. The distribution for ovicaprids and small ungulate is more even, with both extremities and higher quality cuts represented. The butchery pattern for phases 7, 8 and 9 was similar and is described at the end of phase 9.

Phase 8: Mid 3rd to late 4th century

This phase relates to the use of five barrels situated within cuts, their hoops have in some instances remained *in situ*. The barrels are thought to have been used for the storage and settling of brine. A total of 1,362 bones were recovered from this phase, in particular from the disuse fills of structures 16 and 31, which produced 73% of the bone from this phase.

A feature of these later Roman deposits seems to be an increase in the relative importance of ox and large ungulate deposits: 60% in this phase, compared to 18% ovicaprid and small ungulate and 3% pig.

For ox in the structures with the larger samples (S16, S19, and S31) there does seem to be a relatively higher proportion of skull fragments and mandibles compared with other parts of the carcase. However, the samples are small and the numbers of skulls may be affected by increased fragmentation.

The samples for ovicaprid are too small to suggest any emphasis on particular parts of the carcase.

In structure 16, in addition to the usual debris from ox, ovicaprid and pig, a tibia and maxilla of red deer was found and the remains of at least four dogs, the latter in fill 1070. Dogs are discussed in a later section.

Phase 9: Mid 3rd to late 4th century

This phase is concerned with the construction of a large rectangular building (S22), with postholes and cuts south of the building and to the west two clay-lined pits, a cut and other features. The largest quantity of bone comes from this phase, 2,308 bones from nineteen structures.

Two structures in particular, an enclosure (S27) with 1,730 bones and a number of layers (S102) with 228 bones, contributed 85% of the bone in this phase. Some of the contexts which comprise the ditch may include some later material from phase 11, although they have been included here.

The dominance of ox and large ungulate is increased in phase 9 to 70%, ovicaprid and small ungulate is 18%, and pig 3%.

In some contexts from the largest sample, the ditch (559 and 564), concentrations of certain parts of the carcase of ox are apparent. In 559, of a total of 175 ox bones, 24 are scapula, 13 skull, and 13 mandible fragments. The remaining bones represent the limbs, vertebrae and ribs and, apart from 27 rib fragments, no single anatomy numbers more than nine. Similarly in 564, of 439 ox bones, 122 are skull, 25 mandible and 30 scapula fragments. Many of the skull fragments showed evidence of butchery, including cleaving and possibly poleaxing. The high number of skull remains is not only a feature of high fragmentation, since varying stages of completeness were recognised.

No other disposal patterns for ox or any other species were recognised for this phase. Dog/fox and hare were identified from single tooth fragments.

The butchery pattern for phases 7, 8 and 9 echoes that of earlier phases. The larger sample of chopped ox mandibles is dominated by those chopped across the gonion and condyles, with a small proportion chopped across the diastema and under the alveoli. Only one mandible was cut, across the proximal end. As previously mentioned, skulls were chopped and cleaved, with the horn cores chopped away.

Scapulae were most frequently chopped across the neck and sometimes had been cut there. A few had been chopped across the glenoid, separating out a meat joint distally delimited by chopping through the midshaft of the humerus, which was also sometimes chopped across the distal articulation. Radii were chopped and cut proximally, sometimes chopped across the midshaft and split axially. Metacarpals and metatarsals were again less fragmentary than other limb bones, being broken midshaft, proximally chopped and split axially.

On the hind limb, pelves were chopped across and around the acetabulum, and less frequently on the ilium and ischium. The femur was not well represented and was broken at the midshaft, as well as occasional examples of chopping at the proximal and distal ends. The poor representation of this bone may be attributable to heavy fragmentation/butchery, so that it is often not specifically identifiable rather than a true absence. Tibiae were also broken at the midshaft, and sometimes chopped or cut.

Ovicaprid skulls were cleaved in three instances and a few mandibles (nine) were chopped and smashed proximally. Scapulae were chopped across the neck and humeri broken across the midshaft or split axially. Radii were largely broken midshaft or split, and a few metacarpals and metatarsals were split, smashed or sometimes chopped. On the hind limb, pelves were chopped or cut about the acetabulum. The femur was broken midshaft, as was the tibia, which was also chopped midshaft and proximally. The difference between butchery on ox and ovicaprid was that, although the chop marks were found in the same places in both, for the latter it was generally less intensive as one would expect with a smaller carcase.

Since pig only made up 3% in these phases, the incidence of butchery was relatively low. It was most frequently observed on mandibles under the alveoli and occasionally at the proximal and distal end. Humeri and radii were chopped and broken midshaft. On the hind limb, a few pelves were chopped on the acetabulum and ilium, tibiae were broken midshaft, and chopped proximally and midshaft.

Except for one apparently 'split' metatarsal, no butchery was observed on horse in these phases.

Phase 10: 5th to 11th century

Only three bones were recovered from this phase, two ovicaprid and one large ungulate.

Phase 11: 12th to 14th century

In this phase activity is concentrated in the eastern half of the excavation and includes disuse fills from the enclosure ditch (S27) constructed in phase 9. The bone from this structure is 70% of the entire sample from this phase, which totals 958.

In the ditch there were some concentrations of ox skull and mandible fragments, and in context 523 also a relatively high concentration of first phalanges, which are likely to be waste material. None of the other samples were large enough to be significant.

Butchery on ox skulls included axial chopping, and two possible instances of poleaxing. Three horn cores had also been chopped. The main area butchered on mandibles continued to be the proximal end around the condyles. Scapulae were chopped across the neck and on the glenoid cavity as before. Humeri and radii were mainly broken and chopped midshaft, as were metacarpals and metatarsals, some of which had also been split. On the hind limb, pelves were chopped through the acetabulum, femora chopped on the proximal to midshaft surface and tibiae broken or chopped across the midshaft.

Butchery on ovicaprid skulls included axial cleaving, chopping through the base of the horn cores, and chop marks at the back of the skull. The condyles of two mandibles were chopped off and one scapula was chopped through the neck. Six humeri were broken at the midshaft, two radii were split axially and one chopped both proximally and distally. Metacarpals and metatarsals were occasionally split axially and showed cut marks at the proximal end. Pelves were most commonly chopped around or on the acetabulum, femora and tibiae were broken midshaft.

Horse included one scapula chopped across the neck.

The butchery practices do not appear to change from the Roman period to the medieval: rib fragments are still chopped and vertebrae laterally trimmed.

Phase 12: 15th to 18th century

The number of bones is inflated by a cat skeleton from structure 197 and a foetal pig from structure 32, without which only 248 bones would have been recovered. The structures containing bones from this phase are largely ditches and their disuse fills. Ox and large ungulate fragments are almost twice as numerous as ovicaprid and small ungulates.

A goat horn core was identified from context 1546 (S156). The size of the samples is too small for any suggestion of differential disposal of carcase/species to be made. The butchery patterns echo those of phase 11.

Ageing of ox, sheep and pig

Although the epiphyseal fusion of the post cranial skeleton was recorded, tooth eruption and wear is a more precise way to assess age stages. The method of Grant (1975) was used and a numerical value can be calculated for a relatively complete mandible. Where the value is estimated, at least two teeth were present and their wear stages compared with those of Grant. Their values are shown by phase in Table 3.

For ox, the third permanent molar erupts around the value of 30, often two years of age. The largest samples are in the Roman occupation, phases 4 to 9, but are only 38 in all. However, it can be said that the majority (87%) are over three years old, as are two out of three in the Iron Age and all five in the medieval period.

Ovicaprids, suggested to be largely sheep with only a few goats present judging from Boessneck's (1969) metrical separation index on the metapodials, have larger samples. This is particularly true in phase 4 and, to a lesser extent, in phase 5. However, the largest total sample for ovicaprid is from phase 9, suggesting a

Table 3 Ageing of ox, ovicaprid, and pig mandibles (using Grant 1975)

Phase 1	2	3	4	5	7	8	9	11	12
Ox									
16e	41e		21e	16	34	21e	21e	46x3	41
39–42e			22e	41	37	34e	45		47
			39e	45	38e	37	46		
			40———>47		39e	41x2			
			42———>		41x3	44x3			
			44———>			45x4			
			49e———>			46			
			49———>			47			
						49			
						52			
Ovicaprid									
31	17e	40e	11e	8–10e	41	23	24ex2	22e	30–33e
34	33		21e	11–17e		25e	25e	24e	33
	33x2——————>22e			19e		36	27ex2	25e	35
	34		24e	22e		37	48	29	48
	40x2		25e	25–35e				33	
			28e	27e——————————>				34x2	
			33	32e				35———>	
			34	37				36———>	
			34———>40					40	
			35x2	43				45———>	
			36	44					
			38e						
			38———>						
			39						
			41x2——>						
			42x2——>						
			44x2——>						
			48						
Pig									
			13e	6e	23e	5e	10e	19———>10e	
			17e	26	42———>10ex2				22ex2
			28–38e			28e			25e
			39			34–37e			

e = estimated value on range due to incomplete jaw
——> = dated over more than one phase

proportionally higher concentration of mandibles with ageable tooth rows in phases 4 and 5.

The value 20–25 is estimated to be between one and a half to two years. Of the total sample from all phases of 70 mandibles, only four have a value of less than 20 and thirteen less than 25. Including the small samples from phases 1 to 3, most sheep were kept until at least two years and later, and would have been primarily used for wool, milk and breeding, with meat as a secondary product. This applies to all phases. The samples are not large enough to suggest any culling patterns for the mature animals.

The number of pig mandibles that could be aged is very small (nineteen), none earlier than the Roman period. The end of the second year is thought to be indicated by 20–25 and, as is usual for pig, a higher proportion of immature animals is indicated, meat and hide being their only use.

Abnormalities were observed on some jaws of ox and sheep. For ox these were mainly two congenital conditions: the absence of the second premolar and a reduced last cusp on the third molar. These conditions were found from phase 2 to 11. In one mandible from phase 8 the third molar was absent and on another an extra nutrient foramen was observed. As congenital abnormalities these do not imply any nutritional problems, but seem to imply relatively high numbers compared with the number of normal occurrences.

The number of third molars with a reduced third cusp, against the total number for the phase (excluding loose teeth), is shown below:

Phase 2; of 2 mandibles retaining M3, 1 had a reduced cusp.
Phase 4–5; of 6 mandibles retaining M3, 1 had a reduced cusp.
Phase 5; of 5 mandibles retaining M3, none had a reduced cusp.
Phase 7–9; of 7 mandibles retaining M3, 1 had a reduced cusp.
Phase 7; of 4 mandibles retaining M3, 3 had a reduced cusp.
Phase 8; of 17 mandibles retaining M3, 3 had a reduced cusp.
Phase 9–11; of 12 mandibles retaining M3, 4 had a reduced cusp.
Phase 9; of 5 mandibles retaining M3, 1 had a reduced cusp.
Phase 11; of 11 mandibles retaining M3, 1 had a reduced cusp.

Therefore over phases 2 to 11, 22% of permanent third molars had a reduced third cusp. Similarly, a relatively high proportion (some 48%) of mandibles showed the permanent second molar to be absent, the number in each phase is shown below:

Phase 2; of 2 mandibles, none showed PM2 to be absent.
Phase 4–5; of 3 mandibles, 2 showed PM2 to be absent.
Phase 5; of 2 mandibles, 1 showed PM2 to be absent.
Phase 7–9; of 3 mandibles, 2 showed PM2 to be absent.
Phase 8; of 7 mandibles, 4 showed PM2 to be absent.
Phase 9–11; of 6 mandibles, 3 showed PM2 to be absent.
Phase 11; of 6 mandibles, 2 showed PM2 to be absent.

On sheep mandibles the second premolar was absent in two cases, and one mandible had an extra nutrient foramen. One mandible showed resorbtion and exostoses round the alveoli and another had swollen tooth roots with resorbtion round the alveoli. Both may have resulted in feeding difficulties and were of Roman date.

No pathology was observed on pig jaws.

The metrical data

Measurements were taken after Jones *et al* (1981) and are accessible in archival form. From some length measurements various factors can be used to calculate the withers heights, usually using the lengths of whole bones. The withers heights for ox and ovicaprids are shown in Table 4, divided by phase. The largest sample for ox is from phase 8, mostly from metapodials which is some reflection of the completeness of these bones in this phase. The greatest range in size is seen within phases 7 to 9, the late Roman period, where shoulder heights of 100 to 121cms occur.

Despite the less intense fragmentation of a small carcase, the smaller numbers of sheep bones from the site result in far fewer withers heights being available. The four Iron Age examples, ranging from 54 to 55cms, have a smaller upper limit than those of the Roman period, where up to 62cms have been calculated.

It is not possible to calculate withers heights for pig, the small number of pig bones resulted in few measurements being taken. There were not enough proximal radius widths for comparison with those from Friar Street.

Occurrence of other species

Horse

Horse bones were found in small quantities in all phases. Some bones showed evidence of chop marks and knifecuts, which may not necessarily be evidence that horse was eaten, but the result of skinning and dismembering the carcase. Certainly horse was not subjected to the same intensive fragmentation and butchery as ox.

Table 4 Withers heights in cm for ox, ovicaprid, horse and dog (using the factors of Fock, Maltolcsi, Teichart, Kiesewalter and Harcourt respectively)

Phase 1	2	3	4	5	7	8	9	11	12
Ox									
	1089m		1092——>r			1096r	1168m	1188h	1083r
			1052——>m			1075r		1119m	
			1102——>m			1211m		1107m	
			1101——>m			1175m		1183m	
			1230——>m			1113m		1170m	
			1144——>m			997m		1113m	
				1052m		1071m		1101m	
				1095m		1138m		1132m	
					1156——>m			1090m	
						1101x2			
						1022m			
						1128m			
						1062m			
						1073m			
						1177m			
Ovicaprid									
545m									
	543——>r		548——>r		595——>m				
	553r		532——>m		538——>r				
	553m		533——>m		625——>m				
			552——>t			530——>m			
			561m			586m			
			583m				561m		
		514r							
Horse									
	1324r		1343——>t			1417m		1382m	
	1253t		1386——>t			1526t			
						1448t			
						1087m			
Dog									
			531r			440r	270t		
						437r			
						504r			
						449f			
						446f			
						446f			
						292f			

Key
All heights in mm
——> = dated over more than one phase
r = radius h = humerus m = metacarpal or metatarsal f = femur t = tibia

The estimated withers heights are shown in Table 4. In the Iron Age in phase 2, ponies of twelve and thirteen hands are present, and in the early Roman period ponies of thirteen hands. The greatest size differences are in the late Roman period, where ten to fifteen hands (the latter is a horse) was found. The only data for the medieval period is thirteen hands in phase 11.

All the ageing evidence for horse was for mature animals, with all teeth in wear and all epiphyses fused. The exception of a skull fragment from phase 2 in which the presence of deciduous teeth indicated the animal was less than three years old (Thompson 1949, 240).

Exostoses were seen on a metacarpal from phase 5 and on the proximal end of a metatarsal from phase 4–5. This development of 'extra bone' may be a reaction to stress around a joint, which could increase with age.

Red deer

A few red deer bones were found throughout the occupation of the site including a cast antler (from P1), indicating this valuable raw material was collected in the Iron Age, a sawn tine from the late Roman period, a late Roman/medieval skull with the antlers sawn off, a maxilla from phase 11 with all teeth in wear and a few postcranial fragments.

Dog

Dog bones were found as isolated bones and as skeletons. In particular, in structure 16 in phase 8 (2:D14) the remains of at least four dogs were found. Two of these were male and the sizes range between 43 and 50cms at the shoulder (Table 4). Exostoses resulting in the fusion of some lumbar vertebrae, and others in a less severe condition showing extra bone around the articulation, indicated elderly arthritic animals. Eburnation on the glenoid cavity of a scapula, an abrasion on the proximal end of a radius, and a fractured rib, were also evident on one of the skeletons from this feature.

Immature animals were represented by unfused femurs from phases 4 and 5.

Some of the bones of other species from this site showed evidence of dog gnawing, particularly on the articulations.

Cat

A cat skeleton from phase 12 (post-medieval) showed ante-mortem tooth loss in the maxilla and mandible and was a mature animal.

Birds

Apart from the buzzard bone, a number of domestic fowl and goose bones were identified. Domestic fowl was present from phase 4 (early Roman) onwards, totalling 61 bones. All parts of the postcranial skeleton were represented, particularly the tibiotarsus and humerus. Knifecuts were only observed on the humerus, on the proximal and distal surfaces in late Roman deposits, and on the proximal surface in a medieval deposit.

Four goose bones were identified, and a late Roman furcula had a hole of traumatic origin. A carpometacarpus and ulna were of medieval date. Both these species are likely to represent food remains, any cut marks being the result of preparation for the table.

Conclusion

Comparison of the Old Bowling Green and Friar Street animal bone is given at the end of the latter report (Chapter 32).

16 Human bone *Michael Nellist and Simon Woodiwiss*

A near complete articulated skeleton was recovered from the final fill (1686) of a brine tank (S7) assigned to phase 5 (early 2nd to mid 3rd century AD). The position of the limbs suggests that the body had been thrown into the already partially filled pit (Pl 2). A radiograph (Pl 6) of the lower left molar region showed that there was a congenital absence of the lower left third molar. Another radiograph (Pl 7) of the right tibia showed the existence of two distinct, and numerous fainter, Harris or transverse arrested growth lines. Though it was not possible to determine the sex due to the immaturity of the skeleton, the presence of an unerupted upper right third molar crown (Pl 8) and the nearly closed root apices of both the lower left and upper right second molars, suggest an age of fourteen to fifteen years.

This immature skeleton was reasonably preserved and did not appear to show any signs of ante-mortem trauma. A full report is given elsewhere (2:E13–14). The presence of transverse arrested growth lines in the right tibia indicated periods of previous starvation and or illness or psychological stress.

Fragmentary skeletal remains were also recovered from eighteen other contexts, though these were not submitted for specialist comment. The finds were usually of single bones, with the exception of a number of vertebrae and infant bones recovered from ditch fills (S29 and unstratified). Four fragments were from Iron Age phases: the rest (with three exceptions) were from Roman phases, and these were mostly mid 3rd to late 4th century in date.

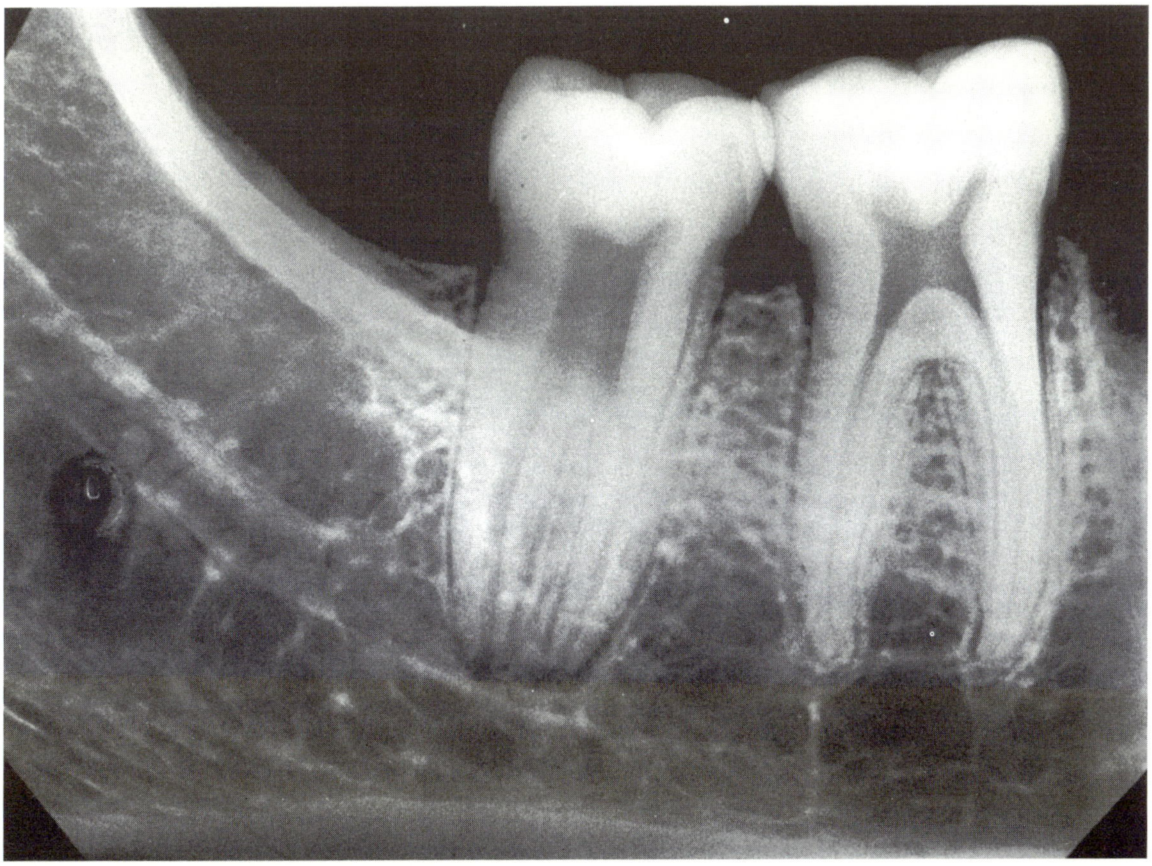

Plate 6 Radiograph of lower left molar region

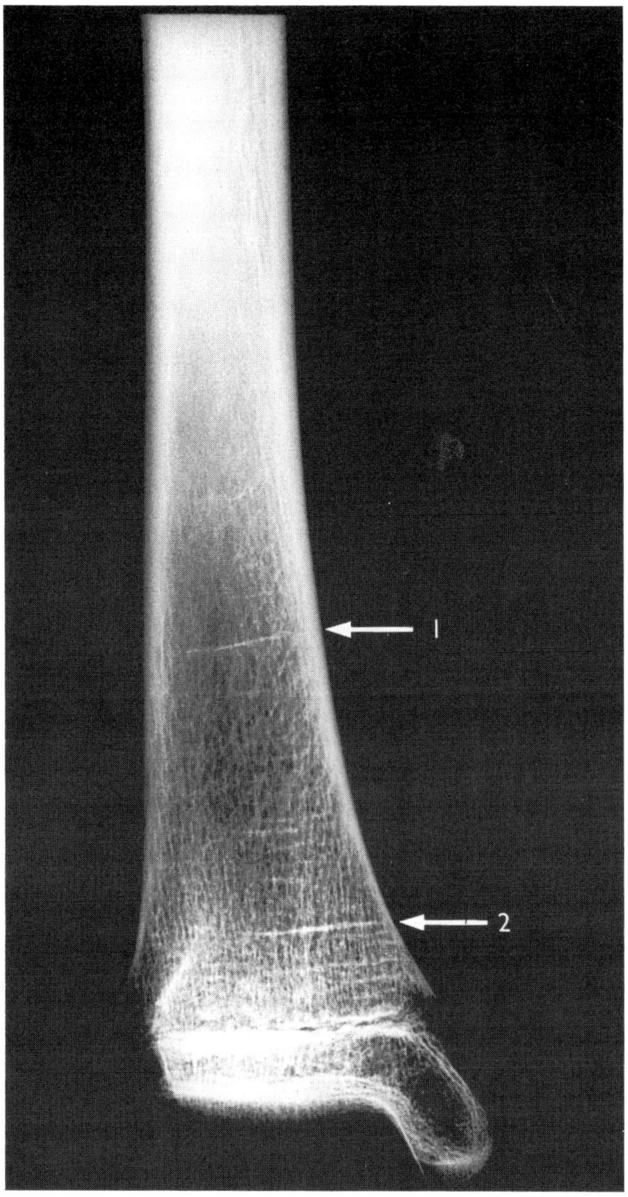

Plate 7 Radiograph of right tibia showing arrested growth lines (1 and 2)

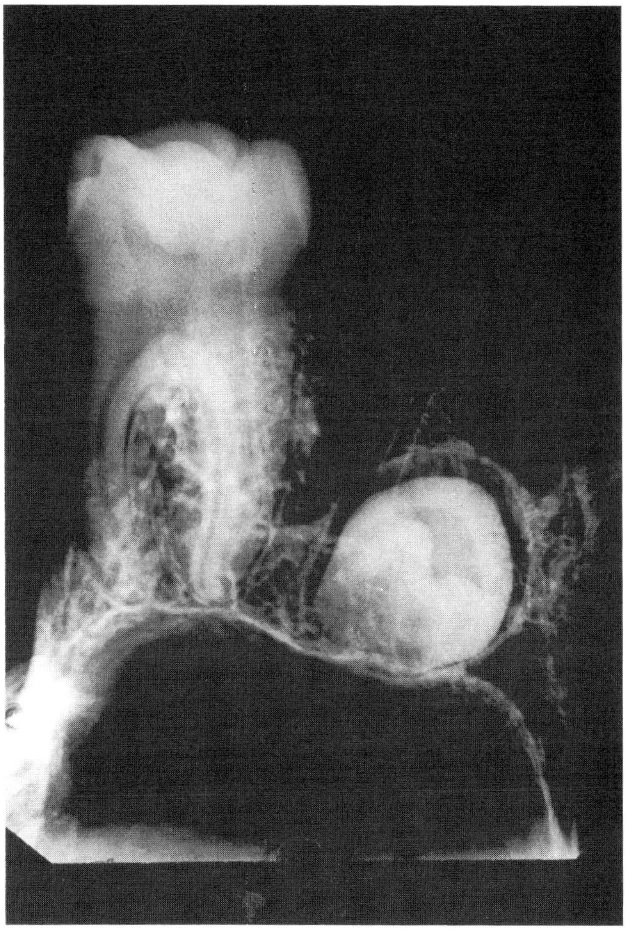

Plate 8 Radiograph of upper right third molar region

17 Worked bone and antler *Anne Crone*

Of the 23 pieces of worked bone and antler, six (and possibly one other) were from pre-medieval contexts. A full catalogue exists in archive.

Fig 54

1 Complete highly polished bone needle with a flattened, perforated head. Context 448, structure 83, phase 2.
2 Bone awl. Context 2183, structure 30, phase 4.
3 Double-sided bone comb. Context 166, structure 30, phase 5?.
4 Bone pin with groove around top. Context 1624, structure 177, phase 5.
5 Bone pin with spherical head. Context 1714, structure 177, phase 5.
6 Perforated bone fragment (burnt). Context 731, structure 94, phase 7–8.
7 Antler, all sides dressed and thin knife marks visible, curved and tapered along length. Context 1153, structure 19, phase 8.

The needle (no 1) is one of very few indicators of Iron Age domestic activity from both this and the Friar Street excavations. Of the post-medieval material, only a tuning key (100, S69, P14) is illustrated (Fig 54 no 8). The remainder included five buttons, two toothbrushes, three handles, a spoon, two needles (all of bone), and an antler fragment.

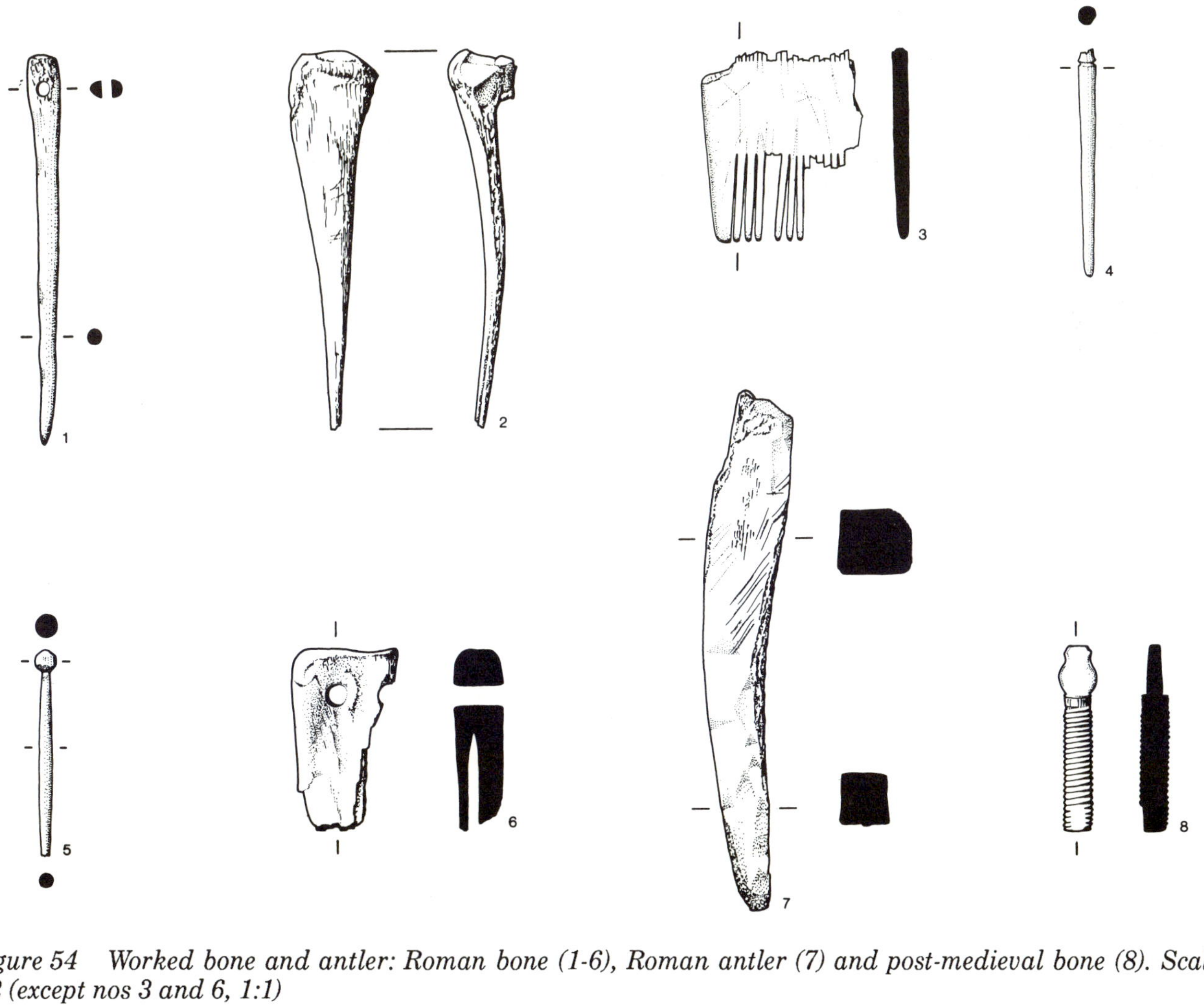

Figure 54 Worked bone and antler: Roman bone (1-6), Roman antler (7) and post-medieval bone (8). Scale 1:2 (except nos 3 and 6, 1:1)

18 Environment *Susan Colledge and James Greig*

Summary

Various suitable contexts were examined for plant remains (pollen, seeds, etc) to try to find out more about the salt working industry which had been on the site. Most of the indications are of weed vegetation, so the plants may well represent phases of abandonment of those particular parts of the site. Remains of cereals with cornfield weeds, and of sedges and other wetland and marsh vegetation, could perhaps represent materials such as corn, straw and sedge that were brought to the site for use as building material, animal fodder or fuel. There is little sign of a salt-tolerant vegetation from the excavation. Compared with more domestic sites there is far less sign of crop plants.

Introduction

Droitwich has plenty of potential for environmental archaeological work because of the sticky clay derived from the Mercian Mudstone (Keuper Marl), which makes the excavation so difficult, but which also provides good waterlogged conditions and hence preservation of a range of plant and animal remains. Part of the interest in the results is that it is mainly an industrial rather than a domestic site and there are few environmental results from sites such as this. A further point of importance is to see whether there is any sign from the flora of the saline conditions caused by the salt springs.

In the following discussion the plants are usually discussed in taxonomic order, which is the same as their order in Table 5. Selected examples of the recovered plant remains are illustrated in Plates 9 and 10.

Wetland vegetation

This is represented by a number of taxa, many of them characteristic of banksides and such wet places. In context 1961 (P5–9) *Ranunculus flammula* (lesser spearwort), *Hydrocotyle vulgaris* (pennywort), *Apium graveolens* (wild celery), *A nodiflorum* (fool's watercress), *Polygonum lapathifolium* (pale persicaria), *Juncus* sp (rush), *Eleocharis uniglumis/palustris* (spike-rush), *Schoenoplectus tabernaemontani* (bulrush) and various species of *Carex* (sedges) are present. Of these, most are to be found nowadays, although pennywort was considered rather uncommon (Amphlett and Rea 1909) and has doubtless not got any more common since. The only plants indicative of standing water are *Ranunculus* subg, *Batrachium* (water crowfoot) and *Glyceria* sp (reedgrass) and there was only one seed of each.

The other botanically rich sample (559, S27, P9(11)) provided a fairly similar wetland flora. The post-medieval material from context 3008 (S34, P12) was very rich in *Ranunculus sceleratus* (celery-leaved crowfoot) which grows in mud, preferably when enriched with animal dung. It also had an aquatic snail fauna and must therefore have contained standing water. Waterlogged samples usually have a wetland element in the plant remains, and when the site is in a damp river valley perhaps it is only surprising that there were not more.

Weeds and ruderals

All samples had floras that were mostly weeds and ruderals, both in numbers of taxa and in abundance of remains. Context 1961 (S29, P5–9) provided a fairly typical flora: most of the taxa were from weeds of one kind or another, the list dominated by the twenty or so weed taxa. Some of

Plate 9 Selected examples of the recovered plant remains

Table 5 Species list

Name	Sample context number								
	1707	1775	1961	132	559	578	1070	1098	3008
Pteridium aquilinum L	–	1	3	1	–	–	–	–	–
Ranunculus cf acris L	–	–	1	–	–	–	–	–	–
Ranunculus subg Ran	–	2	4	–	8	2	–	21	2
Ranunculus parviflorus L	–	–	1	–	–	–	–	–	–
Ranunculus flammula L	–	1	3	–	–	–	–	–	1
Ranunculus sceleratus L	–	1	–	–	9	–	–	–	600+
R subg *Batrachium*	–	–	1	2	–	–	–	–	–
RANUNCULUS type	3	–	–	–	5	2	3	1	5
Papaver argemone L	–	–	5	1	3	–	–	–	–
Papaver sp	–	–	–	–	1	–	–	–	–
Fumaria sp	–	–	1	–	–	–	–	–	3
Brassica sp	–	–	1	–	–	–	–	–	1
Thlaspi arvense L	–	1	–	–	–	–	–	–	–
CRUCIFERAE	1	–	–	–	3	1	4	–	4
Viola sp	–	1	1	1	1	–	–	–	–
Stellaria media group	–	2	3	–	–	–	–	–	31
Stellaria palustris/graminea	–	–	1	–	1	–	–	–	–
Arenaria sp	–	–	1	–	–	–	–	–	–
Scleranthus annuus L	–	1	–	–	–	–	–	–	–
CARYOPHYLLACEAE	–	–	–	–	3	–	–	–	1
Monia fontana ssp *chondrosperma* (Fenzl) S M Walters	–	1	2	–	–	–	–	–	4
Chenopodium cf vulvaria L	–	–	–	–	–	–	–	–	1
Chenopodium cf album L	–	44	37	9	31	–	–	1	2
Chenopodium ficifolium Sm	–	–	–	–	–	1	–	–	–
Chenopodium murale L	–	–	–	–	3	–	–	–	–
Chenopodium sect *Pseudoblitum* (Gren) *Aschers*	–	–	32	–	–	–	–	–	1
Atriplex sp	–	12	9	–	259	–	–	–	218
CHENOPODIACEAE	5	–	–	–	13	2	15	1	15
TILIA	1	–	–	–	–	–	–	43	1
cf *Vicia*	–	1*	1*	–	–	1*	–	–	–
Vicia sativa L	–	1*	–	–	–	–	–	–	–
TRIFOLIUM REPENS	1	–	–	–	24	–	5	–	–
TRIFOLIUM PRATENSE	–	–	–	–	7	–	3	–	–
Trifolium sp *calyx*	–	–	4	–	–	–	–	–	–
Trifolium sp *corolla*	–	1	4	–	–	–	–	–	–
Lotus tp	–	–	–	–	4	–	–	–	–
LEGUMINOSAE	–	–	–	–	3	–	–	–	–
FILIPENDULA ULMARIA	–	–	–	–	–	–	2	–	–
Rubus cf idaeus L	–	–	–	–	4	–	–	–	–
Rubus cf fruiticosus agg	–	2	5	–	24	2	–	–	–
Rosa or Rubus (thorns)	–	1	1	–	–	–	–	–	–
Potentilla anserina L	–	1	–	–	–	–	–	–	–
Potentilla erecta (L) *Rauschel*	–	–	2	–	–	–	–	–	–
Potentilla cf *Reptans* L	–	–	5	–	8	–	–	–	–
POTENTILLA tp	1	–	–	–	9	–	–	–	1

Table 5 (continued)

Name	Sample context number 1707	1775	1961	132	559	578	1070	1098	3008
Aphenes arvensis L	–	1	2	–	–	–	–	–	–
Aphanes microcarpa (Boiss and Reuter)	–	1	5	–	–	–	–	–	–
Prunus cf *institia* (L) C K Schneid fr	–	1	–	–	–	–	–	–	–
Crataegus sp	–	–	–	–	2	–	–	–	–
Crataegus/Prunus (thorn)	–	–	1*	–	–	–	–	–	–
ROSACEAE	1	–	–	–	–	–	–	–	–
Hydrocotyle vulgaris L	–	–	3	–	–	–	–	–	–
Torilis japonica (L) Houtt	–	–	1	–	3	1	–	–	–
Conicum maculatum L	–	–	–	–	101	3	–	–	30
Apium graveolens L	–	–	1	–	1	–	–	?1	37
Apium nodiflorum (L) Lag	–	2	4	–	–	–	–	–	–
Petroselinum crispum (Mill) Nyman	–	–	–	–	5	–	–	–	–
Aethusa cynapium L	–	2	–	–	1	–	–	–	1
? *Foeniculum vulgare* L	–	–	–	–	–	–	–	–	1
Pastinaca sativa L or *Heracleum sphondylium* L	–	–	2	–	2	–	–	–	–
Daucus carota L	–	–	–	1	–	–	–	–	–
UMBELLIFERAE	10	–	–	–	6	2	5	–	8
Polygonum aviculare L	–	3	27	5	7	–	–	6	–
POLGONUM BISTORTA type	5	–	–	–	–	–	9	–	2
Polygonum lapathifolium L	–	1	3	–	–	–	–	–	–
Polygonum convolvulus L	–	–	1	2	–	–	–	–	–
Rumex acetosella L	–	4	8	–	15	–	–	–	–
Rumex sp	–	4	5	–	9	–	–	–	4
RUMEX tp	3	–	–	–	3	–	2	–	–
Urtica urens L	–	–	11	–	1	–	–	–	3
Urtica dioica L	–	11	15	–	4	–	–	1	15
URTICA	–	–	–	–	4	–	–	–	–
ULMUS	–	–	–	–	–	–	–	–	88
Betula sp	–	1	–	–	–	–	–	1	–
BETULA	2	–	–	–	1	–	1	1	3
ALNUS	1	–	–	–	1	–	4	103	6
Alnus glutinosa (L) Gaertn	–	–	1	–	–	–	–	1	–
Corylus avellana L	–	–	1	–	–	–	–	–	–
CORYLUS	4	–	–	–	4	–	15	63	5
FAGUS	–	–	–	–	–	–	–	–	1
QUERCU	4	–	–	–	1	2	7	29	7
SALIX	–	–	–	–	–	–	1	–	–
ERICALES	–	–	–	–	6	–	2	–	–
cf *Fraxinus excelsior* L	–	1tw	–	–	–	–	–	–	–
FRAXINUS	–	–	–	–	1	–	–	–	–
Atropa bella-donna L	–	–	–	–	35	–	–	–	–
Hyoscyamus niger L	–	–	2	–	42	–	–	–	–
Solanum dulcamara L	–	–	–	–	–	–	–	–	–
Solanum nigrum	–	–	–	–	–	–	–	–	2
Verbascum cf *pulverentulum* Vill	–	–	–	–	–	–	–	–	1

Table 5 (continued)

Name	Sample context number 1707	1775	1961	132	559	578	1070	1098	3008
RHINANTHUS tp	1	–	–	–	2	–	–	–	3
Mentha cf *arvensis* L	–	–	–	–	2	–	–	–	–
Mentha sp	–	–	3	–	–	–	–	–	–
MENTHA tp	–	–	–	–	1	–	–	–	–
PRUNELLA tp	2	–	–	–	–	–	–	–	–
Prunella vulgaris L	–	2	4	–	–	–	–	–	–
? *Stachys palustris* L	–	–	–	–	–	–	–	2	–
? *Stachys sylvatica* L	–	–	–	–	1	–	–	–	–
Ballota nigra L	–	–	–	–	11	–	–	–	–
Lamium sp	–	–	–	–	1	–	–	–	51
LAMIUM tp	2	–	–	–	?1	–	–	–	–
Galeopsis sp	–	1	1	–	–	–	–	–	–
Ajuga reptans L	–	–	1	–	–	–	–	10	–
Plantago media L	–	–	3	–	–	–	–	–	–
PLANTAGO MEDIA	–	–	–	–	1	–	–	–	–
PLANTAGO LANCEOLATA	6	–	–	–	39	8	26	–	4
CAMPANULACEAE	–	–	–	–	1	–	–	–	–
cf *Galium* sp	–	1	–	–	–	–	–	–	–
GALLIUM tp	–	–	–	–	–	–	3	–	1
Sambucus nigra	–	1	2	–	130	5	–	106	8
SAMBUCUS NIGRA	–	–	–	–	1	–	–	1	2
Valerianella locusta (L) Laterrade	–	1	–	–	–	–	–	–	–
Valerianella rimosa Bast	–	–	–	1	–	–	–	–	–
Valerianella dentata (L) Pollich	–	1	4	–	–	–	–	–	–
Senecio jacobea L	–	–	–	–	–	–	–	–	–
Anthemis cotula L	–	–	–	–	–	–	–	–	34
Tripleurospermum inodorum (L) Schulz Bip	–	–	3	–	–	–	–	–	–
ANTHEMIS tp	6	–	–	–	–	1	5	–	10
ASTER tp	–	–	–	–	–	–	1	–	–
ARTEMISIA tp	–	–	–	–	6	–	–	–	2
ARCTIUM tp	1	–	–	–	–	–	–	–	–
Cirsium cf *palustre* (L) Scop	–	–	–	–	–	–	–	–	3
Cirsium cf *arvense* (L) Scop	–	–	1	–	–	–	–	–	–
Cirsium sp	–	–	–	–	1	–	–	–	–
CIRSIUM tp	–	–	–	–	–	2	–	–	–
CENTAUREA CYANUS	–	–	–	–	–	–	–	–	1
CENTAUREA NIGRA tp	–	–	–	–	5	15	3	–	1
TUBULIFLORAE	–	–	–	–	12	–	–	–	10
Lapsana communis L	–	–	–	–	–	–	–	–	1
Leontodon sp	–	–	–	–	–	–	–	–	2
Sonchus oleraceus L	–	–	–	–	–	–	–	–	29
Sonchus asper (L) Hill	–	–	1	–	–	–	–	–	6
Taraxacum sp	–	–	–	–	–	–	–	–	2
LIGULIFLORAE	37	–	–	–	42	9	17	2	8
Alisma plantago-aquatica L	1	–	–	–	–	–	–	–	–
? POTAMOGETONACEAE	–	–	–	–	1	–	1	1	–

Table 5 (continued)

Name	**Sample context number**								
	1707	**1775**	**1961**	**132**	**559**	**578**	**1070**	**1098**	**3008**
Juncus sp	–	1	19	–	4	–	–	–	2
Luzula sp	–	–	1	–	–	–	–	–	–
TYPHA LATIFOLIA tp	–	–	–	–	–	–	2	–	–
Eleocharis uniglamis/palustris	–	2	6	–	71	1	–	–	2
Schoenoplectus tabernaemontani (C C Gmelin) Palla	–	5	3	–	–	–	–	–	–
Isolepsis setacea L	–	1*	1	–	–	–	–	–	–
Carex cf *flava* agg	–	–	–	–	24	–	–	15	1
Carex cf *riparia* Curt	–	–	–	–	5	–	–	–	–
Carex cf *limosa* L	–	–	–	–	–	5	–	–	–
Carex cf *hirta* L	–	–	–	–	7	–	–	–	–
Carex cf *otrubae* Podp	–	–	–	–	7	–	–	–	–
Carex cf *divisa* Huds?	–	–	–	–	1	–	–	–	1
Carex cf *divulsa* Stokes	–	–	16	–	2	–	–	–	–
Carex cf *ovalis* Good	–	–	–	–	–	–	–	–	1
Carex sp	–	1,1*	19	1	5	2	–	–	–
CYPERACEAE	1	–	–	–	14	–	8	–	4
Bromus sp	–	1*	3*,1	–	–	–	–	–	–
Glyceria sp	–	1	1	–	–	–	–	–	–
Gramineae (mostly Poa)	–	1	11	–	–	–	–	–	–
GRAMINEAE	55	–	–	–	164	13	84	14	35
Triticum sp (grain)	–	–	4*	–	1*	–	–	–	–
Triticum spelta gl/b	–	1	4*,2	–	1*	–	–	–	–
Cerealia (grain)	–	1*	2	–	–	1*	–	–	–
Cerealia (culm nodes)	–	1,1*	1	–	–	–	–	–	–
CEREALIA	–	2	–	6	1	1	–	8	–
Total macroscopic remains	–	129	335	25	901	24	–	160	1111
Total pollen	396	–	–	–	58	230	291	226	2004
Other remains									
wood charcoal	–	+++	+++	+	–	–	–	–	–
beetle remains	–	+	+	+					
fly puparia	–	+	+	–					
cadis cases	–	–	+	–					
twigs	–	+	+++	–					
coal	–	–	+						

All remains waterlogged except when marked *. Remains other than seed are fruitstone fragments (fr), twigs (tw), and glume bases (gl/b), and flower parts calyx and corolla. Order and taxonomy from Clapham *et al* 1962

these, such as *Chenopodium* (goosefoot), *Atriplex* (orache) and *Urtica urens* (lesser nettle), *Tripleurospermum inodorum* (mayweed) and *Sonchus asper* (sow thistle), are very widespread and provide little information beyond the presence of disturbed soil. *Chenopodium* and *Atriplex* were especially abundant in context 559 (S27, P9(11)), and *Atriplex* in 3008 (S34, P12). These weeds could also have grown in cornfields, but there is no actual evidence that this was the case.

Hyoscyamus niger (henbane) was present in two samples, and quite abundant in one of them (559). It is a rare plant now according to the *Botany of Worcestershire* (Amphlett and Rea 1909), although recorded from Droitwich, and it is somewhat unsuccessful when grown outdoors now. There has been discussion (Greig 1988, 373) as to why it seems to have been common on Roman sites such as Droitwich, although without a clear answer.

There is also a small number of plants whose habitat is ruderal or rough grassland, the sort that are found now on the verges of roads. *Potentilla erecta* (common tormentil) and *P reptans* (creeping tormentil), *Torilis japonica* (upright hedge-parsley) and *Ranunculus parviflorus* (small-flowered buttercup) are the most typical examples. *R parviflorus* was described in the *Botany of Worcestershire* (Amphlett and Rea 1909) as local and rather rare, although Dodderhill and the canalside below Salwarpe were given as places where it had been found (Amphlett and Rea 1909) and, like henbane, it is one of the weeds which seem to have been commoner in Roman times.

A range of weeds is also evident from the pollen spectrum from context 1707 (S6, P4), such as Chenopodiaceae (goosefoot, etc), *Rumex* type (docks), *Anthemis* type (which includes *Anthemis* and *Tripleurospermum*) and there are other pollen types that probably represent weeds, such as Cruciferae. The weeds are mainly insect pollinated, and may therefore be less well represented by pollen than by their seeds.

Weeds of specific habitats

Cornfield weeds

In context 1961 (S29, P5–9) there was a large group, including *Papaver argemone* (poppy), *Fumaria* sp (fumitory), *Arenaria* sp (sandwort), *Scleranthus annuus* (knawel), *Aphanes* species (parsley piert), *Polygonum convolvulus* (bindweed), *Rumex acetosella* (sorrel), and *Valerianella dentata* (cornsalad), which are especially characteristic of light sandy soils and many of them are traditionally weeds in cornfields on these soils. The other samples show much less sign of such a group. Much of Droitwich seems to have heavy clay soil and therefore a rather unsuitable habitat for these weeds, which raises the question of whether they might have grown elsewhere and been brought in, perhaps with straw, eventually being deposited with rubbish. This is discussed further in connection with the cereal remains. Context 3008 from structure 34 (P12) contained *Centaurea cyanus* (cornflower) pollen and seeds of *Anthemis cotula* (stinking mayweed), which are both characteristic of medieval, but not Roman, cornfield weed floras.

Plate 10 Selected examples of the recovered plant remains

Perennial weeds, ruderals and scrub plants

Context 559 (S27, P9(11)) has signs of a more perennial weed flora than the other samples. There were 101 seeds of *Conium maculatum* (hemlock), and 35 of *Atropa bella-donna* (deadly nightshade). The latter grows where there is limestone and Droitwich seems a rather unlikely site for it, there being no nearby records in the *Botany of Worcestershire*. There were also 24 seeds of *Rubus fruticosus* (bramble) and 130 seeds of *Sambucus nigra* (elder). Hemlock and elder grow best where human occupation has enriched the soil and such vegetation springs up on abandoned settlements. It is surprising that there were not more nettle seeds, unless the overgrowth was too dense. This flora seems to be a sign that the surrounding area was overgrown with weeds and scrub growing on the good soil, and that it was probably not in use at the time. The alternative theory that plants were brought to the site for use seems less attractive, as there are so many different scrub and weed taxa present.

Grassland plants

Possible grassland plants are the best represented group in the pollen spectrum from context 1707 (S6, P4) that can be identified to a particular habitat. There are Gramineae (grasses), *Trifolium repens* (white clover), *Prunella* type (self-heal), *Plantago lanceolata* (ribwort plantain), and perhaps Compositae (Liguliflorae, dandelions, hawkweeds, etc), which all generally speaking grow in grassland, although many of these plants can spring up on wasteland together with weeds, without fully- developed meadowland needing to be present. Some such as *Ranunculus* type (buttercup) can represent either weed or grassland taxa. These plants are well represented in the pollen record, particularly the wind-pollinated ones such as grasses and plantain, as one would expect. In the macrofossil floras of contexts 1961 (S29, P5–9) and 559 (S27, P9(11)), grassland plants are represented by small numbers of taxa (less than ten) and some of these would also grow up on wasteland as well as in proper grassland. One of the *Ranunculus* seeds was well-enough preserved to be identified as *R acris* (meadow buttercup), and there were seeds of *Pastinaca/Heracleum* (wild parsnip or hogweed), *Prunella vulgaris* (self-heal), *Ajuga reptans* (ground ivy), and probably *Poa* (a grass family with weed and grassland species). There were also floral remains of *Trifolium* (clover) in context 1961 and pollen of both red and white clover in 559. Bracken frond was present. In the post-medieval possible canal (S34) there was also a good record of *Centaures nigra* (knapweed), a fairly good example of a grassland plant.

Trees and shrubs

Trees, normally well represented when they are present, are scarce in most of the pollen results. Only context 1098 (S31, P8) has a substantial amount of tree pollen, mainly *Alnus* (alder) *Corylus* (hazel), and *Quercus* (oak). The post-medieval sample (3008, S34, P12) had an unusually large amount of *Ulmus* (elm) pollen which seems to be very local in origin, perhaps from flowering elm boughs brought to Droitwich for the production of water pipes: some of which have been found in the town. In the macrofossil records there is just the occasional birch or alder seed, or piece of hazelnut shell, or thorn of sloe/haw, but trees do not usually show up well from macrofossils unless they were actually overhanging the find-site. The shrub *Sambucus nigra* (elder) has been mentioned as part of scrub vegetation (see above). Large amounts of charcoal lumps, as well as the finer particles that washed through the sieves in a number of samples may represent the residue of wood or charcoal burning on a huge scale that was used to evaporate the brine.

Cultivated plants

These were only represented by *Triticum spelta* (spelt wheat) which was identified from the glume bases (part of the chaff) and the other cereal remains of rather nondescript charred grains might then be from spelt as well. There were a few charred seeds, such as *Vicia* (vetch) and *Bromus* (brome grass) which may be from plants charred along with the straw that was burnt. The possible post-medieval pit, structure 34 (3008, P12), had the pollen of *Linum usitatissimum* (flax) and *Vicia faba* (broad bean). These are typical crop plants for this period and presumably they, together with the cereal pollen, owe their presence there to general rubbish.

Deposit formation

In the case of some spectra, such as that from context 1707 (P4), which came from the fill of a brine tank (S6), and context 3008 (P12), possibly part of a canal (S34), pollen could have been deposited from the atmosphere, thus to some extent representing the general surroundings. In the case of other samples such as from context 559 (S27, P9(11)), where the organic content is high, pollen could also have come from whole plant material. The landscape around the site can be interpreted from this grassy/open land pollen spectrum to have been mainly of grassy and weedy wasteland vegetation. Such a pollen spectrum is commonly obtained from Roman wells and ditches. The plant seeds probably come from the very local vegetation, as seeds are not usually dispersed as far as pollen. Some seeds could have come from vegetation containing them (such as straw, reeds, etc), which was brought in to the site for some reason.

General interpretation

This assemblage of plants appears to represent a mixture. Some of the weeds and grassland plants could well have grown on the site itself and their remains would thus represent the local vegetation at the time. The cornfield weeds and some of the cereal remains represent grain, straw and threshing waste. These would have been brought to the site from cornfields and barns. They might have had a number of purposes apart from nutrition: straw can be used as animal fodder, building and roofing material, and as fuel, and there is not much indication of which is the case here. The wetland plants could have grown on the site, perhaps in the pits and ditches which were wet enough to preserve the evidence. They might also have been brought into the site for use as roofing, flooring, etc.

Plate 11 Fragment of post-medieval 'barrow' or conical wicker basket for drying salt crystals

The large post-medieval pit or canal 3008 (S34, P12) was particularly interesting; the clay fill preserved pollen well at most levels and one of the authors (Sue Colledge) analysed enough samples for a pollen diagram (Fig 55) to be prepared, as well as the plant macrofossils and the molluscs. Wetland plants identified from the seeds and pollen show that it probably remained at least damp throughout the year. Fine clay sediment and the aquatic snails seem to suggest that it was waterfilled, although there were few signs of true aquatic plants. *Ranunculus sceleratus*, the most abundant seed, would have colonised the damp edges of the pit, where wild celery *Apium graveolens* and hemlock *Conium maculatum* would also have found a suitable habitat. Large numbers of weed seeds show that the surroundings may have been overgrown with weeds such as *Atriplex* (orache). The comparatively large amounts of tree pollen and particularly of *Ulmus* (elm) have already been mentioned.

A pollen analysis such as this provides evidence not possible from the macrofossil record, and the site does seem to have enough tree pollen to show that the surroundings were probably fairly overgrown, which is in contrast to the almost treeless Roman pollen spectra already mentioned. Crop plants such as the cereals are well represented here, as well as *Vicia faba* (field bean) and *Linum usitatissumum* (flax), which produce much less pollen. Perhaps these crops were stored or processed in the area, and the last two might even have been grown nearby, as they were often grown as garden rather than field crops. Another interesting facet of the pit was the retrieval of the remains of a basket made of thin wooden strips in lattice work. It was photographed as soon as it was found (Pl 11) although conservation later on proved to be unsuccessful. There is little indication as to the original function of this feature; most places with waterlogged plant remains such as pits, ditches, and wells tend to preserve plant remains that give little indication of what the original function was. Their fills usually represent subsequent disuse, since they filled up only after the original use had finished. It is therefore not really possible to give any botanical evidence about what this pit was used for. The late date of the deposit is interesting as there are very few such results from the post-medieval period. The Birmingham Moat (Greig 1978–9), similarly post-medieval, showed some sign of fully aquatic plants such as waterlilies and probably therefore remained open long enough for this to establish while it gradually silted up. There the crop plants were cereals, as at Droitwich, and in addition *Cannabiaceae*, perhaps from hemp being processed or even retted in the moat itself. Environmental archaeology needs to be carried out on more post-medieval sites to obtain continuity of information, as well as to compare with the more abundant evidence from earlier periods.

Comparison with other finds from Droitwich

Plant remains have been studied from a number of other Roman excavations in Droitwich. The villa at Bays Meadow (HWCM 678) produced a well, although the fill with so many hemlock and nettle seeds seems to represent a phase of abandonment rather than the surroundings of Roman Droitwich as such (Greig 1988). The main part of the site produced charred remains with large amounts of spelt wheat, together with some rye and barley (Straker 1979), which parallels spelt in the cereal remains from the Old Bowling Green. At the Hanbury Street excavation (HWCM 681) there was a large amount of charred wheat chaff, some grain and a range of charred weed seeds: *Ranunculus* sp (buttercup), *Agrostemma githago* (corn cockle), *Stellaria* sp (chickweed), *Chenopodium* sp (goosefoot), *Trifolium* sp (clover), *Polygonum convolvulus* (bindweed), *Rumex* sp (dock), and *Tripleurospermum inodorum* (scentless mayweed; Vaughan 1982), which shows that these may have been the main cornfield weeds. The Old Bowling Green seeds were mainly waterlogged (even though there were very large amounts of charcoal) and

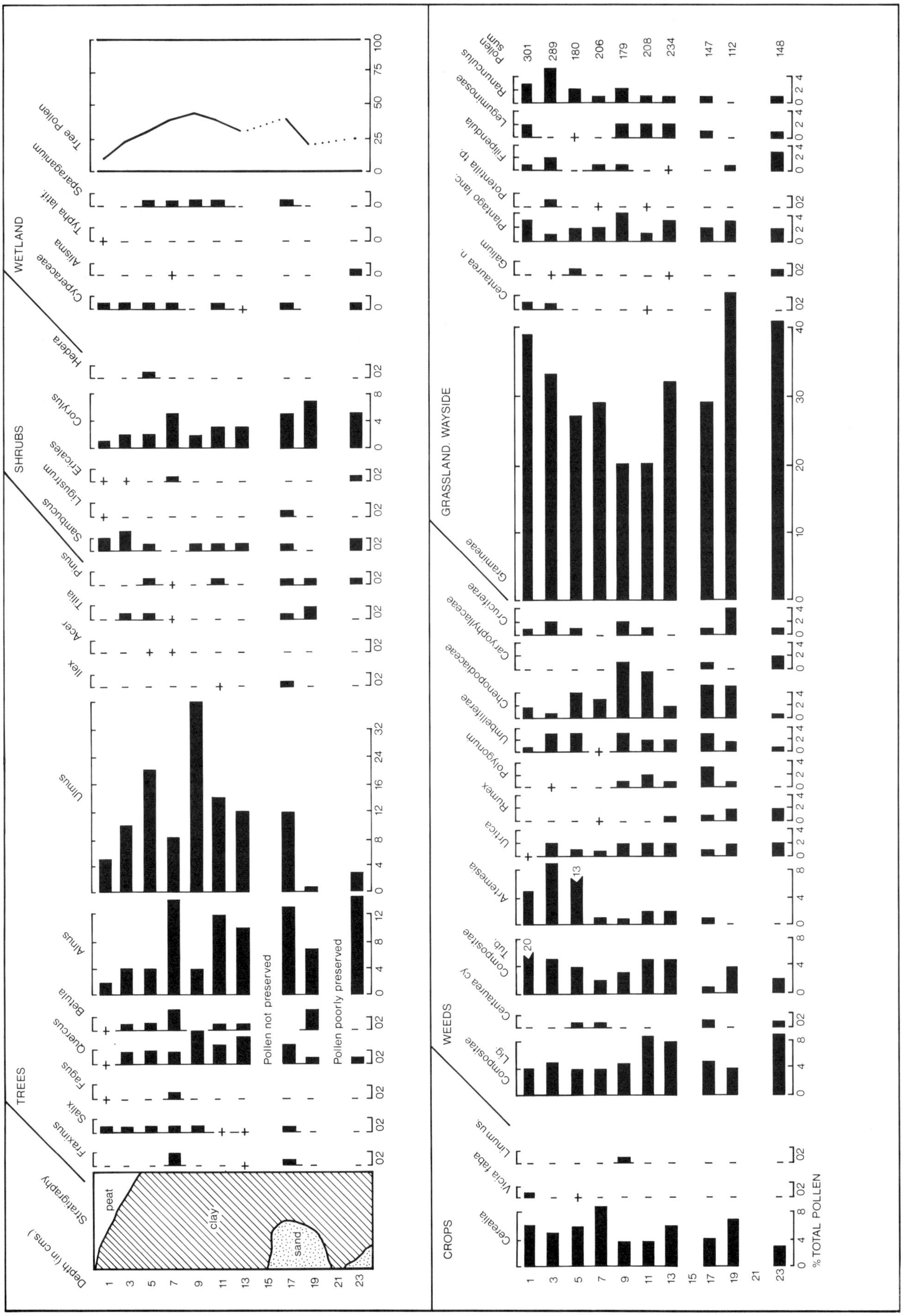

Figure 55 Pollen from the fill of a canal or large pit (S34), phase 12

these deposits thus differ in nature from the rich charred finds, mainly of cereals, mentioned above. The Old Bowling Green flora does, however, have some similarity to the finds from the well at Bays Meadow in the preponderance of weed seeds, and the sample from context 559 (S27, P9(11)) especially so. An industrial site for salt extraction such as this would probably be a place with little vegetation because of the trampling, burning fires, piles of salt, etc, but if areas became disused, weeds such as those that were found would have moved in. The remains may have become preserved if the plants either grew by pits or ditches themselves, or if their seeds (together with topsoil) were deposited in such wet places when the site was cleared for re-use.

Signs of halophytes

The brine springs at Droitwich may have had a flora of salt tolerant plants in the past. These might originally have come from seeds carried in the guts of grazing birds, such as geese. It is interesting to see if there is any sign of this in the plant record. The only possible halophyte (salt tolerant plant) found so far is *Apium graveolens* (celery). In the recent past it was 'very abundant' along the Droitwich canal (Amphlett and Rea 1909), but the authors have not seen nor heard of it there now. In recent years, wild celery seems to have retreated to a few coastal sites in Britain, the only locality the authors could find being near Portsmouth. As well as growing wild, celery was also cultivated in Roman times, it is believed for the aromatic seeds rather than for the stems that are eaten today. The seeds from context 1961 (S29, P5–9) even still smelt of celery. Celery seed has been found at other Roman sites far from the salt springs, such as at Alcester (Moffett pers comm) and Berinsfield, Oxfordshire (Robinson pers comm). The presence of celery seed at Droitwich cannot then be taken as an indication of salinity, since it seems to have been cultivated, although the scarcity of other cultivated plants in this Droitwich material might offer at least the suggestion of such salt-tolerant natural vegetation.

19 Wood *Anne Crone*

The excavation produced a total of 423 wood samples, of which 372 (88%) came from four of the excavated brine tanks, structures 6, 7, 8 and 9. The bulk of the wood was derived, either from the boards lining the pit sides or from debris within the pits. Further pits to the north (S3, S4 and S5) were excavated and recorded as being lined in a similar manner but the linings were not totally retrieved. However, small samples of the wood debris were recovered from structures 3 and 5, brine tanks and a large pit (S34). A small sample of the wattle lining of structure 1 was also retrieved. Two well preserved barrels, split along their axes, were lifted, one (S16) of which has been conserved by the National Maritime Museum (2:F6). Six wooden posts, found *in situ*, were recovered. The remainder of the sampled material derived from debris from miscellaneous features, layers, pits, ditches and postholes. Carbonised material and that from post-medieval contexts is not considered in this discussion.

Phase 3: Late Iron Age

Brine tanks (S3, S5, S6, S7, S8 and S9)

The species associated with the brine tanks are summarised in Table 6. One hundred and ninety-three pieces of wood were recovered from the construction of the brine tanks (S6, S7, S8 and S9), consisting of boards, board splinters, scantling, worked and unworked branchwood. The boards had mostly been cut to form a point with one or two oblique cuts. Often, especially in structure 8, the boards are buckled just above the base, presumably as a result of hammering the boards into the base of the tanks. The survival of toolmarks is variable, although those from structure 6 are particularly clear. This occurs, most frequently, as a series of ripples or stepped ridges down the side facets, varying in length from 15 to 50mm and in width from 30 to 50mm. Oak (*Quercus* sp) was by far the most common wood used in the construction of the brine tanks (Table 6).

Of the wood recovered from the disuse of the brine tanks of this phase (S3 and S5), most was of boards but also included a handful of small oak and hazel (*Corylus avellana*) twigs (162, S3), a split oak board with a rough oval hole just above the tip, and two unworked branches of hazel and field maple (*Acer campestre*; 494, S5).

Phase 4: Mid 1st to early 2nd century AD

Disuse of the brine tanks (S3, S6, S7 and S8)

Most (134 pieces) of the wood was either unworked or simply worked branches, board splinters and scantling. However sixteen artefacts were retrieved from the fill (3116) of structure 6 and four boards from the fill (3118) of structure 8 displayed carpentary details indicative of a previous function.

Sliding mechanism

Of the artefacts from structure 6 a rectangular piece of oak measured 215 x 80 x 20mm, the full length of which did not survive (Fig 56 no 1). For nearly 140mm of its surviving length it is divided into two rectangular prongs and tapers slightly from the upper end to the pronged end. Three grooves had been cut across one face. The grooves had very thin ridges along their sides, reminiscent of sawmarks. When found *in situ* the two prongs were slotted into two rectangular holes in the middle of a long square beam. A peg entered the beam at right-angles to the prongs and penetrated between them but did not penetrate the opposite side of the beam. This would have enabled the beam to slide up and down along the prongs. Its function is not at all clear. It may be the locking mechanism of a barred door but its dimensions suggest something smaller than a door; a chest for instance.

Cutting blocks

A chord off a large oak trunk was recovered, on which sapwood and bark still remain in places. Another sample was a roughly 'L' shaped segment, also an oak trunk, one arm of the L crossing the trunk and the other arm going up the trunk. Almost every surface on both pieces of wood is covered in thin cutmarks running mainly across the grain of the wood. Within the curve the cutmarks are so profuse that the wood appears 'feathered' in texture. The cutmarks are short, very thin and no more than 2–3mm deep, suggesting cuts produced by a sharp edge such as a knife.

The most obvious explanation for the function of these two pieces is as cutting blocks, though why some of the more awkwardly shaped surfaces are also covered in cutmarks may need further explanation.

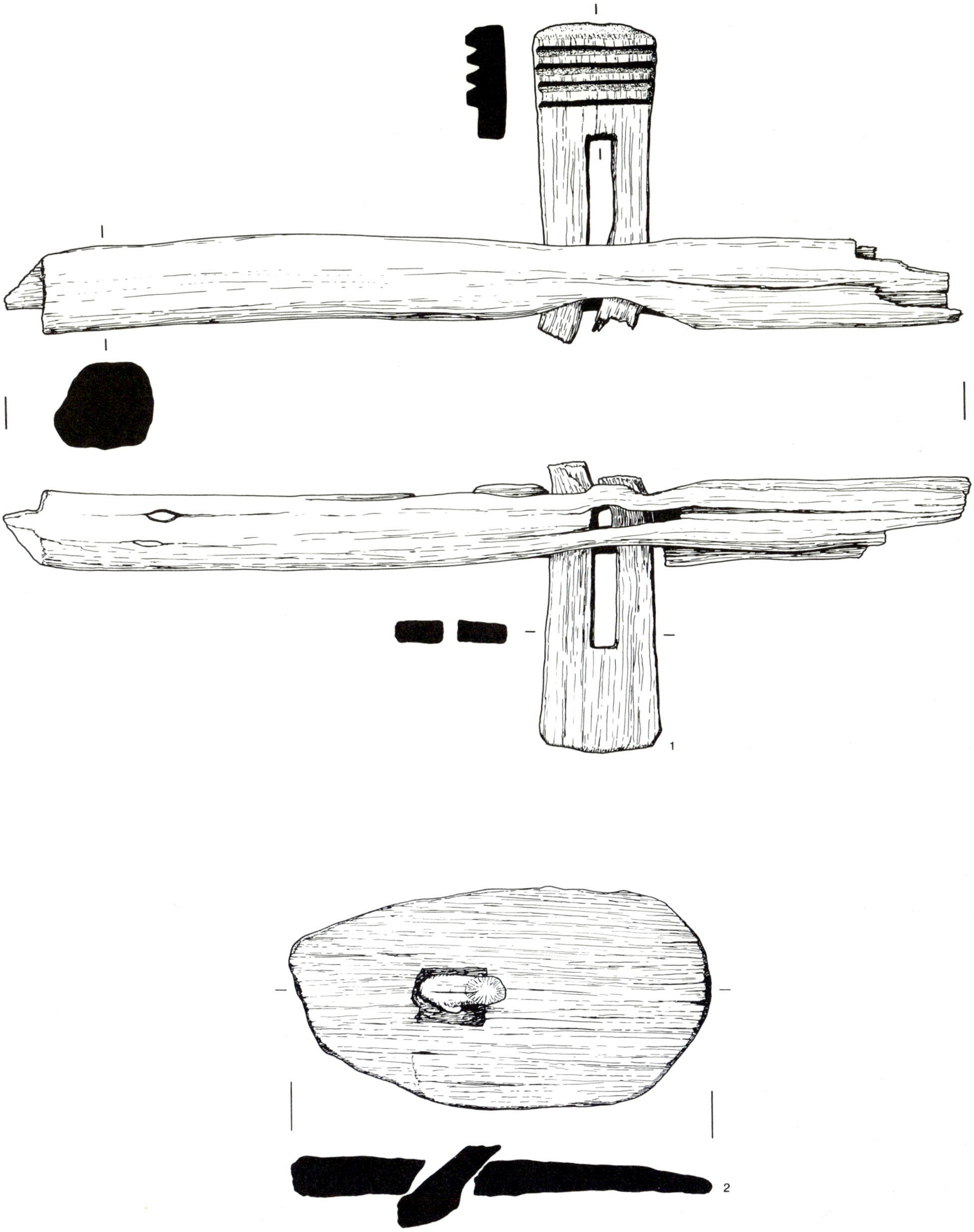

Figure 56 *Wood. Scale 1:4*

Table 6 Wood from the brine tanks

Phase Structure Context	Wood Category	S	T	B	I	Pr	Al	Po	Ac	F	C	Q	Non ident	Total
2–4	Branchwood											10		10
1														
28D														
3	Board											2		
3	Amorphous											1		
162	Miscellaneous										4	3		10
4	Amorphous												1	
3	Board splinter									1				2
129														
3	Board											1		
5	Board splinter											1		
494	Branchwood							1			1			4
3	Board											36		
6	Board splinter											13		
1745	Scantling											8		57
4	Board											1		
6	Board splinter											5		
3116	Branchwood		1			1					4	5		
	Scantling											1		
	Worked branch											1		
	Artefacts			1	1				1			13		35
3	Board											7		
7	Board splinter											2		
1773	Worked branch									1				10
4	Board											7		
7	Board splinter									1	1	14	1	
3117	Branchwood						2		2	1	6		1	
	Worked branch							3	2	2	6	1		
	Amorphous							3		1		3		58
3	Board											45		
8	Board splinter											21		66
1693														
4	Board				1							10		
8	Board splinter								1	3		9		
3118	Branchwood	1				2	1	1	1	3	2	11	2	
	Worked branch							1	1	1	2	1		
	Amorphous		1			1	1		2	1		4	1	65
3	Board											69		
9	Branchwood									2		2		73
2702														
TOTALS		1	1	3	2	3	5	8	11	17	26	307	6	390

Key: S (*Salix*, willow), T (*Tilia*, lime), B (*Betula*, birch) I (*Ilex*, holly), Pr (*Prunus*, wild cherry), Al (*Alnus*, alder), Po (*Pomoideae*, apple/pear/quince group), Ac (*Acer*, field maple), F (*Fraxinus*, ash), C (*Corylus*, hazel), Q (*Quercus*, oak)

Shovel blade

Of interest was a roughly oval piece of birch (*Betula* sp), 300mm long and 150mm at its widest point. It narrowed to 100mm at one end where it became slightly 'waisted'. It was 15mm thick and slightly concave along its length. At the wider end the very edge of the piece was burnt. At the narrower end, just below the waist, was a rough triangular slot 45 x 30mm penetrating the piece. The upper and lower sides of this slot were parallel to each other but angled down the length of the piece. When first found, a small piece of roundwood, chopped at one end, was sitting within the hole (Fig 56 no 2).

This artefact displayed characteristics identical to the blade of the composite shovel, described in detail most recently by Carole Morris (1981). Briefly, the shovel blade is generally oval to sub-rectangular in shape, waisted at one end, with pegholes above the waist and a rectangular slot just below it. A separate shaft was fitted into the slot from the concave, or upper, face and secured by means of pegs through the pegholes (Morris 1981, 52). It lacks a peghole but as the piece is broken above the waist, a peghole may have existed. Alternatively, another means of securing the shaft may have been used. Morris describes two complete blades which do not have pegholes (1981, 56, nos 24 and 26) and suggests that they must have been secured by rope or cord bound around the shaft and blade shoulders.

The shovel blades that have been recovered from secure contexts during excavation have been dated to the 9th to 13th centuries AD, the earliest known in this country coming from 10th century levels at York and Durham (Morris 1981, 59). However, a similar blade was found in the 19th century in the lead-mines at Shelve, Shropshire and attributed to the Roman period. Rastel (1678, 1062) describes a 'loote', which is a board fixed slope-ways on a haft about 1.00m long and was used to ladle out salt crystals from the boiling pans.

Forked stakes

Some eleven forked stakes appear to have been dumped as a bundle into structure 6 (Pl 1). With the exception of one stake of field maple all were of oak. All have been fashioned from straight lengths of roundwood which fork at one end. The branches vary in diameter from 45 to 90mm. The most complete example (Fig 57 no 3), is 1,750mm long, while another (Fig 57 no 4) measured 1,830mm although incomplete. The subsidiary branches have been trimmed by oblique chops to lengths varying from 140 to 330mm. A third (Fig 57 no 5) had long thin facets along its length, presumably, as a result of stripping off the bark, but all the others were still covered in bark.

At the opposite end to the forks the branch had been reduced to a long rounded tip by a series of facets. There are no signs of compression on those stakes where the tip still exists. Five of the forked stakes, had a halo of surface burning on the stem above the stake tip, and one was also burnt in the tips of both forks. The toolmarking was well preserved on a number of the stakes. It occurred most commonly as a series of shallow ripples or stepped ridges down the length of the facets at the stake tip, but varied greatly in length from 5–30mm.

The most obvious explanation for the function of these forked stakes is that they were designed to be set in the ground, supporting a horizontal member between the forks. The halo of burning noted on some of the stakes suggests that they may have been set in the ground near open fires, allowing the stem to scorch. A clue to their function comes from a more recent source. In Book XII of his treatise on metallurgical practices, *De Re Metallica*, Georgius Agricola describes the various processes of salt extraction that he observed being practised in Germany during the 16th century (Hoover and Hoover 1912). An illustration of a contemporary salt works shows two pairs of men crossing a yard carrying a cask of brine suspended on a pole between them. Each man carries a forked stake identical in appearance to those from structure 6 and Agricola describes them in a note to accompany the illustration as 'Forked sticks in which the porters rest the pole when they are tired' (Hoover and Hoover 1912, 549).

Discussion

In addition one piece of scantling (SRBB) was found in the debris (3116). One end had broken across a rectangular hole 50mm wide, of which 60mm of its length remained. Sawmarks were visible in the corner of the hole. A tenon, 30mm diameter, extended from the centre of the opposite end. It had been roughly fashioned by a number of small facets which reduced the end of the scantling, leaving the tenon protruding. The tenon survived to a length of 45mm.

Of the boards from structure 8 (3118), four display features indicative of a previous function. One board of oak is 163mm in length and 23mm wide, with a series of four holes cut down the middle of the board. Two holes are square with angled edges, both measuring 30 x 45mm while the other two are oval, one measuring 30 x 75mm and the other 40 x 90mm. A similar line of holes are found on a second oak board measuring 1780 x 180mm. The holes are off-centre along the board and are very irregular in size and shape, suggesting that the wood has eroded away along the line of the holes. One end of the third sample has broken across a roughly square hole 55mm wide. The fourth board, a sample of holly (Ilex aquifolium) is a very unusual piece, not least because of its species. It was very fragile and fragmented when lifted. It appears to be the chord from a trunk, with the bark still adhering in places. A groove 30mm wide and 4mm deep is cut across the width of the plank on the flat face and the underside is burnt in patches.

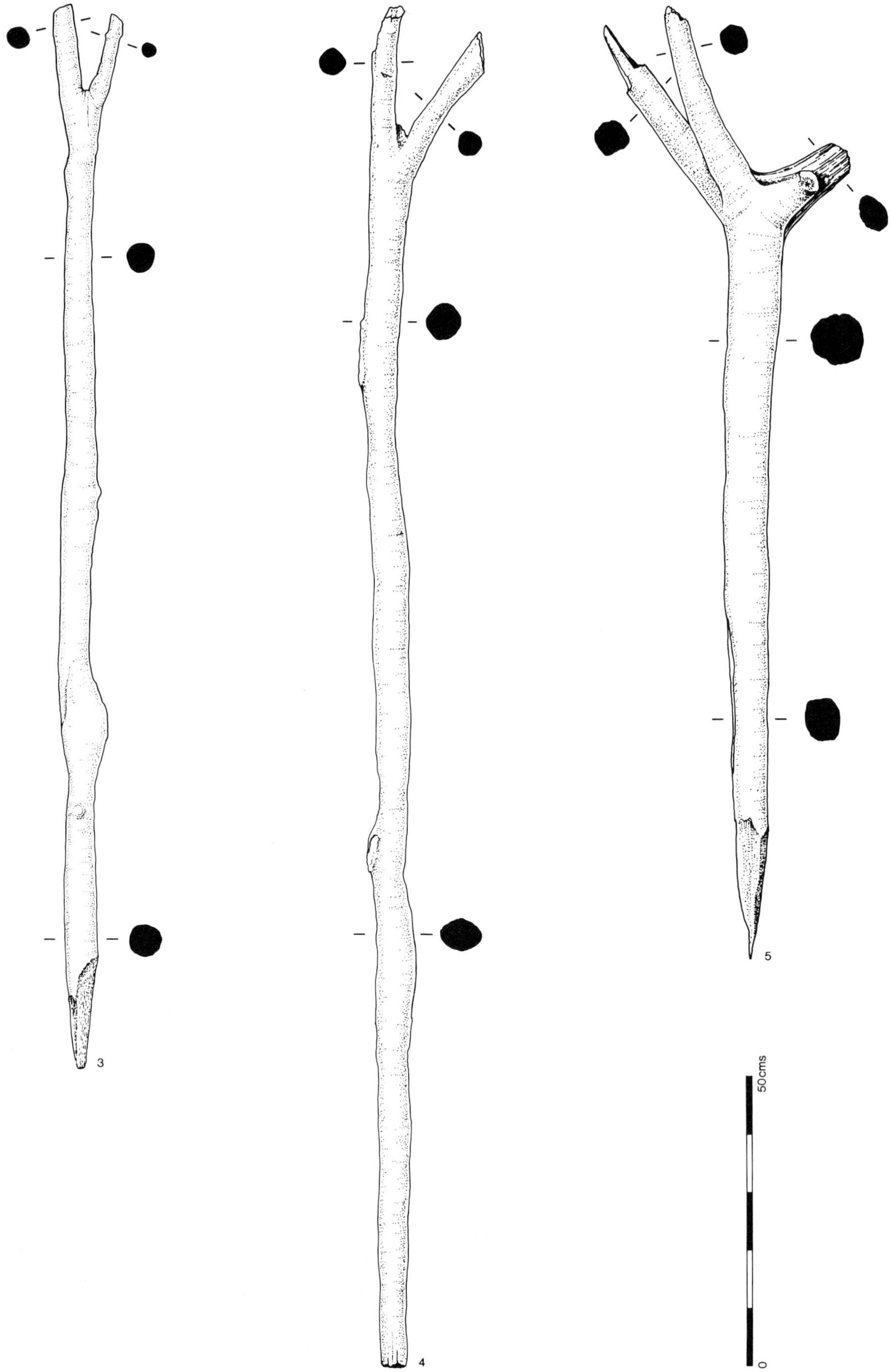

Figure 57   Wood (continued). Scale 1:10

Table 7 The barrels (dimensions in m)

	Height	Max diam	Min diam	No of staves	Stave width	Stave specie	Hoop specie
Old Bowling Green S16, 1102	1.78	1.25	1.00	10	0.11–0.20	Abies	Quercus
Old Bowling Green S19, 3060	*c* 1.80	1.30	1.20	11	0.13–0.18	Abies	Quercus
Old Bowling Green S21, 1727	1.90	–	–	3	0.16–0.18	Larix	–
Old Bowling Green S15, 1228*	1.65	1.10	1.00	–	–	–	Quercus
Old Bowling Green S20, 1229*	1.70	1.00	1.00	–	–	–	Quercus
Silchester Insula XVIII	1.95	0.86	0.72	19	0.15	Abies	Corylus
Silchester Insula XVII	1.97	0.84	0.68	18	0.165	Abies	Corylus
Harelbeke	1.60+	1.20	0.80	18	0.17	Abies x 12 Larix x 6	Corylus

* dimensions estimated from size of pits

Disuse of a possible leat (S30)

The fills of a possible leat contained four splinters of bark covered branch (168), probably debris from the squaring of a post. Another two splinters possibly came from a plank (476). All were of oak.

Phase 5: Early 2nd to mid 3rd century AD

Posthole alignment (S146)

This structure included an oak post (1776) which remained in situ. Its top was tapered and hollow, due to rotting. The base was flat and radial, and rotting was in its early stages (see phase 9 for full discussion).

Phase 8: Mid 3rd to late 4th century AD

Barrels (S15, S16, S19, S20 and S21)

Of the best surviving barrels, one (1102, S16, Fig 17) was sent to the national Maritime Museum, where it was examined and then conserved for display in Droitwich (2:F6). The other (3060, S19, Fig 18) was stored with the rest of the wood assemblage and finally dismantled in the course of examination at Droitwich. Unfortunately, the upper staves had collapsed and only the lower staves were still well preserved. Both the examination at Greenwich and that at Droitwich produced comparable data. The staves of both barrels were of silver fir (*Abies alba*) and the hoops were of oak.

The method of conversion of the staves was examined closely. Traditionally, the staves are fashioned from radially split wedges so that the medullary rays remain largely intact and form an impervious membrane along the circumference of the barrel (Kilby 1977, 30). Many of the staves had not been converted so carefully and were cut at oblique angles across the rays. It is suggested that, instead of radial splitting, the trunk was quartered and then the staves were sawn off parallel to one of the radii.

The hoops were chords about 20mm wide and 5mm thick, cut from small bark covered branches, so that the bark constituted the greater part of the hoop. The hoops only survived in small strips and, in many cases, only their shadow was visible on the stave. No means of securing the hoops were observed.

Structure 21 was much disturbed and consisted of three barrel staves (1727) found *in situ* in the base of structure 31. These collapsed during lifting but the fragments were identified as radially split pieces of European larch (*Larix decidua*).

The barrel hoops of structures 15 and 20 were of oak (1228 and 1229).

Neither larch nor silver fir is native to this country, both being introduced in the 19th century, and up until then their nearest natural habitats were the mountain ranges of Central Europe (Godwin 1975, 1–02–3). The literature on Roman trade describes goods such as wine and samian ware being transported in barrels and there have been numerous finds of barrels in Roman contexts across Europe and in Britain.

The barrels from the Old Bowling Green are compared with a selection of these in Table 7. The Old Bowling Green barrels fall within the standard range of dimensions and are manufactured from the same range of species. The larch staves of a barrel from Harelbeke were positioned at irregular intervals in groups of one or two around the circumference, similar to the barrel from the Old Bowling Green (1727). A barrel from Aardenburg, Zealand also used both species (cited in Frison 1961).

The only feature that appears to be quite different is the nature and species of the hoops. All other samples cited have used split hazel wands, a more suitable choice, as hazel is known for its elasticity. At Harelbeke the hoops were secured by simple knots (Frison 1961, 772).

As the hoops of the barrels were not fastened to the staves, except as a complete belt keeping the staves *in situ* under pressure when cut, the barrel would probably have fallen apart. It would therefore seem probable that whole barrels were laid in the ground and then cut longitudinally, or even left whole. This last suggestion would however, make access difficult and there was very little in the way of barrel parts in their fills as might be expected if they had remained whole. Removal of the staves and end boards would not necessarily have greatly disturbed the hoops, and these would have remained *in situ* (S15 and S20).

Phase 9: Later mid 3rd to late 4th century AD

The rectangular building and posthole alignment (S22 and S161)

Very little of the single remaining oak post (2465) survived as it was very much decayed. It was located *in situ* on top of a large stone, acting as a post-pad (2464). Again this oak post from an alignment (3115, S161) was *in situ* and was much decayed.

The remains of six wooden posts were recovered *in situ* from the excavation as a whole. One (2465, P9) was found sitting on its padstone within a posthole (Pl 4) while the others were set directly into postholes. Despite being found *in situ* only two of the posts had well preserved bases. The base of 1171 (P5?) was completely flat whilst that of context 1776 (P5) had been shaped by two short downward chops on opposing sides, leaving a flattened ridge across the width of the post. With the exception of context 3115 (P9) all the posts displayed similar patterns of decay, although at differing stages on the process. In all cases the centre of the post had rotted away (Pl 4) and the decay was spreading out along the medullary rays. In the case of context 307 this had resulted in total fragmentation of the post so that only wedges of the original trunk remained. Only half of post 2465 remained and decay was very well advanced on this piece, creating a scalloped appearance on both inner and outer surfaces. Context 1171 still retained its complete circumference although it tapered considerably from base to top. At the top the interior was completely hollowed out, leaving only an outer shell, and the very centre had rotted away down to the base. Context 1776 was similarly tapered and hollow at the top but the centre had not completely rotted away at the bottom. Instead, the centre and some of the rays were still filled with the soft, spongy partially decayed wood.

Posts found during excavations at the broch of Buchlyvie, Stirlingshire and displaying similar characteristics to those described above, were the subject of a recent study into this type of decay (Barber pers comm). Acetolysis and microscopic examination of thin sections taken along the length of the posts, indicated that fungi were the main agency of decay but the species of fungi involved could not be identified.

The fungi attack the wood via the wood-vessels or pores, which provide vertical access through the wood. They also provide access for oxygen and water, both necessary for fungal activity. It is estimated that the number of pores on the cross-sectional surface is four times as many at the centre of the trunk as it is at the circumference (John Barber pers comm). Therefore, decay will occur more rapidly in the centre of the post. As the centre decays, access is enlarged and the fungi begin to attack via the smaller horizontal cells, ie the medullary rays. This process is much slower on the exterior of the post, partly because the necessary oxygen decreases with increased soil depth. Therefore the rate of decay will also decrease with depth, thus producing the tapered profiles seen in contexts 1171 and 1776. This type of decay will only stop with the onset of anaerobic conditions which arose, in this case, when the water table was reached.

Uncertainly phased

Only two samples (28D) were recovered from structure 1 (P2–4), a brine tank and were derived from the wattle lining.

Of the *in situ* posts which were uncertainly phased, the samples from contexts 303 and 307 (S36, P4–5) were very decayed. Decay had occurred along the rays. Another *in situ* oak post (1265, S108, P4–5) again showed advanced decay along the rays (see phase 9 for full discussion).

A layer (1023, S114, P7–9) produced a fragment of a small vessel stave, measuring 65 x 45 x <5mm. The full length of the stave has not survived. It was neatly squared at one end, and 10mm above that end a groove, 5mm wide and 2mm deep, had been cut across the stave. The dimensions suggest a small, finely made vessel; possibly a bucket. The choice of species (yew; *Taxus baccata*) is unusual. Two other fragments of planking may have been barrel staves of silver fir, but were not well preserved.

Of the material recovered from post-Roman contexts, a fragment of wickerwork (3008, S34, P12; Pl 11) is worthy of note. Conical wicker baskets were used for the draining of salt crystals after they had been removed from the boiling pans (see Chapter 1).

Discussion

With the exception of the roundwood used to rebuild one corner of structure 8, all the brine tanks were lined entirely with oak. Although the inner- and outermost rings are missing on all the boards so that the number of rings measured during tree-ring analysis represents only the minimum age of the tree many of them clearly came from long-lived trees (Table M15, 1:E6). Tree-ring analysis also indicates that the boards were fashioned from relatively slow-grown oaks.

The lining boards had been split radially from the trunk, presumably using wedges, and varied from 15 to 60mm in thickness. Practical experiments with seasoned and unseasoned oak have demonstrated that it is much more difficult to split seasoned oak into halves, let alone into thin boards (Darrah 1982). Furthermore, the sapwood had not been trimmed and remained virtually intact on many samples. It therefore seems most likely that freshly-felled oak trunks were brought to the site and converted into planks for immediate use in the tanks. The use of unseasoned wood is also suggested by the buckling observed near the bases of many of the boards, a feature characteristic of green timber hammered into a hard surface. Under similar duress, seasoned wood tends to become merely compressed at the tip.

Apart from the basic shaping of the tips (see above), there was no additional trimming or dressing of the faces. As described above, the toolmarks on the shaped tips are either a series of ripples or stepped ridges. Experimental work in the Somerset Levels has demonstrated that the axe leaves abrupt steps where the blade has bitten in and the chip broken off, similar to those at Droitwich. The toolmarks are rarely larger than 50mm and although subsequent chops will remove part of the toolmark it seems likely that a relatively small axe was used to fashion the lining boards.

There is evidence for a more varied toolkit in the debris thrown into the tanks. Both the knife and the saw were used. Sawmarks are visible in the corner of the rectangular hole at one end of SRBB and in the grooves on the sliding mechanism. The Romans used both a frame saw and a bow-saw but the latter is probably more suitable for the finer work in evidence here (Liversidge 1976). The knife seems to have been used for finishing work. The bark was stripped off on the the forked stakes with a knife and the tenon at the other end of SRBB had been finished using a knife. The knife would also have been the most suitable tool of shaping the square holes found on planks amongst the debris (see above). The Romans used both spoon-shaped augurs and bow-drills for boring holes (Goodman 1962) but the holes on the Droitwich wood are too eroded to determine the method of their manufacture.

A wide range of species was found in the debris thrown into the tanks, much of it twigs, branches and some woodworking debris. The species represented are commonly found today in the environs of Droitwich. Poplar, willow, and alder line the banks of the Salwarpe, while field maple and ash appear as hedgerow trees. The fruits (Pomoideae), apple and pear, and the wild cherry (*Prunus* sp) may have been deliberately sought out for their value as good fuel (Taylor 1981).

The artefacts found amongst the debris, particularly the shovel blade and forked stakes, are of interest in that they parallel some of the equipment used in salt-extraction some 1,600 years later (see above).

20 Miscellaneous *Simon Woodiwiss*

Leather artefacts were sporadically collected and processed. No conservation was undertaken and the assemblage was discarded. A single unidentifiable decayed fragment was retrieved from a Roman context (192, S214, P4(5)) and one other from a layer of phase 10 (1205, S110). The rest of the assemblage was derived from post-medieval or unstratified contexts.

A number of fragments of mortar were recovered from Roman contexts but these formed too small an assemblage for worthwhile comment.

21 Friar Street (HWCM 605), the excavation

Justin Hughes and Alan Hunt

The site was located towards the western end of Friar Street, on the northern side, at National Grid Reference SO 8974 6349 (Fig 1). Excavation was carried out between August 1973 and July 1975, with a break between March and August 1974.

The project was directed by Alan Hunt (County Archaeological Officer) and supervised by David Freezer from August 1974 to the end. It was funded throughout by the Historic Buildings and Monuments Commission for England, formerly the Ancient Monuments Division of the Department of the Environment, and by Hereford and Worcester County Council.

Archaeological problems and questions were defined as follows:

1 There was a strong possibility of prehistoric settlement on the site, perhaps connected with salt working in that part of the Salwarpe valley later occupied by the town. Excavations had previously been concentrated at the Roman sites at Dodderhill (St Joseph 1938; Whitehouse 1962; HWCM 603), and at Bays Meadow (Gelling 1957; HWCM 678). Little prehistoric material was recovered from these sites on the higher ground north of the valley. However, an early focus of activity beside the Salwarpe or on the lower valley slopes remained a probability.

2 The early Roman fort at Dodderhill and the villa at Bays Meadow might be interpreted as administrative centres for a Romano-British salt industry at Droitwich, attested by the name *Salinae* given in Ptolemy's *Geography* and *Salinis* in the *Ravenna Cosmography* (Rivet and Smith 1979, 451). This industry was presumably sited in the Salwarpe Valley or on its lower slopes and a domestic settlement (or settlements) might be associated with it. Indeed, the pottery found by Hodgkinson (1933) in the town, including the Friar Street area (HWCM 671), might reasonably have been associated with a salt working settlement.

3 The Saxon documentary evidence indicated salt working at Droitwich in this period and raised the possibility of continuous brine exploitation from Roman times onwards.

4 The medieval urban topography of Droitwich has been summarised by Freezer (1977). The earliest morphological unit seemed likely, on topographical grounds, to be close to the point where the Roman roads crossed the Salwarpe. Perhaps significantly, the road aligned north to south here used to be known as Market Street (now Queen Street). The High Street/Friar Street alignment was postulated as a later development, perhaps of a planned or regulated nature. A very slight variation of alignment where High Street joins Friar Street, by the parish church of St Andrew (HWCM 607), seemed to indicate that this secondary development took place in two stages. Friar Street seemed more likely to be the later of the two. The date and character of these developments needed elucidation. An earlier excavation in a burgage plot on the south side of High Street (David Peacock pers comm; HWCM 4167) was frustrated by shallow deposits and considerable disturbance.

It was clearly important to explore early settlement in the area occupied by the town and to investigate the origins, chronology and character of the urban development. A high value was placed on potential development sites in the historic centre of Droitwich, particularly where Roman material had been recovered. An initial survey found that developments were planned on several sites of potential archaeological importance. The Friar Street site was of particular interest, because it lay within the medieval town, close to the site of the demolished parish church of St Nicholas (Fig 1. HWCM 255) and opposite the late medieval timber-framed Priory House (HWCM 609). Roman material had also been found in this area (Hodgkinson 1933), although findspots could not be identified with any certainty.

The site therefore presented an opportunity to explore an area occupied in Roman and medieval times and possibly shed some light on the intervening period. Plans for building a new fire station were well advanced. The west part of the site, formerly occupied by a church hall, was in County Council ownership and had already been cleared; to the east was a hall still in use. Development was scheduled to begin in 1975, which allowed time to carry out extensive excavations.

The subsequent investigation produced an almost continuous sequence of occupation from the mid to late Iron Age, to the 20th century. An area of 400 square metres was examined, yielding up to 3m of stratified deposits. One thousand one hundred and eighty contexts were recorded, and these were subsequently divided into 174 structures and ordered into fourteen phases. A full description of

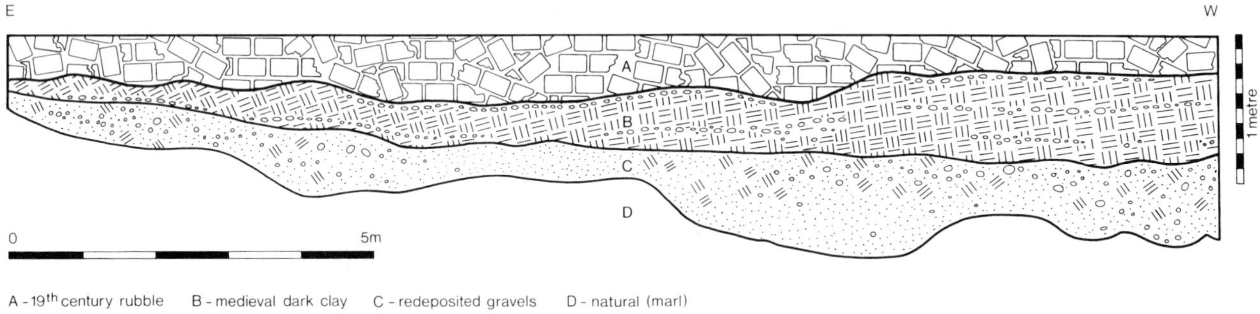

Figure 58 Simplified section on southern edge of excavation

the stratigraphical analysis is given in fiche (2:F7–3:A8).

The depositional sequence was for the most part well defined (Fig 58). Iron Age and Roman features (P1 to P3ii) were cut into the marl subsoil and were sealed in the sub-Roman period by a compacted pebbly clay. In the 12th century the beginnings of an extensive build-up of dark, highly organic clay loam were detected, sealing drainage features in phase 4i and tanning pits in phase 4ii. This rich soil was cultivated during the sequence of domestic occupation described in phases 5 to 9. The post-medieval sequence of brick buildings (P10 and P11) was sealed by rubble from their demolition in the early 19th century.

Phase 1: Iron Age (Fig 59)

The earliest features on the site were associated with mid to late Iron Age salt production.

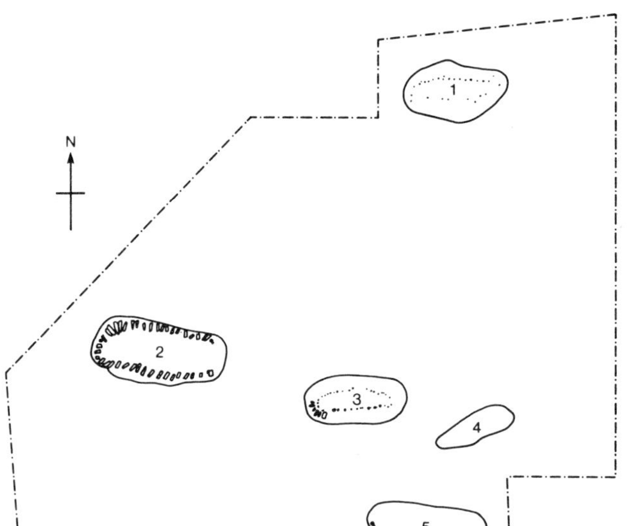

Figure 59 Phase 1 features

Brine tanks (S1, S2, S3 and S5)

All but one of the five structures of this phase were identified as brine tanks. Two (S1 and S3) were badly eroded but each had a series of postholes cut through its base and sides, demonstrating that both had been lined with timber. Structures 2 (Pl 12) and 5 were identical in profile and lined with clay. The lining of the former was also revetted by wooden planks and wattles and had a dendrochronological date of AD 19 (Table M24, 3:A14). Wood stains were recorded in the base and sides of structure 5, but the pit was truncated at its northern end. Figures 60 and 61 illustrate the contrasting construction techniques of structures 2 and 3.

Hearth (S4)

To the east of structure 3 an elliptical pit (Fig 62), with almost vertical sides to a depth of 0.78m, was interpreted as an open hearth. It was filled with large quantities of briquetage, charcoal and ash, in a complex series of deposits of green and grey clays, dark brown and grey loams, and silt. The lower fills may have been associated with the use of the hearth, as the concentration of fuel waste was higher than in the upper fills.

Discussion

The brine tanks and the hearth contained large quantities of briquetage and were aligned east to west. Two of the four tanks (S1 and S3) were poorly preserved and significantly smaller than the others, but all four were clearly related functionally. The great quantity of fragmented and crushed sherds of briquetage in the fills of the brine tanks and the hearth also indicates a related function. The tanks were probably used for the storage and settling of brine (see Chapter 2 for a more detailed discussion of the function of these features). The fuel waste and the large quantity of base sherds recovered from the hearth indicate the close association of briquetage vessels with a process of controlled heating. Alternatively, it is

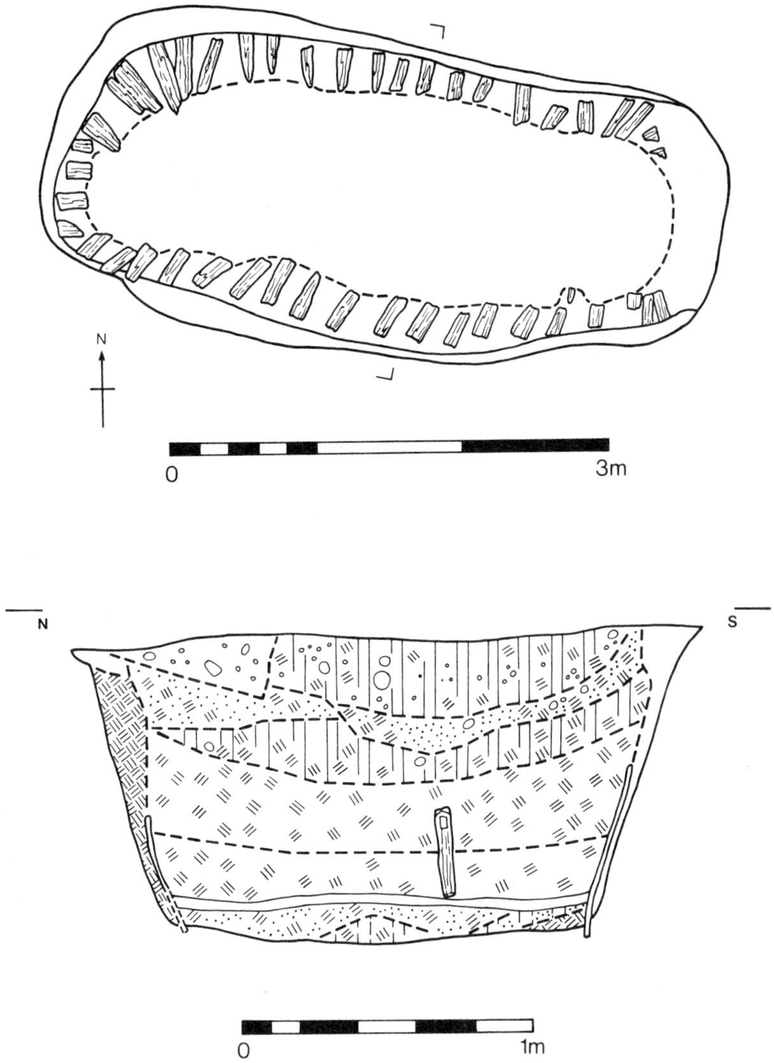

Figure 60 Brine tank (S2)

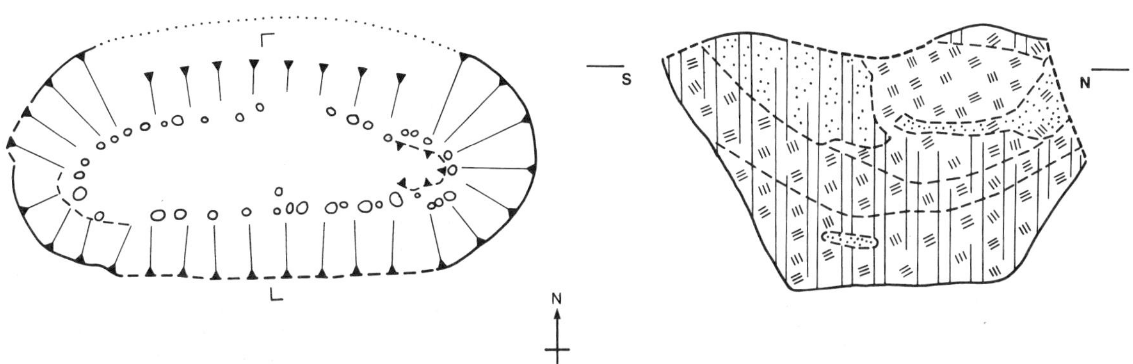

Figure 61 Brine tank (S3)

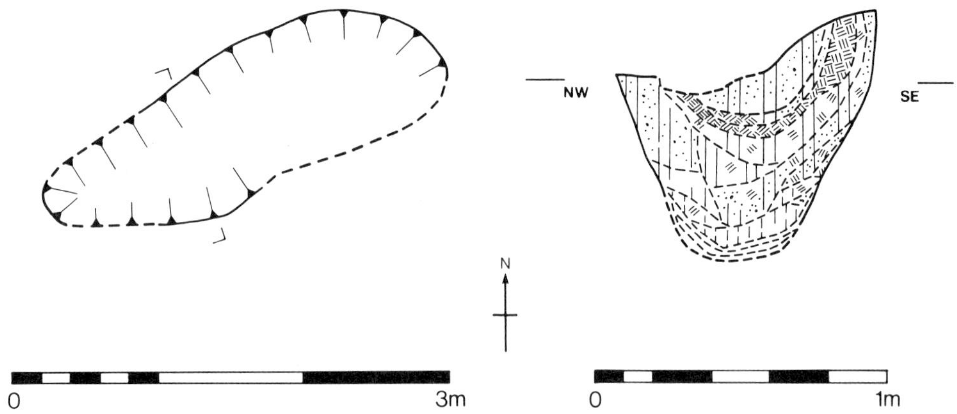

Figure 62 Hearth (S4)

possible that the hearth was used for the firing of briquetage vessels.

Whether briquetage was used as the boiling vessel and or for the draining of wet salt crystals is not clear. Petrological analysis has identified two general fabric types, organic and sandy (Morris 1985, 342–4). Some of the sandy, flat-based vessels had narrower bases (eg Fig 82 nos 5–6). This variation may have been related to some special function or may merely indicate smaller containers.

The remainder of the pottery assemblage from this period was small, but it included fabric types representing a wide range of trade links (Chapter 22).

Plate 12 Brine tank (S2, P1)

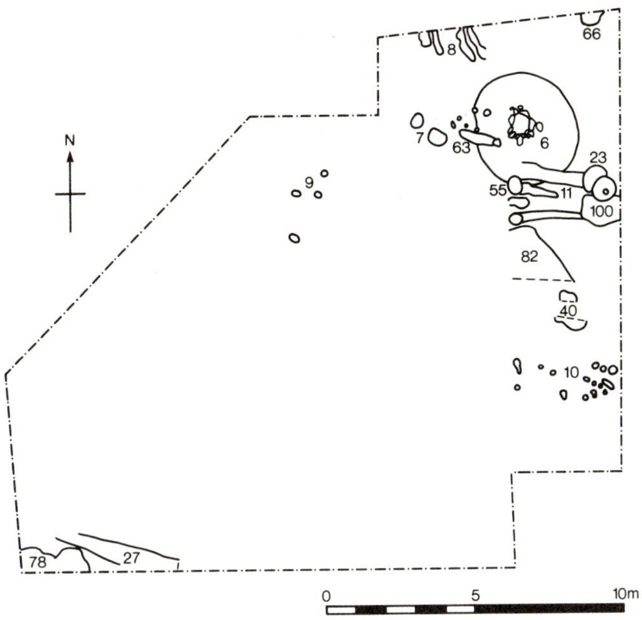

Figure 63 Phase 2 features

Phase 2: Early to mid Roman (Fig 63)

The construction elements of phase 2 dated from the mid 1st to the 2nd century, when salt production ceased. Features were mainly in the form of pits and postholes, the main area of activity lying in the north-eastern corner of the site where a cluster of features formed a possible area of domestic occupation.

Disuse of the brine tanks (S2, S3 and S5)

The fills of structures 2, 3 and 5 were mainly composed of clay and loam, and contained large quantities of briquetage and charcoal. Structure 2 contained an assemblage of wood, including a forked stake (Pl 13) similar to examples from the Old Bowling Green (see Chapter 19). A *terminus post quem* of the 2nd century AD, derived from pottery, was given for their disuse. In addition a dendrochronological date of AD 45 was obtained for this material (Table M24, 3:A14). Radiocarbon dates were obtained from three charcoal samples taken from the fills of structure 3. These were 400–100 Cal BC, 366–40 Cal BC, and 331 Cal BC–Cal AD 60 (Table M28, 3:B5). However the samples were taken from redeposited soils and the *terminus post quem* provided by the pottery is more reliable.

Plate 13 Wood debris in the fill of a brine tank (S2, P2)

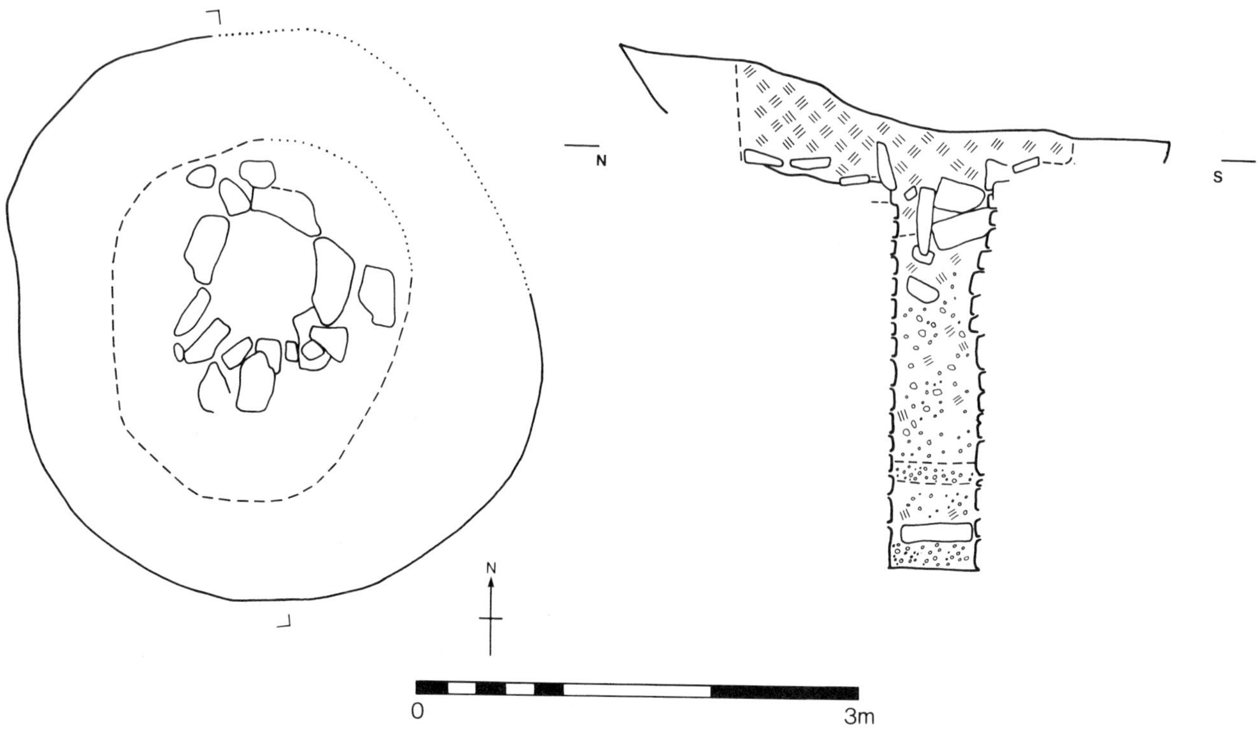

Figure 64 Well (S6)

Well (S6)

Structure 6 had two construction elements, a pit and a sandstone shaft (Fig 64). The shaft was 3.25m deep and varied in diameter from 0.73m at its top to 0.59m at its base. It was built with dressed grey and green sandstone blocks. Some unevenly worked blocks of green sandstone were retrieved from the upper fills of the well. The primary fill of the shaft was composed of a clay silt, while the upper fills had a similar composition of clay and sandstone. Fragments of green sandstone lay over the well-shaft.

Pits, postholes and stakeholes (S7, S9, S10, S23, S40, S55, S66, S78, S82 and S100)

Three of the ten structures (S7, S10 and S55) formed post or stakehole alignments in the eastern half of the excavated area. Structure 7 consisted of four postholes and three stakeholes on an east to west alignment; structure 10 comprised a cluster of sixteen stakeholes, six of which were aligned north-west to south-east; and structure 55 was formed by four stakeholes. All three groups were filled with clay and loam soils.

The other seven groups (S9, S23, S40, S66, S82 and S100), with the exception of structure 78, were also constructed in the eastern sector around the well (S6). However, these pits had no clear association with the features previously described, except that they were cut from a similar level and were filled with clay loam. Structure 78 consisted of two pits adjacent to structure 27 but of a slightly later date.

Ditches (S8, S11, S27 and S63)

Three of the four ditch groups were located close to the well. Structure 8 consisted of four drainage gullies aligned north-west to south-east and lying to the north of structure 7. Structure 11 comprised two parallel ditches aligned east to west which post-dated the disuse of the well (S6). Structure 63 consisted of a ditch and three pits, possibly associated with structure 55. The fourth group (S27) lay approximately on an east to west alignment in the south-western corner of the excavation, and consisted of a ditch punctuated along its southern side by six stakeholes, from which some timber survived.

Layers (S45, S57, S77 and S84)

The features described in this phase were partially sealed by deposits of gravelly clay soils.

Discussion

The brine tanks went out of use prior to the construction of the late 1st to early 2nd century well and its associated features. The main area of activity was in the north-eastern area of the excavation, with peripheral activity in the south-west. The well probably provided fresh water. Despite the presence of brine in groundwater in the area of Droitwich, later records suggest that fresh

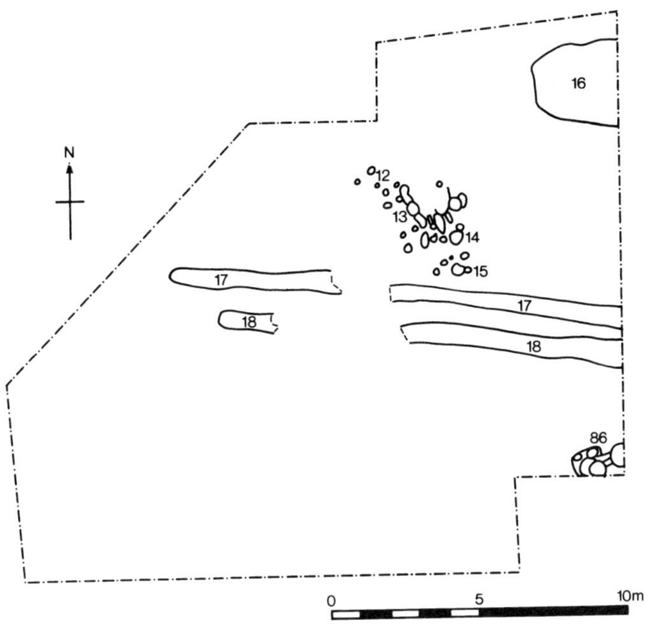

Figure 65 Phase 3i features

or mildly saline water was obtainable from surface wells (Richardson 1930, 104). The well presumably lay within an area of domestic settlement. It contained a primary silt probably formed during its operational life, while the upper fills contained pottery suggesting that it was backfilled in the second half of the 2nd century.

Structures 7 and 8 probably provided drainage for the area. Structure 8 had a gradient running down to the north-west, away from the well. Structure 55 also appeared to have been associated with the well, perhaps serving as the foundation for a superstructure to draw water from the shaft. The function of structures 9, 10, 11, 23 and 100 are unclear.

In the south-west a palisade trench (S27) probably marked a land boundary. The six stakeholes driven into its upper slope may have supported a fence. The boundary was presumably associated with the activity described in the north-east of the site, although the proximity of the backfilled phase 1 brine tanks in the central area may have made the construction of buildings difficult.

Traces of building materials (a small assemblage of tile) were recovered from the fill of the well. The pottery indicated that there was in the 2nd century occupation in the vicinity, probably of a domestic character. It has been suggested that, when the Romans occupied the area in the second half of the 1st century AD, they quickly acquired control over the salt industry (Freezer 1977, 8). A manifestation of this control may possibly be seen in the construction of the fort at Dodderhill (HWCM 603). Some reorganisation would then seem likely, possibly involving relocation or contraction of the salt producing area. With the evidence of salt production continuing into the early Roman period at the Old Bowling Green excavation (Chapter 2), it is possible to view this as indicating contraction of the salt producing area.

Phase 3i: 3rd to 4th century (Fig 65)

Evidence of small-scale settlement was apparent in the late 3rd century from a number of structures to the north of two linear features (S17 and S18). An apparent absence of archaeological features to the south of structure 18 was misleading because the area was heavily disturbed in phase 5ii.

Pits, postholes and stakeholes (S12, S13, S14, S15 and S86)

Structures 12, 13, 14 and 15 formed four posthole and stakehole groups clustered along a north-west to south-east axis. Several peripheral features within structure 12 were probably linked with this alignment.

Structure 86 consisted of a series of inter-connecting pits and a ditch in the south-eastern corner of the excavated area.

Ditches (S16, S17 and S18)

Structures 16, 17 and 18 were constructed on an east to west alignment. Structure 16 was a ditch with a steep profile and a narrow flat base, which terminated 4m west of the eastern excavation baulk. Its primary fill of clay contained charcoal and sandstone fragments. Structures 17 and 18 consisted of two narrow, shallow gullies which lay parallel to structure 16. The western end of structure 17 was clearly delimited but the eastern edge merged into structure 174. A slight downward gradient was noted as both structures ran westward. Their edges in the central area were not clearly defined and, although no physical evidence of disturbance was recorded, such disturbance is more than likely given the position and depth of structure 34 (see P 4ii).

Layers (S54, S64, and S68)

These deposits were composed of gravels and clays.

Discussion

The structures in the late Roman period were irregular, but they included a series of drainage channels and small enclosures (S13, S14, S15, S16, S17 and S18). A drainage function was implied particularly in structures 17 and 18, which ran down the natural valley slope. To judge by its size, structure 16 may have been part of an enclosure dividing two plots of land. The cluster of pits (S13, S14 and S15) to the south of the ditch was difficult

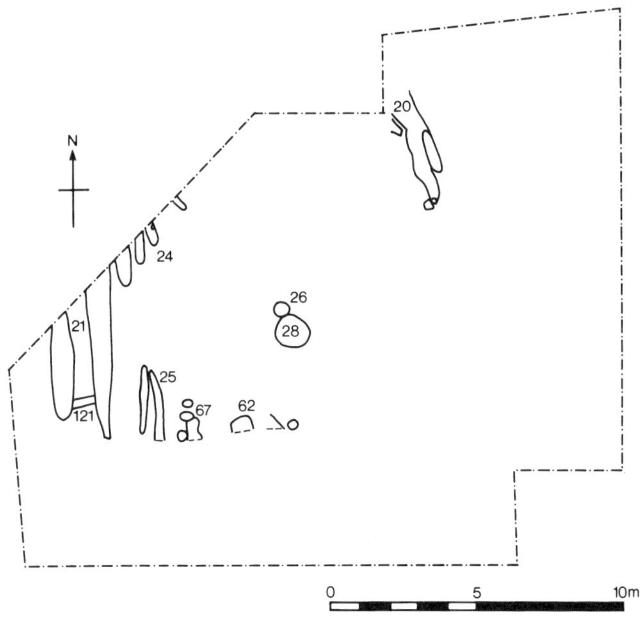

Figure 66 *Phase 3ii features*

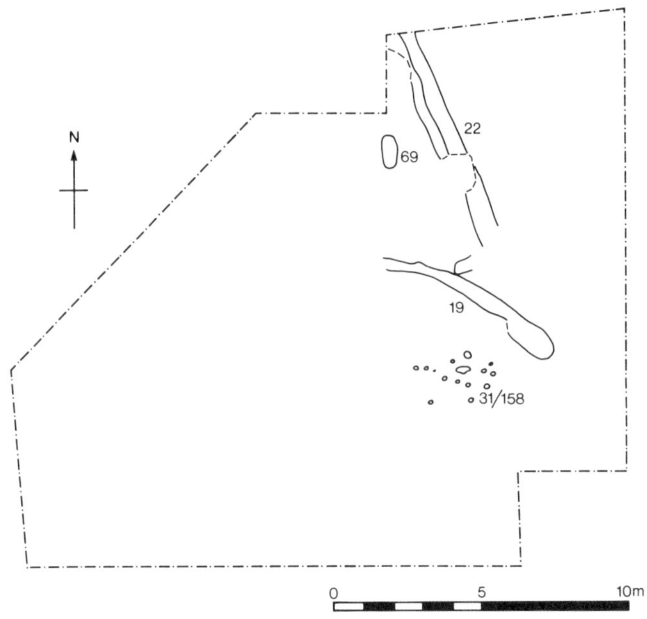

Figure 67 *Phase 4i features*

to interpret within the limited area of excavation, but they may have formed parts of substantial timber structures.

The composition of structures 54 and 68 and the general character of adjacent features may suggest that soil cultivation was taking place next to a settlement area. The two drainage ditches (S17 and S18) were constructed on the same east to west course as the enclosure ditch, indicating that the settlement limits were clearly defined. The domestic nature of that settlement was implied by the ceramic assemblage, which included tablewares and *mortaria* (Chapter 22). Small quantities of glass were also present.

Phase 3ii: Sub-Roman and Anglo-Saxon (Fig 66)

A series of linear features were cut into the layers comprising structure 174.

Pits, postholes and stakeholes (S26, S28 and S67)

Three poorly preserved groups of features were constructed in the central area. Structures 26 and 28 consisted of a possible posthole and a pit; and structure 67 was an incomplete line of stakeholes and a ditch.

Gullies (S20, S21, S24, S25, S62 and S121)

Five of the six groups were constructed on a north to south alignment. Structures 21 and 24 consisted of a parallel series of shallow gullies, but to the east of the former, structures 25 and 62 were badly disturbed. A ditch which was recut twice (S20) lay in an apparently isolated position to the north-east. Finally, a narrow east to west gully (S121) was cut by structure 21.

Layers (S65, S71, S171 and S174)

These deposits were largely composed of clay loam and pebbles, and were compacted to level the ground surface which had previously sloped on a slight gradient from north-east to south-west.

Discussion

Six of the structures identified above formed a series of shallow linear features, generally on a north-south alignment. Structures 20, 21, 24 and 25 probably represented plough furrows; the few abraded Romano-British pottery sherds from these features were consistent with deposition through field manuring. Soil within the furrows was mainly composed of loam, which further suggested that the site was being cultivated. However the activity was difficult to date as only abraded residual Roman pottery was recovered. A date somewhere in the sub-Roman or Anglo-Saxon periods is suggested on the basis of three 'grass' tempered sherds which were recovered from these deposits (Chapter 22).

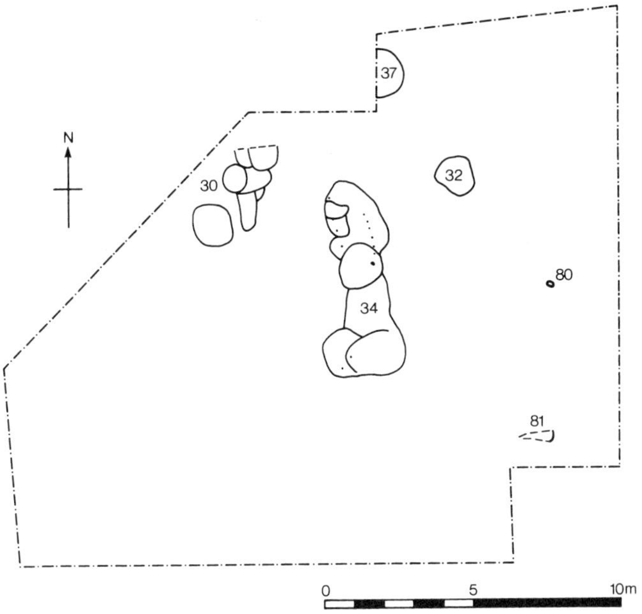

Figure 68 Phase 4ii features

Phase 4i: Earlier Saxo-Norman (Fig 67)

Phase 4i consisted of a series of pits, ditches and layers upon which little functional interpretation could be placed.

Pits, postholes and stakeholes (S31, S69 and S158)

Structures 31 and 158 consisted of a line of stakeholes running parallel to structure 19. Structure 69 was an isolated pit to the west of structure 22.

Ditches (S19 and S22)

Structure 19 consisted of a ditch constructed on a curving north-west to south-east line. Its north-western end was not located, as it had been disturbed by structure 34. The ditch widened and deepened towards its south-eastern terminal. Structure 22 was cut on a similar alignment and deepened to the south-east.

Layers (S93 and S170)

These two layers of sandy clay and loam were disturbed during the construction of the large boundary ditch in the 13th century (S36, P5ii).

Discussion

Phase 4i comprised structures associated with drainage (S19 and S22) and rubbish disposal (S69). Structure 31 possibly formed a fenceline to the south of structure 19. There seems to have been

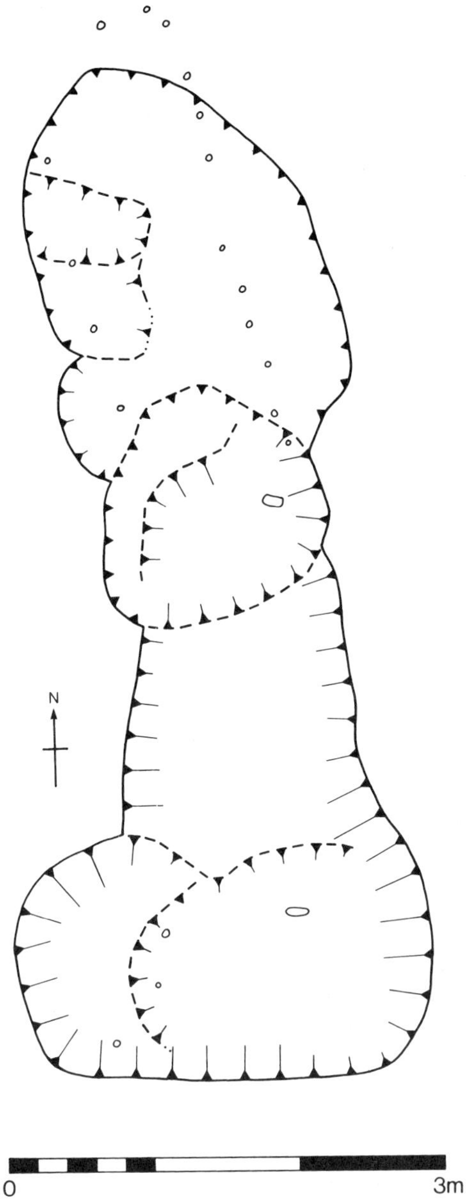

Figure 69 Pits possibly used in the processing of animal hides (S34)

little activity on the site in this period, but the distribution of the features suggested more intense occupation to the north and east. Surviving features were generally poorly preserved and the finds assemblage was too small to aid interpretation.

Phase 4ii: Later Saxo-Norman (Fig 68)

Phase 4ii comprised a complex of pits in the central and northern parts of the excavated area.

Pits (S30, S32, S34, S37, S80 and S81)

Structure 34 formed the most substantial feature (Fig 69). It comprised six pits and a wide shallow ditch. The bases of the interconnecting pits were

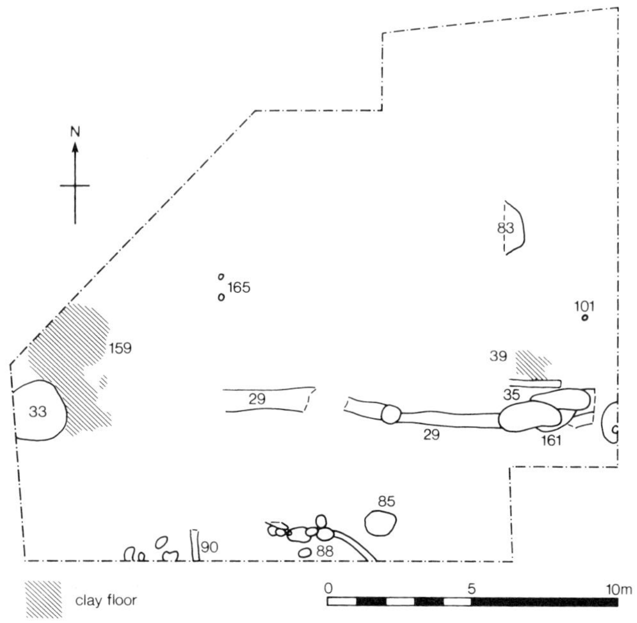

Figure 70 Phase 5i features

punctuated by a series of stakeholes. A second structure (S30) consisted of four pits, two postholes with posts and a connecting ditch. Structures 32 and 37 were similar, and a small assemblage of horn cores was recovered from the latter. The fills of these groups contained organic deposits and fragmented, decayed wood and charcoal.

The two remaining structures (S80 and S81) were partially disturbed and were not clearly associated with the main activity.

Layers (S70)

Two layers of black clay and loam lay over much of the excavated area and were notably deeper in the eastern area, in a slightly lower part of the site. The deposit was rich in humic material and its accumulation and cultivation continued from phases 4ii to 9.

Discussion

Structure 34 may have been used for the processing of animal hides, and the parallel lines of stakes were probably used as a rack to dry the skins. A comparable structure of similar date was discovered at Lower Bridge Street in Chester (Mason 1985, 26). The horn cores and fragments of skulls indicate the processing of animal products, possibly tanning (where the untanned hide may be acquired with the horns still attached; see Chapter 32) or hornworking. The central feature (S34) also contained large quantities of decayed wood and organic material, which are characteristic of the waste products formed during the processing of animal hides.

The pottery assemblage indicated a wide range of trade contacts with, for example, the Cotswolds area and Lincolnshire (Chapter 22). This confirms the commercial importance (founded on the salt trade) of the early medieval town across a wide geographical area, extensively referred to in Domesday.

Phase 5i: 12th century (Fig 70)

Activity included the construction of ditches and pits. The phase was chiefly characterised by the continued deposition of clay loam over the site.

Pits, postholes, stakeholes and ditches (S29, S33, S35, S39, S83, S85, S88, S90, S101, S161 and S165)

Three of the twelve structures (S29, S33 and S35) were directly associated by their alignment. Structure 29 consisted of a ditch (probably a drain) of indeterminate length, running east to west. A pit (S33) lay to the west, and structure 35 comprised three pits, connected to the eastern terminal of ditch 29. Structure 39 consisted of a bowl-shaped pit filled with clay and large quantities of charcoal and ash, a pebble and clay surface, a ditch containing traces of a horizontal timber floor, two possible postholes and the western edge of two further postholes. The timber setting probably represented the impression made by a cill beam, which might suggest that a timber-framed building stood to the north of the rubbish pits of structure 35.

The function of the remaining pits and small postholes within structures 83, 85, 88, 90, 101, 161 and 165 was difficult to determine because of disturbance in phase 5ii.

Layers (S70 and S159)

The soils of structure 70 were first deposited during phase 4ii and eventually reached a maximum depth of one metre in this phase. They were composed of black, clay loam, sealing structures 29, 33 and 35. Structure 159 was a compacted surface adjacent to the rubbish pit (S33).

Discussion

The main structural developments in phase 5i point to domestic activity, with a sequence of drains and rubbish pits. Associated buildings were not identified to the north of these features, where the build up of black organic clay loam was recorded (S70). Structures 29, 33 and 35 were contemporary and were for drainage. Structure 35 possibly provided a sump for substances dumped and poured into the ditch (S29). The timber setting and adjacent clay floor of structure 39 suggested that a timber building existed in this area, but its poor preservation prevented a more precise definition.

A higher percentage of locally manufactured pottery was noted during the 12th century,

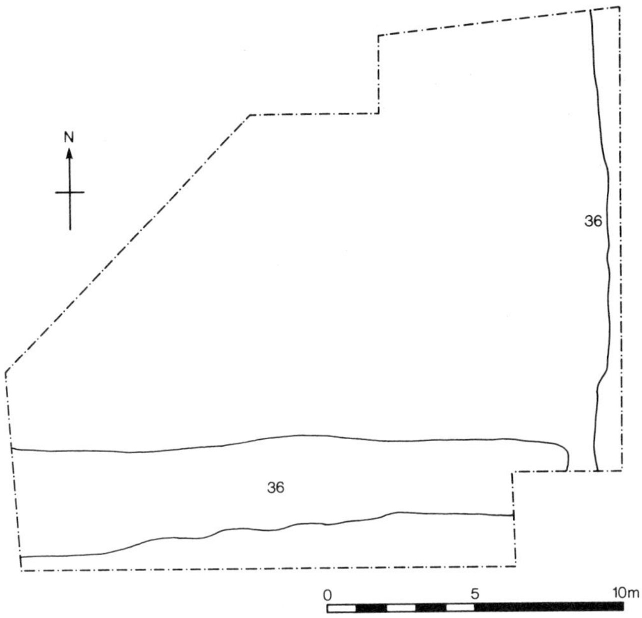

Figure 71 Phase 5ii features

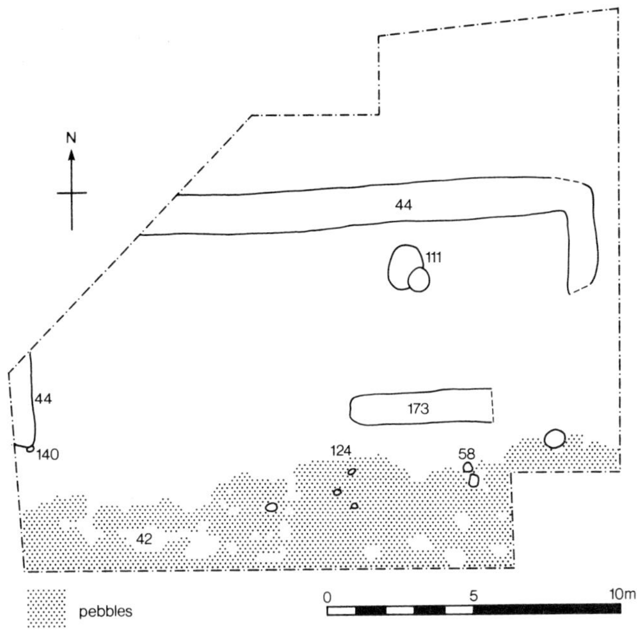

Figure 72 Phase 6 features

especially the Worcester-type ware, with a relative decline in the presence of pottery from more distant sources such as the Cotswolds (Chapter 22).

Phase 5ii: Early 13th century (Fig 71)

Constructional activity was confined to the cutting of one large and one slighter ditch.

Ditches (S36)

Structure 36 consisted of a large ditch running east to west and a smaller one running north to south. The eastern end of the former terminated within the excavated area. It had steep sides, but was shallower and narrower towards the east. The sides of the lesser ditch were steeper but its width could not be determined. Primary fills of silt possibly related to the use of the larger feature as a drainage ditch, after which it was filled with tipped layers of clay and loam. The upper fills of the slighter ditch contained ash, plaster and rubble, indicating the destruction of a building by fire. This is the only archaeological evidence for the existence of this building. The quantity of domestic pottery found in these deposits suggest that the building was a house, probably of middling or fairly high status (see below).

Layers (S70)

The substantial deposit of black clay loam originally observed in phase 4ii continued to accumulate in the northern area of the excavation.

Discussion

The form and size of the east to west ditch suggested that structure 36 was constructed to define the southern and eastern boundaries of a substantial plot. The narrower, shallower ditch, which was cut at a later date, probably subdivided a larger original plot. It is notable that the plot at Priory House (HWCM 609), to the south of the excavation, had its longest axis running east to west, and a similar layout may be suggested here. Contemporary internal features could not be identified; if any existed they could have been removed by later activity.

The presence of a substantial house, in a strongly delineated and probably large enclosure, suggests that domestic occupation of a higher status was established by this time. It seems likely that Friar Street served as a wealthier residential area to the west of the commercial centre (High Street).

The ceramic assemblage recovered from these levels largely comprised wares of cooking pot type, with some pitchers, probably manufactured at Worcester. A decorated lead spoon was found in the upper fill of the slighter ditch (Chapter 30), perhaps confirming the relatively high status of this newly established residential area of the town.

Phase 6: 13th to 14th century (Fig 72)

The boundaries identified in phase 5ii were superseded by a smaller enclosure. A cobbled road was also laid along the south end of the excavated area, forming the medieval course of Friar Street.

Road (S42 and S98)

The road was constructed with cobbles and it traversed the southern end of the excavated area.

Wall foundation (S44 and S173)

The main foundation trench consisted of a ditch running east to west, which turned to the south approximately 1.00m from the excavation edge. The southern extent of the cut respected the line of the road. Another trench running north to south was not physically linked with the first but it was probably contemporary. The depths of both cuts were similar, with steep-sided profiles and flat even bases, and there was evidence of packing within the loam and rubble fills, which also contained compacted clay.

Pits (S58, S111, S124 and S140)

Three posthole structures (S58, S124 and S140) and two pits (S111) seemed to be linked with the line of the road, as they were established along its edge. It can be assumed that a fenceline existed between the plot and the road.

Layers (S70, S99, S108, S113 and S116)

These deposits were largely composed of the dark organic clays within structure 70, but some variations were noted, particularly in the ashy content of structures 99, 113 and 116.

Discussion

The road, the first excavated manifestation of Friar Street as a metalled street, was represented by structure 42. It is perhaps relevant that Droitwich received its first documented grant of pavage in 1316 (VCH ii, 90). Structure 44, contemporary with the road surface, represents the foundation of three sides of a building, whose internal dimensions were 17.50m east to west and 7.50m north to south. Structure 173 probably provided a foundation for the main south wall. Unfortunately any internal evidence was removed by activity during phase 7.

The accumulation of dark soil deposits (S70 in P 5i and P5ii) continued into the 14th century. The high ash content within structures 99, 113 and 116 indicate that a fire caused some damage in the eastern section of the excavated area. A fire is documented in 1297, which caused widespread damage, and probably started at St Andrew's church, at the junction of Friar Street and High Street (VCH iii, 75).

Worcester-type wares continued to dominate the ceramic assemblage but a significant quantity of Malvernian ware was also represented. Domestic vessels were recovered in large quantities which, combined with a range of domestic items of metalwork (especially copper alloy pins and thimbles, and structural ironwork fittings), pointed to the domestic character of occupation. A small quantity of smithing waste was also recovered from 14th century deposits. Four fragments of smithing slag, including a hearth bottom, were found. However they were associated with the layers identified as road make-up (S98) and adjacent soils (S108), and so could have been brought onto the site for road construction.

Figure 73 Phase 7 features

Phase 7: 15th to 16th century (Fig 73)

In the 15th century the character of the road remained the same, and in the mid 16th century two substantial timber and sandstone buildings were erected.

Road (S42)

Four metres of the total width of the road was recorded, the southern edge lying beyond the excavated area. The surface was not uniform and the differing patches of cobbling indicated that repairs had been made. Its location in this phase was almost identical to that in phase 6.

Buildings (S41 and S43)

A building (S41) occupied the eastern half of the excavation, but was largely obscured by later activity. The main surviving feature was a sandstone masonry construction interpreted as a chimney stack, probably attached to the external wall of a timber-framed house in the unexcavated area at the eastern end of the site. A north to south alignment (S49) of four postholes lay parallel to the chimney. These may have been part of the

Plate 14 Cellar (part of S43, P7)

postulated building or, more probably, of scaffolding associated with its construction. To the north of the chimney, a drain with a sandstone lining appeared to lead away from the north-western end of the building, and a cobbled surface lay adjacent to the west side of the chimney.

A second building (S43) with sandstone walls, aligned at right-angles to the street, lay to the west of structure 41. Its four walls enclosed a cellar (Pl 14) with a partially preserved floor of sandstone flags, at a level lower than the local ground surface. A small hearth made of tile occupied the centre of the cellar, which was possibly used as a brine boiling house. A line of five padstones lay to the south, presumably the foundations for the wall fronting the street. A north to south posthole alignment, parallel to the east wall, probably provided support for an outshut attached to the conjectured central bay of the building.

Postholes (S59, S60, S122, S138, S163 and S164)

Structures 59, 163 and 164 formed three posthole groups between the buildings, but their precise functions could not be determined. Structures 60, 122 and 138 were parallel to the road, and may have formed a boundary.

Layers (S70 and S135)

The black clay loams identified in previous phases were cultivated during the 16th century. Large quantities of flat roofing tile in structure 135, which also contained other building rubble, may have been used in the construction of the buildings (S41 and S43).

Discussion

Structures 41 and 43 were constructed in the mid 16th century, partly of timber framing on sandstone sills or pads and partly of sandstone masonry walls, with tiled roofs. Post-medieval activity removed much of the eastern building, but its external chimney stack survived with an adjacent cobbled yard. The remains of the western building (S43) were more substantial, with a recognisable central room. Rooms to the north and south were indicated by the slight continuation of the walls forming the central bay and a further room to the west was conjectured. Of the western structure (S43) two bays of a possible three-bay building were excavated, but the neighbouring building to the

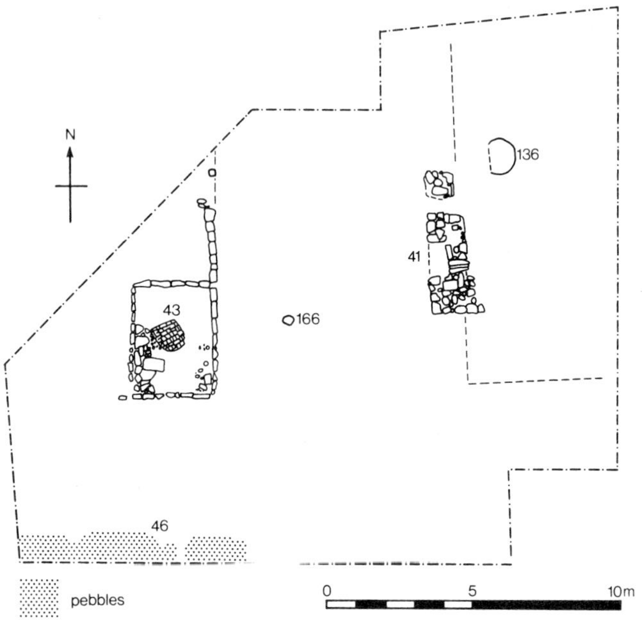

Figure 74 Phase 8 features

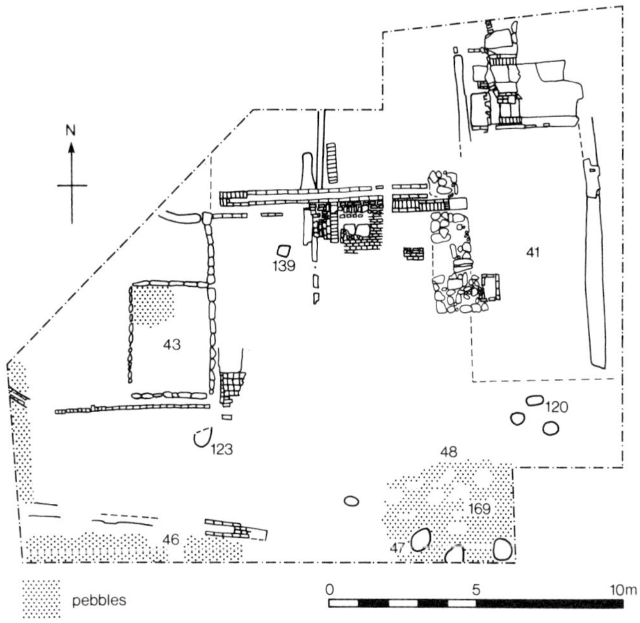

Figure 75 Phase 9 features

east (S41) was more difficult to interpret because only the masonry foundations of a chimney stack survived. This was interpreted as an external stack of a timber-framed house whose conjectured outline is shown on Figure 73.

The domestic nature of the buildings was identified from the range of pottery forms in use, including a high proportion of Malvernian products, in particular glazed jugs and jars (Chapter 22). Phase 7 also produced a wide range of copper alloy (Chapter 28) and iron (Chapter 29) artefacts, mainly in the form of personal and household items.

Although there was no sign of a boundary between the two buildings it is likely that they occupied separate plots. If this was so it seems that these plots were much smaller than those described in phases 5ii and 6. The plan of structure 43 may suggest that it stood in a conventional burgage plot, but structure 41 presumably lay in a larger rectangular unit, whose proportions and dimensions cannot be estimated. Unfortunately the evidence does not allow us to explore the relationship between these two units. However, it seems clear that by the 16th century a range of buildings reflecting a variety of status (and perhaps function) were in place in this area of the town.

Phase 8: 16th to mid 17th century (Fig 74)

The buildings continued in use into the 17th century, while the line of the street running along the southern edge of the excavation was slightly modified.

Road (S46)

The northern edge of the cobbled road was repaired and now lay slightly to the south of its line in the previous phase. This indicates either a reduction in width or a movement of the road southwards.

Buildings (S41 and S43)

The character of the two buildings constructed in phase 7 remained the same, but the outshut attached to the central bay of structure 43 was demolished.

Pits and postholes (S136 and S166)

Only one pit and one posthole survived later disturbance. The pit (S136) appeared to be internal to the eastern property, and may have formed the base of a hearth. The posthole (S166) was isolated, beyond the east of the central bay of the western building (S43), and its function was not determined.

Discussion

Much of the activity detected in this period was disturbed by the reconstruction and extensions associated with structures 41 and 43 in phase 9. Later building work particularly affected the line of the road (S46), the eastern edge of which was lost by the cutting of a modern pipe trench. Much of structure 41 was disturbed by the later erection of brick-built features. However the chimney was still

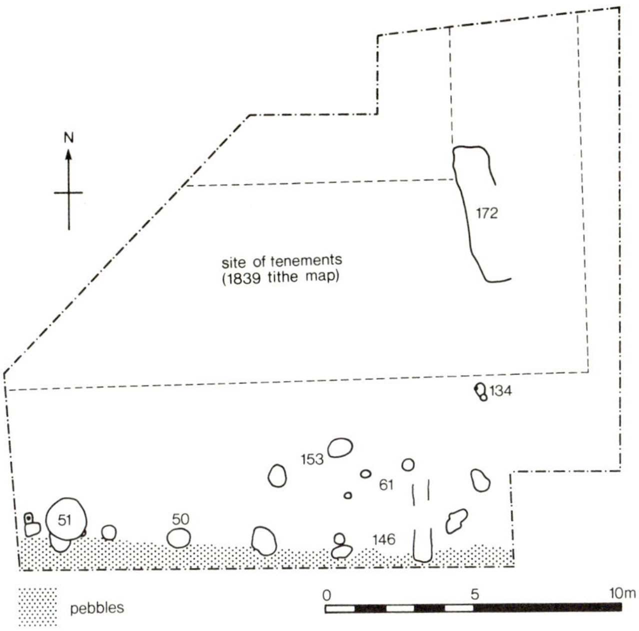

Figure 76 Phase 10 features

in use. The artefactual evidence was similar to that recovered from 16th century levels, demonstrating continuity of domestic occupation from *c* 1550 to 1650. Phases 7 and 8 were divided only on the basis of the demolition of the outshut and the repairs to the street.

Phase 9: Mid 17th to 18th century (Fig 75)

In the early 18th century structures 41 and 43 underwent large-scale change. Structure 43 was partially rebuilt with a number of brick walls, while structure 41 was refurbished in brick, with an extension to the north. Between the two buildings, and connecting them, a brick extension was constructed with a system of drains in its floor.

Road (S46, S48 and S169)

The road (S46) was disturbed in the south-eastern corner of the excavation by a cobble spread (S48 and S169), which possibly formed a yard, with a pit cut into the surface (see S47 below).

Buildings (S41 and S43)

To the north of the chimney stack foundation a series of internal features connected with a kitchen and cellar or sunken-floored room was added to the eastern building (S41). A brick floor outside the entrance to the cellar led to three steps descending into it, and onto a floor made of sandstone flags edged with brick. A tiled floor lay adjacent to the steps, and a brick wall lay to the west of the two floors, forming the western side of the cellar. Its parallel east wall had been largely robbed. The rear wall was constructed with brick and tile, and a partition wall within the cellar also survived. To the south of the cellar, but still within the building, a rectangular brick feature was built into the side of the chimney.

The east wall of the western building (S43) was rebuilt in the early 18th century. The sandstone base was retained but the upper structure was of brick. The south wall had been partially robbed at its western end and only one course of bricks survived. Two largely robbed floor surfaces lay adjacent to it, and were constructed with brick, and re-used malthouse tiles. They were enclosed by a brick wall and were therefore probably external to the house, which had a front garden bordered by a brick-built wall along the street frontage.

A brick and cobble-surfaced area with two brick-lined drains was constructed between the two buildings. The function of this area was not clear but it seems likely to have been an outbuilding.

Pits (S47, S120, S123 and S139)

Structures 120, 123 and 139 were associated with the buildings. They consisted of five pits which were probably used for rubbish disposal. Structure 47 comprised three postholes on an east to west alignment, which seem to have been linked to the alterations made to the course of the road. However it may be that the line of the road was disturbed when the buildings were demolished (see P10).

Discussion

The main activity associated with phase 9 was the rebuilding and extension in brickwork of the late medieval sandstone and timber buildings. The central drains seem to have served both buildings. The main brick-built extension associated with the east building (S41) seems to have been a cellar below a possible kitchen area. The street frontage was separated from the site by a further brick-built boundary wall, which had been robbed to the south of the eastern property.

Phase 10: 18th to early 19th century (Fig 76)

The two buildings were demolished in the 18th century and the road was rebuilt. In the early 19th century two tenement buildings within a block of eight were erected on the site.

Road (S146)

The road described in phases 8 and 9 (S46) was replaced by structure 146, which was constructed with more closely packed cobbles and a more clearly defined northern edge.

Pits and ditches (S50, S51, S61, S134, S153 and S172)

A boundary line (S50), consisting of seven evenly spaced postholes, separated the road from the tenements. Other pit groups included a clay-lined well (S51) and six pits, forming structures 61, 134 and 153.

A robber trench (S172) provided evidence of the removal of internal building features from the eastern plot. It is possible that it was excavated to remove a boiler or fireplace, as the disused phase 9 chimney stack foundation lay adjacent to it.

Layers (S142, S147, S149, S150, S152, S154 and S155)

Twenty-five layers of clay and rubble were recorded and were clearly associated with the demolition of the phase 9 buildings. Some of the debris also sealed the pits described above.

Discussion

Structures 41 and 43 had been demolished by the mid 18th century, according to the dating of pottery recovered from the rubble of structure 149 (an early 19th century penny also found in this rubble may well have been intrusive). The road had been slightly realigned further to the south, narrowed and resurfaced. Several small pits were dug near the northern edge of the street surface in this period, possibly marking a boundary between the read and the tenements. The remains of these brick-built tenements were removed during machine clearance early in the excavation but their limits, shown on the Tithe Map of 1839 (CRO R760/256/1572), are included on Figure 76. The tenements were demolished at some time between the preparation of the Tithe Map and the first edition Ordnance Survey 1:2500 map in 1884.

Phase 11: 19th to 20th century (Fig 77)

Phase 11 represented the final period of activity on Friar Street prior to the construction of the Fire Station in the late 1970s. It comprised the building of two halls used by church and community organisations.

Boundary wall (S53)

A brick-built wall bordering the west and south of the excavated area was constructed prior to the first edition Ordnance Survey map of 1884.

Church hall (S52)

The foundations of a late 19th or early 20th century building (S52) consisted of brick walls which were 14.00m from east to west and approximately 6.00m from north to south. An interior north to south wall also survived. Three demolition layers (S151, S156 and S157) sealed the building. A second church/community hall was built at about the same time, immediately to the south of structure 52. Its original function is unknown but latterly it was used by the Scouts and Guides organisations. Both halls were demolished in recent times, structure 52 in the 1960s and the hall to the south in 1975.

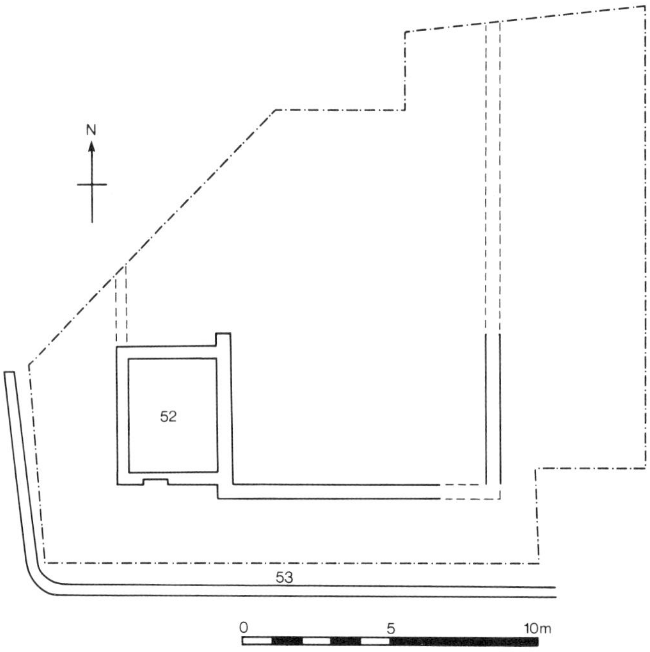

Figure 77 Phase 11 features

Discussion

Structure 53 was constructed as a boundary wall to the south of the site in the second half of the 19th century. Its earliest dated appearance was on the 1884 Ordnance Survey plan, which also indicated that the excavated area was unoccupied at this time.

After the excavation the site was acquired by Hereford and Worcester County Fire Service and, at the time of writing, is still occupied by a fire station.

Conclusion

From the late Iron Age to the early Roman period, Droitwich brine was being exploited and salt was being produced. Production involved the clay- and stake-lined tanks and hearths, used respectively for the storage and or settling, and boiling, of brine. This industrial production ended in the early 2nd century AD and the area was then used for domestic occupation characterised by the construction of a well. Only sporadic activity was evident in the 3rd and 4th centuries, by which time salt production seems to have ceased in this area.

Sub-Roman and Saxon occupation was not well represented, although the linear features within phases 3ii and 4i may have been related to light agricultural activity in this period. Their consistent orientation implies that field boundaries were well established.

By the 10th and 11th centuries the site was again probably exploited for industrial purposes. The central phase 4ii structure was probably used in the processing of animal hides, and nearby pits were dug for the disposal of waste products, probably from tanning.

From the 12th to early 13th century onwards the site was continuously occupied and laid out in a succession of land and property divisions varying in size, arrangement and status. Within these land divisions, domestic and possibly industrial, functions alternated.

The wider archaeological significance of the excavation, in relation to the general development of Droitwich, is discussed in the concluding section of this volume (see chapter 34).

22 Pottery *Derek Hurst*

The pottery assemblage ranged in date from Iron Age through to modern in an almost continuous sequence. A total of 14,867 sherds weighing 214.09kg was recovered, including industrial ceramic vessels, or briquetage, used as salt containers in the earlier period. This assemblage complemented that from the nearby Old Bowling Green excavation, which contained a particularly large component of Roman pottery though lacked a well stratified sequence for the later periods. In particular, the late Saxon and early medieval periods were well represented at Friar Street (Figs 78 and 79). It was therefore possible, by considering both of these sites together, to present an overall picture of the development of pottery usage at Droitwich over at least 2,000 years.

Method

The pottery was macroscopically sorted using the fabric reference series (Chapter 35), with occasional use of a x10 binocular microscope. Forms were classified using non-functional terms into broad vessel types, and sub-types if applicable. Evidence of vessel use, for example sooting and charred food residues was noted separately. In order to avoid unnecessary duplication, reference to vessel form typology for the Roman period is largely to illustrated vessels from the Old Bowling Green. Decorative motifs were recorded using a coding system covering 32 different categories of decoration and surface treatment. Quantification is by weight unless specified otherwise.

Phase discussion

Phase 1: **Iron Age** (Figs 80 and 81)

Fabrics present: 1, 2, 3, 4.2, 5.1, 5.2, 5.3, 6, 9 and 10

Much of the pottery of this phase (95%) consisted of briquetage vessels used to dry and transport salt, with the organic fabric (fabric 2, 67%) being more common than the sandy variety (fabric 1, 28%). Definite rims could not be recognised, as the tendency to break along coil junctions observed by Rees (1986, 48) produced many broken pieces of briquetage which could not be certainly distinguished from rims. In contrast a number of bases were present in both fabrics, as at the Old Bowling Green excavation (Rees 1986, 49). At Friar Street these seemed to be of two types (Morris 1985, fig 3), either flat (Fig 82 nos 2–6) or indented (Fig 82 no 1). The base diameters of the two types

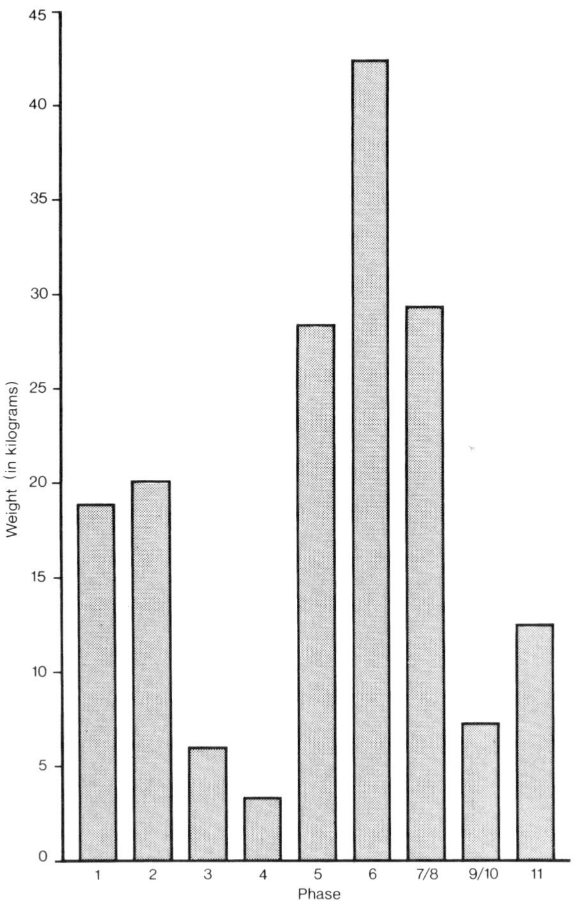

Figure 78 Pottery quantification by weight

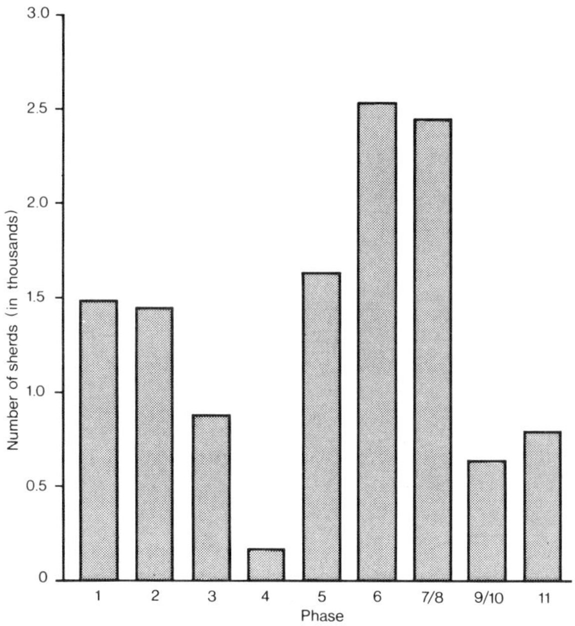

Figure 79 Pottery quantification by sherd count

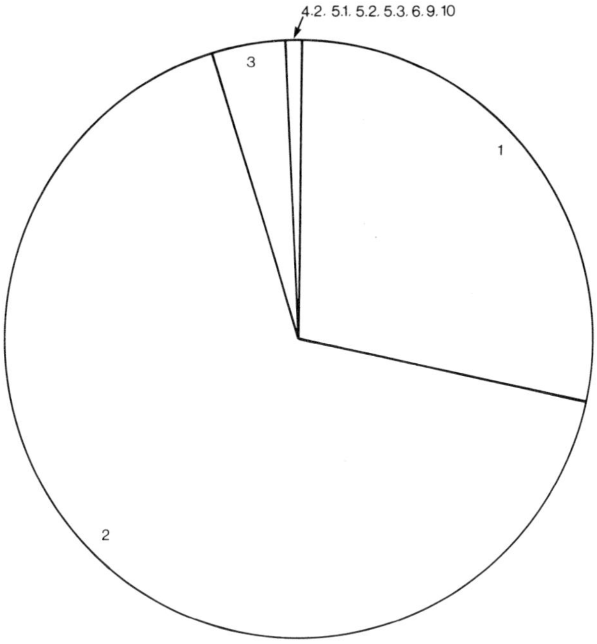

Figure 80 Fabric proportions by weight in phase 1

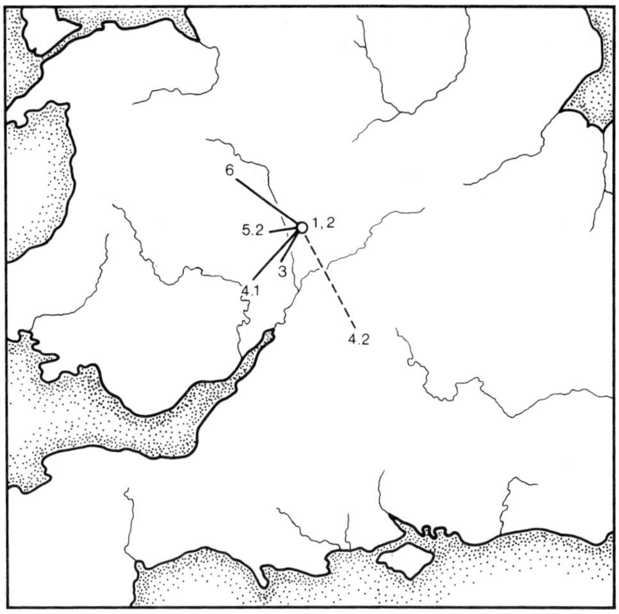

Figure 81 Principal pottery sources in phase 1

seemed similar (*c* 60–100mm), but the flat base only occurred in the sandy fabric. Bases of both types were recovered from contexts probably relating to the use of a hearth (S4), and the unusual number of complete bases from this feature suggested that it may have had some association with briquetage production or utilization (see Rees 1986 for further discussion). The indented type has previously been recognised in Droitwich (Rees 1986, 49, fig 2), and also at other sites such as Sutton Walls (Kenyon 1953, fig 12 no 6) and Beckford, but the flat type has so far only been distinguished in the Friar Street briquetage assemblage.

Other fabrics of this phase were poorly represented, except for the handmade Malvernian ware (fabric 3). This included several rim sherds with linear tooled decoration (eg Fig 82 no 9), which were mostly associated with structure 4. These belonged to jars with plain unbeaded rims, while the linear tooling just beneath the rim was of either horizontal or oblique type. Similar pottery has been found at several other sites in the region, for example Midsummer Hill (Stanford 1981, 144–7) and Beckford (Ford and Rees pers comm). Since petrological analysis conducted by Peacock (1968, 415–21), this type of pottery fabric (Group A) has been known to contain Malvernian rock inclusions and its source is therefore broadly indicated.

Fabrics 4.2, 5.1, 5.2, 5.3, 6, 9 and 10 were present in very small quantities, but no information about their forms could be established. However, petrological analysis (3:B6–5) suggested that they represented, together with the more common fabrics, a wide range of different sources supplying Droitwich with pottery during this phase. These comprised sources to the north-west (fabric 6), south-west (fabrics 3, 5.2 and 9) and south (fabric 4.2). Sandy wares (fabrics 5.1 and 5.3) were considered to be possible local products. This wide range of sources for the pottery, though based on a small assemblage, is observed by Morris (3:B9) to be unique in the region and may well be a result of the salt-making industry functioning as a focus for ceramic exchange (Morris 1983, 355–6).

Dating

The dating of the pottery assemblage of this phase is rather problematical, as the pottery of this period is generally not closely datable. However, linear tooled Malvernian (fabric 3) pottery has been represented at many sites of middle to later Iron Age date in the region. At Beckford linear tooled Malvernian pottery constituted a diagnostic feature of the middle Iron Age ceramic phase but was regarded as probably late in this sequence (Ford pers comm).

Other pottery information pertinent to the dating of this phase was less specific. However, all the fabrics present were of Iron Age date, except perhaps a single decorated sherd in fabric 10 (Fig 83) for which a pre-Iron Age date was suggested by Morris (3:B9). The briquetage also supported an Iron Age date, though the period of its manufacture and use has yet to be determined, and it may have continued in use beyond the Iron Age at least at the production centre (Morris 1983; Rees 1986, 51). However, the massive quantities of briquetage present at Droitwich have made the detection of residuality difficult, and on both the Old Bowling Green and Friar Street (Fig 84) excavations it remained a significant component of the ceramic assemblage throughout the Roman period.

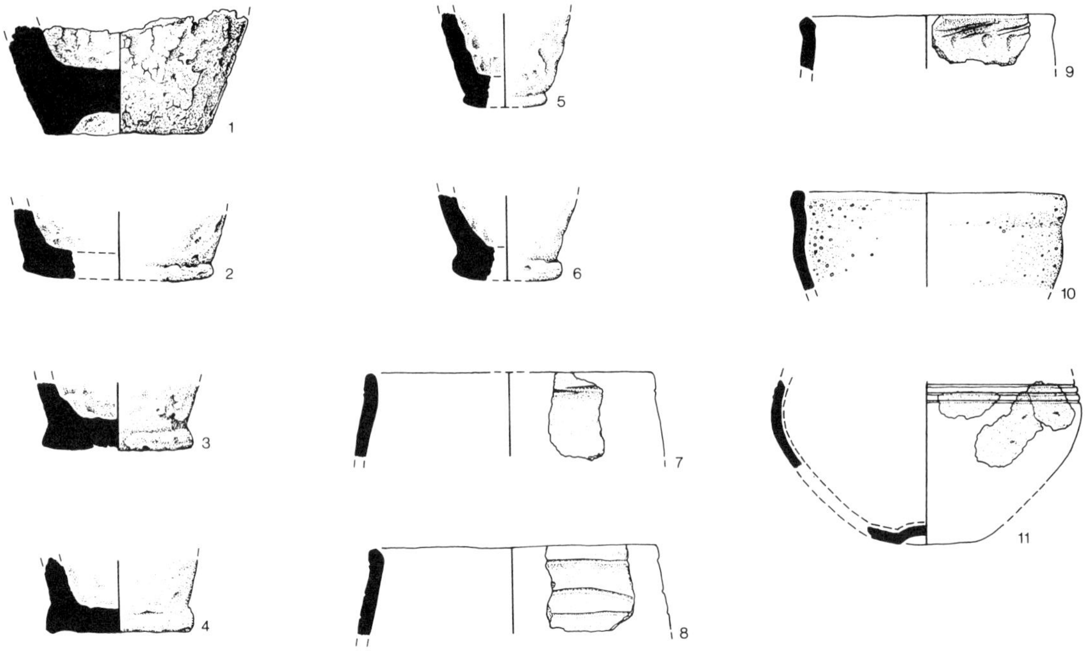

Figure 82 Prehistoric pottery and briquetage: fabrics 2 (1), 1(2–6), 3 (7–9), 4.1 (10) and 5.1 (11). Scale 1:4

Illustrated pottery

Prehistoric pottery and briquetage (Fig 82)

Organic briquetage (fabric 2; no 1), sandy briquetage (fabric 1; nos 2–6), Malvernian (fabric 3; nos 7–9), limestone tempered (fabric 4.1; no 10) and fabric 5.1 (no 11)

1 Briquetage vessel with indented base (1182, S3, P1–2)
2 Briquetage vessel with flat base (469, S4, P1)
3 Briquetage vessel with irregular flat base (1161, S4, P1)
4 Briquetage vessel which is mottled purplish grey throughout (1161, S4, P1)
5 Briquetage vessel with flat base (469, S4, P1)
6 Briquetage vessel with flat base and reduced internal surfaces (469, S4, P1)
7 Malvernian (Group A) jar with linear tooling below rim (469, S4, P1)
8 Malvernian (Group A) jar with linear tooled decoration (469, S4, P1)
9 Malvernian (Group A) jar with oblique linear tooled decoration (885, S1, P1)
10 Limestone tempered (Group B1) bowl (415, S5, P1–2)
11 Burnished bowl with *omphalos* base (1182, S3, P1–2)

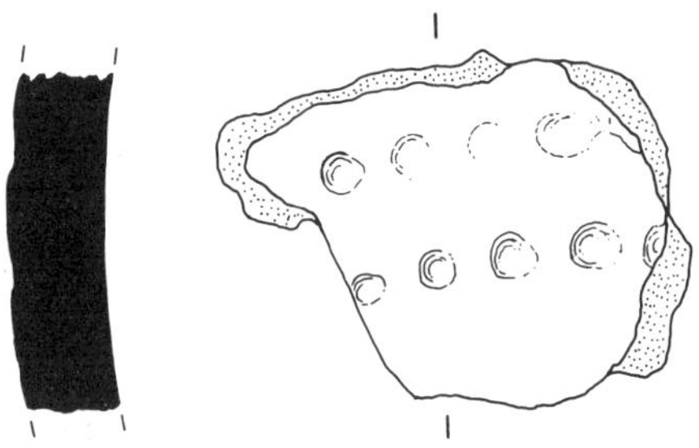

Figure 83 Possible pre-Iron Age decorated sherd (fabric 10). Scale 1:1

Phase 2: Early to mid Roman (Fig 85)

Fabrics present: 1, 2, 3, 4.1, 5.3, 12, 12R, 12.2, 12.4, 13, 14, 15, 21, 22, 41, 42.1, 43, 45 and 98

The amount of Roman pottery in phases 2 and 3 was relatively small compared with the amounts of pottery associated with phase 5 onwards. Only structures 2, 3, 5 and 6 produced much pottery, while the disuse fills of two clay and stake-lined

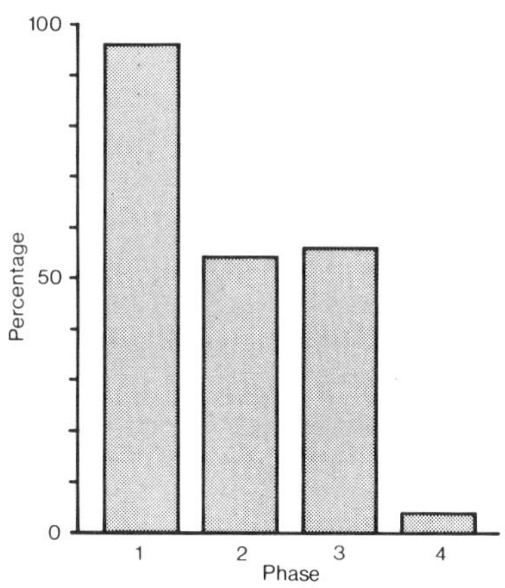

Figure 84 Percentage by weight of briquetage in the pottery assemblage

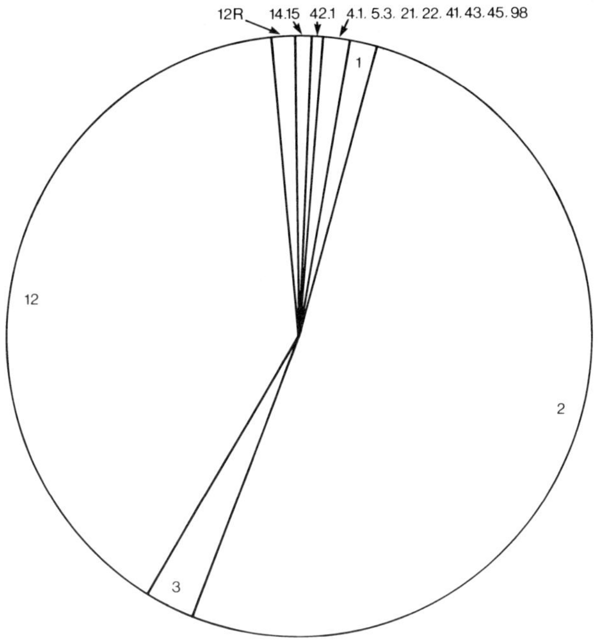

Figure 85 Fabric proportions by weight in phase 2

pits (S2 and S3) accounted for much of the briquetage. The concentration of briquetage in their fills (but not in other probably slightly later features of this phase) perhaps supports the contention that the briquetage went out of use at the same time as the brine tanks, at least in this area of Droitwich. In general the briquetage of this phase was more fragmentary and degraded than in the earlier period, with fewer recognizable base sherds. However, this may have been due to the above average condition of the briquetage from structure 4 in phase 1 rather than any inherent residuality in the phase 2 briquetage assemblage. It was noted that the organic fabric (fabric 2) was even more dominant than in phase 1. In quantity, briquetage was comparable to that of phase 1 and it still formed a major component of the ceramic assemblage (*c* 53%). Handmade Malvernian wares were again featured (<3%), though only structure 27 (a ditch) produced very many sherds. This fabric was of a type described by Peacock (1968, 15–18) which was characterized by tubby cooking pot forms and vertically burnished bands or surfaces. This was in contemporary use with Severn Valley ware and lasted well into the 2nd century (Peacock 1968, 16).

One of the clay- and stake-lined pits (S3), still being infilled during phase 2, produced the fragmentary remains of a single vessel in sandy fabric 5.1 (Fig 82 no 11) associated with Severn Valley ware (fabric 12). The precise context of this vessel within the feature was, however, uncertain. The possibility remained, therefore, that it belonged to phase 1 rather than phase 2. The vessel was very unusual in form, but unfortunately only survived in a very poor condition and had to be consolidated before lifting. It was a bowl with an outside burnished surface (the inside surface was missing) and an *omphalos* base, and it was decorated above the maximum diameter with three horizontal grooves. No comparable vessel is known from Droitwich, despite the petrological analysis suggesting the possibility of a local source (3:B7). The only other Iron Age pottery fabric in this phase was a small quantity of Palaeozoic limestone tempered ware (fabric 4.1).

The Roman pottery assemblage consisted mainly of Severn Valley ware (fabric 12) and included three complete vessels. These were a dish with a slight footring (Fig 86 no 4) and two narrow necked jars (Fig 86 nos 1–2), which were both derived from the bottom fill of the well (S6). Each displayed on their undersides traces of secondary turning, subsequent to throwing and drying to the leather-hard stage. In this way the shallow groove on the underside was equivalent to the footring on the bowl, as both features represented a secondary finishing operation on the wheel. A turned base was also detected on jars from other contexts of this period. Another detail noticed on some Severn Valley ware was a dark grey banding apparently applied when burnishing took place. Other Severn Valley ware vessel types present were the early tankard type (cf Fig 29 nos 11–13) and carinated beaker (cf Fig 29 nos 21–26). Associated with the Severn Valley ware of this phase were grey ware bowls and jars in fabrics 14 and 15, the latter including several with rusticated surfaces. A small quantity of Black Burnished ware type 1 consisted only of cooking pots, except for a single beaker of cooking pot form (Gillam 1976, 64–6). Another beaker (fabric 45), possibly an import, was colour coated and rough-cast with clay. Other imports were Dressel 20 *amphorae* and samian (3:B11–12).

Dating

There were several Severn Valley forms of early type in this phase. These included the dish with footring (Fig 86 no 4) for which Rawes (1982, 45) has indicated a mid 1st to 2nd century date, and another dish from context 1143 (cf Fig 30 no 26) of a type dated by Webster (1976, 35, nos 65–6) to the 2nd century. Fragments of carinated beaker from contexts 647 and 731 may also be of mid 1st to 2nd century date (Webster 1976, 33, nos 59–60). The earlier tankard type with straight fairly vertical sides was also present (1136) and seemed identical to an example published by Rawes (1982, 44, no 143) which was being produced in the Malvern area in the 2nd century. The 'Belgic' influence evident in the production of Severn Valley carinated beakers may also be reflected in an unusual pedestal base in this fabric from context 416 (S5), probably from a jar (Fig 86 no 5). It was associated with a samian sherd dated to *c* AD 50–65 (3:B11) and this assemblage provided the most tightly defined dating evidence for the abandonment of one of the brine tanks, suggesting that, at least in this case, it

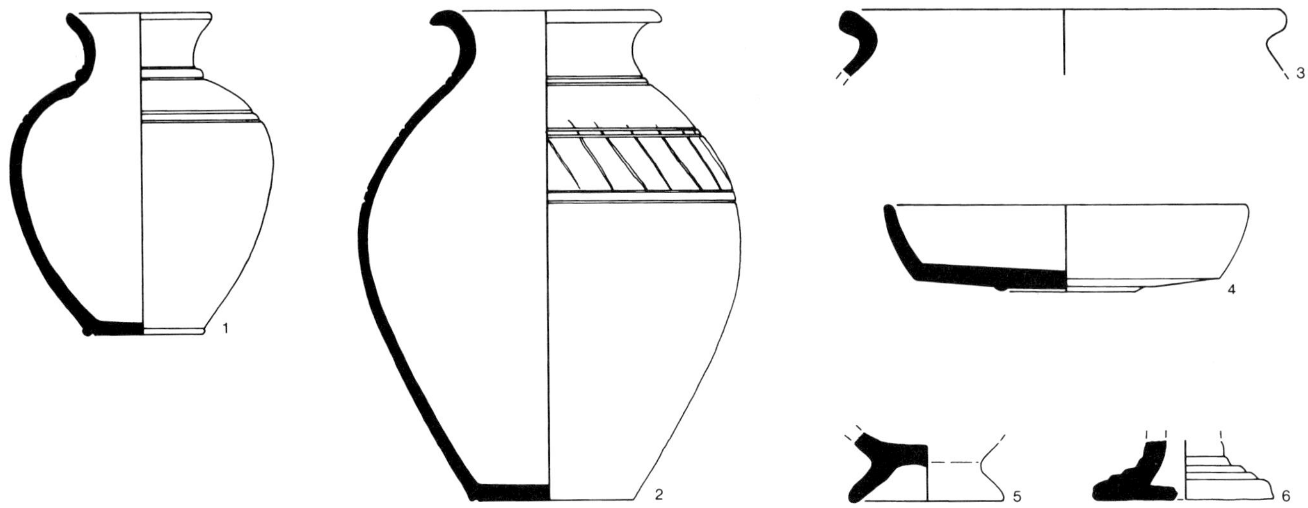

Figure 86 Roman pottery: fabrics 12 (1, 2, and 4–6) and 23 (3). Scale 1:4

could have occurred in the second half of the 1st century.

Other dating evidence was obtained from a small quantity of Black Burnished ware type 1 sherds, none of which needed to have been later than the mid 3rd century. This included a possible handled drinking vessel in a cooking pot form, probably not produced beyond the 2nd century (Gillam 1976, 64), and cooking pots with upright rims and acute angled lattice decoration. Sherds of rusticated grey ware, comparable with examples given by Gillam (1970, 12, nos 95–8; dated to later 1st to 2nd centuries), were derived from the fill of the well (1134 and 1142, S6). The use of this feature (S6) was also associated with a rough-cast beaker datable to this early Roman phase. A small amount of samian from the disuse fills (1134, 1136 and 1142) dated from the Trajanic to the (early) Antonine periods (3:B12).

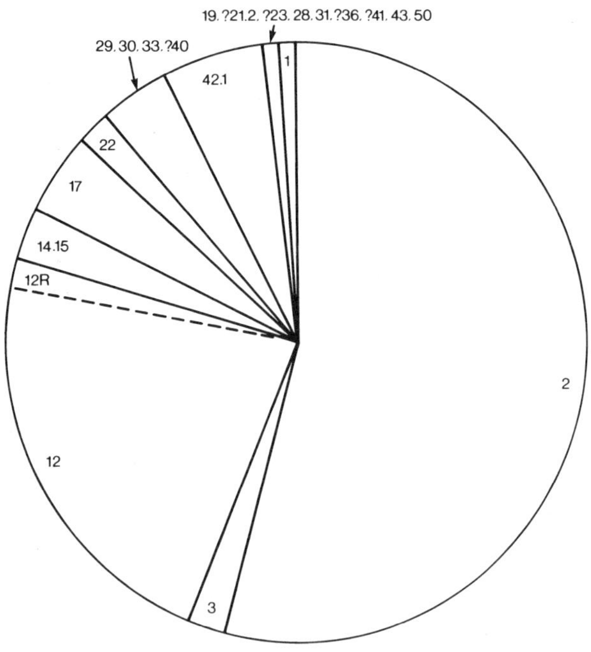

Figure 87 Fabric proportions by weight in phase 3

Illustrated pottery

Roman pottery (Fig 86)

Severn Valley ware (fabric 12; nos 1–2 and 4–6) and shell gritted ware (fabric 23; no 3)

1 Narrow necked jar (1143, S6, P2)
2 Jar with burnished decoration (1143, S6, P2)
3 Jar with internal lime-scaling (308, S74, P5)
4 Dish with footring (413, S5, P1–2)
5 Pedestal base (416, S5, P1–2)
6 Candlestick base (232, S42, P7)

Phase 3: 3rd century to Anglo-Saxon (Figs 87 and 88)

Fabrics present: 1, 2, 3, 12, 12R, (?)12.2, 12.4, 13, 14, 15, 17, 19, (?)21.2, 22, (?)23, 28, 29, 30, 31, 32, 33, (?)36, (?)40, (?)41, 42.1, 43 and 50

The pottery assemblage of this phase was relatively small and degraded. The average sherd size was reduced to 6.8g (5.9g for briquetage), compared with 13.8g for phase 2 and 17.9g for phase 4. This quantity of smaller sherds resulted in less information being available about forms and hence dating was less secure. Two structures (S64 and S68) produced the most pottery in this phase and these formed deposits laid down on the east side of the site.

Briquetage still dominated the pottery assemblage, with over 50% by both weight and sherd number. However, its residual character seemed to be assured by its degraded condition and the absence of any recognisable rims or, more especially, bases. The few identifiable Severn Valley ware forms were generally contemporary with this phase, ie the later type of tankard and the jar with bifid or 'pulley' rim (see below). A grooved rim bowl with internal overhang (cf Fig 30 nos 9, 12 and 13) could not however be closely dated, as it seemed to be a type in use throughout the Roman period. A tendency to high firing of Severn Valley ware was sometimes apparent at this period, leading to a dark purplish surface finish. The purpose of this was uncertain. The final result was reduced but, since other grey ware in a reduced Severn Valley fabric was not so highly fired, there seemed to be no reason to suggest that it was done to effect a dark finish. Possibly the vessels at the front of the kiln during firing were affected in this way and so this may have been accidental. A number of bases, as in phase 2, exhibited a turned groove on the underside.

Cooking wares consisted mainly of Black Burnished ware type 1, though it was not present in any quantity. There was also a small amount of grey ware, for example a plain dish (cf Fig 31 no 29) and single sherds of both late Roman shelly ware (fabric 23) and wheel-thrown Malvernian (fabric 19) cooking pot (cf Fig 32 no 12). Other kitchen wares were Mancetter/Hartshill and Oxfordshire (both white and white slipped) *mortaria*, and large storage jars in Severn Valley ware (cf Fig 27 no 19) and mudstone tempered ware (fabric 17).

Finer tablewares were supplied by the Oxfordshire and Nene Valley industries, with the former predominating, with both its colour-coated and parchment wares being represented. There was also a small amount of fabric 31, a colour-coated ware probably derived from a south-western source (see below).

As in phase 2, the principal imported pottery was Dressel 20 *amphorae* and samian, the latter certainly being residual by this period.

Dating

The dating of this phase was based more on fabric date ranges than form, as the assemblage was generally so degraded. No doubt associated with this, much of the pottery was also very abraded, especially from structures 17, 20 and 45. Several fabrics were present which suggested a 3rd to 4th century *terminus post quem*. These included Oxfordshire and Nene Valley wares, and fabrics 17 and 31. Mudstone tempered ware (fabric 17) has been attested at other Midland sites in the later Roman period (Booth and Green 1989) and may have been associated with the transportation of some special commodity. Fabric 31 has been dated to probably the late 3rd and 4th centuries at Andoversford (Young 1980, 44). Wheelthrown Malvernian pottery (fabric 19) was dated to the 3rd

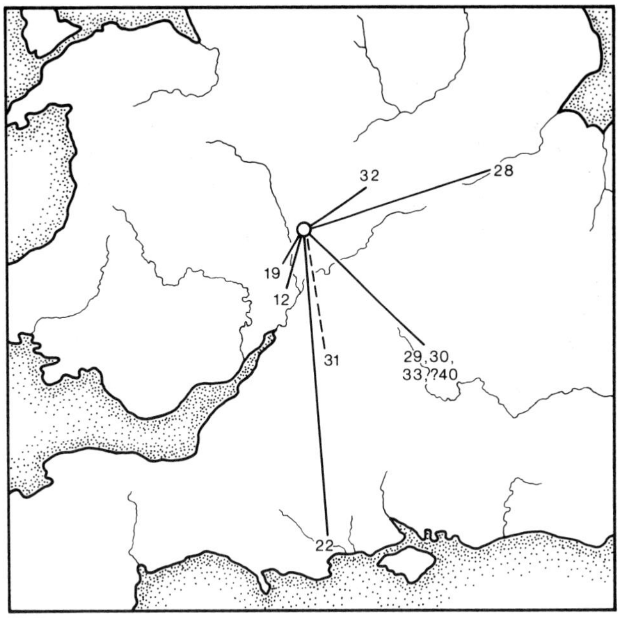

Figure 88 Principal pottery sources in phase 3

to 4th centuries at Beckford (Ford and Rees pers comm), where the evidence suggested a similar date was in order for the handmade slab-built vessel rims in a Malvernian fabric (see Chapter 23). The latter were described by Peacock (1965–7, 24, nos 80–2) and similar material has been detected at the Old Bowling Green excavation (Chapter 5) and a number of other sites in the region. Shelly fabric 23 may also be dated to the later Roman period (Chapter 3).

Fabrics used throughout the Roman period at Droitwich were present in phase 3 in late forms. They included for example, Severn Valley ware late tankard type (cf Fig 29 nos 14–18) and jars (cf Fig 28 no 9) dated by Webster (1976, fig 3 nos 10–11) to the mid 3rd to 4th century. Black Burnished ware type 1 included cooking pots with oversailing rims and obtuse angled lattice decoration (cf Fig 33 no 1) also of this date range.

Three sherds of 'grass' tempered ware (fabric 50), probably from a single vessel, were associated with the later part of this phase (P3ii). They had patchy oxidized and reduced surfaces, and no rim or base survival was evident. Unfortunately these sherds cannot now be located and no detailed description or record of their context is available. However, their most likely context was 682 (S174). The dating of this pottery, though problematical, presented the distinct possibility that phase 3 extended into the Anglo-Saxon period. 'Grass' tempered pottery from the 1983–4 excavations on the site of the Great Upwich brine pit, Droitwich (HWCM 4575) was associated with Anglo-Saxon stamp decorated sherds. A single sherd of 'grass' tempered pottery was also noted by David Freezer from a watching brief at Covercroft Road,

Droitwich (HWCM 4147), though the date of the context could not be fully established from the stratigraphic and artefactual evidence. Elsewhere in Worcestershire, 'grass' tempered pottery has only been observed at Pershore (Vince and Whitehead 1979, 22). In a survey of 'grass' tempered wares in Gloucestershire by Vince (1984a, 240–1) this fabric type was assigned a 5th to 8th century date range and a similar date range may be applicable in Worcestershire.

The evidence of the 'grass' tempered sherds and the degraded and abraded condition of the Roman pottery may be taken as confirming that phase 3 continued into the Anglo-Saxon period. However, the extent of the continuation remains unknown. Heighway (1984, 226–7) has also observed the difficulties of dating the archaeological evidence in Gloucestershire during this period. However, the amount of early to middle Saxon pottery from the Friar Street excavation tends to imply that this area had become rather peripheral by the end of phase 3. In contrast the much greater quantity of pottery of this period from the Upwich excavation, where it was associated with *in situ* industrial structures, suggests that this area instead had assumed greater importance.

Phase 4: Saxo-Norman (Figs 89 and 90)

Fabrics present: 1, 2, 12, 13, 29, 30, 46, 49, 55, 57 and 58

Pit fills (S30, S32 and S37) accounted for most of the pottery of this phase, but the fill of a ditch (S19) that was stratigraphically early was also productive. Overall the quantity of residual pottery was low (*c* 10%) and this consisted mainly of briquetage and Severn Valley wares. The average sherd size increased to 18g because much of the pottery was from pit fills rather than layers.

The two commonest fabric types were St Neots-type ware (fabric 49) and Cotswolds ware (fabric 57). The first was represented by jars (eg Fig 91 no 2), inturned rim bowls (eg Fig 91 nos 7–8) and shallow dishes (eg Fig 91 no 11). The jars occurred most frequently and an almost complete example came from structure 19 (Fig 91 no 2). This exhibited the distinctive internal corrugated surface which was a feature of many body sherds from jars. Slight lid seating on the inside of jar rims in this fabric was also observed in some cases. Droitwich is on the western periphery of the main distribution area of this pottery type (Hunter 1979, fig 104), and it has only very rarely been found at Hereford (Vince 1985, 63).

The Cotswolds wares were exclusively cooking pots, and both straight (eg Fig 95 no 15) and more globular vessels (eg Fig 95 no 12) were present. The straight-sided vessels could either have a flat topped out-turned rim, or were club rimmed with a slight internal overhang (cf Fig 95 no 17, from P5). This material could be related to fabrics identified elsewhere in the region, for example Gloucester (fabric TF41b, Vince 1984b, 262–3), Hereford (fabric D2, Vince 1985, 55–6) and the Oxford calcareous gravel tempered fabric (Maureen Mellor pers comm). It would seem, therefore, to be a very widely spread early medieval type (Vince 1984b, 254). Vince (1984b, 253–4) has suggested that production may have taken place at Haresfield to the south of Gloucester, where the appropriate geological requirements exist.

Sandy, probably local, cooking pots in fabric 55 also became available during phase 4. This fabric has conventionally been referred to as Worcester-type (eg Vince 1979b, 176). It was less common

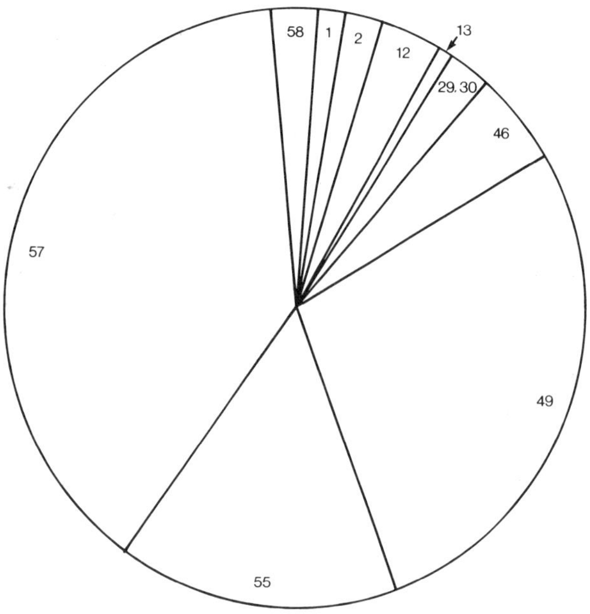

Figure 89 Fabric proportions by weight in phase 4

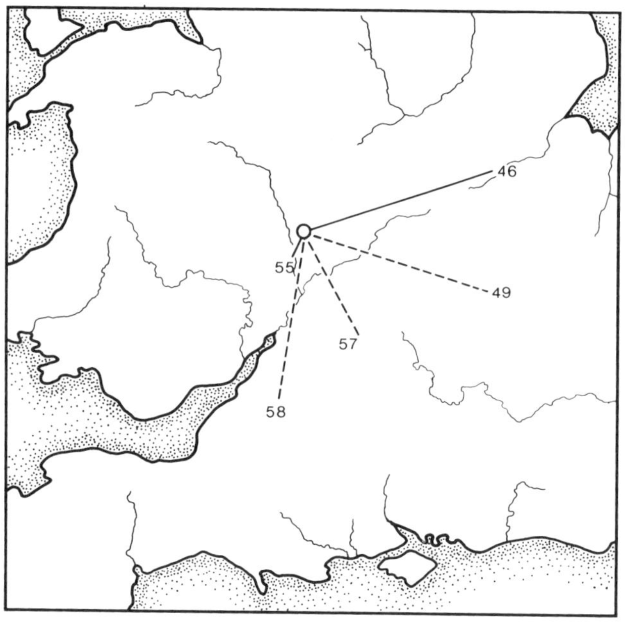

Figure 90 Principal pottery sources in phase 4

than the St Neots-type and Cotswolds wares in phase 4 and, although its form later became very standardized, at its inception greater variation seemed apparent. Thus, one vessel type (Fig 96 no 7) seemed very reminiscent of the Cotswolds form, and another (Fig 97 no 5) was rather unusual. One example (Fig 97 no 3), however, bore a general resemblance to the later standardized form of a more sharply everted and thickened rim (eg Fig 97 no 1, from P5). Context 518 (S31) included two small fabric 55 rims of this standardised type, though these could not be definitely phased as 4 and may have belonged to phase 5.

There were a number of glazed Stamford ware sherds in this phase. However there were no rims and only one handle which derived from a spouted pitcher (Kilmurray 1980, fig 3, form 5). Fabrics fell into the Kilmurray A/B/G range (Howard Leech pers comm), except for one small sherd of Kilmurray fabric D (Howard Leech pers comm) for which Kilmurray (1980, 132–3) has suggested a 10th to early 11th century date. As with the St Neots-type ware, Droitwich was towards the western extreme of the distribution of Stamford wares. In the county of Hereford and Worcester this fabric type has been found at Droitwich, Worcester (Morris 1978, 75; 1980, 223–4) and Hereford (Vince 1985, 58). There was also a small quantity of other glazed sherds, which were not entirely dissimilar to Stamford ware in fabric and glaze colour. However, they contained coarser inclusions than Stamford ware (Howard Leech pers comm) and their provenance remained undetermined.

Fabric 58 was represented by a few body sherds and a single upright rim with thumbing on the outside edge. The precise provenance of this group was uncertain, but it continued into phase 5 when some sherds were decorated with rosette stamps (see below).

There were only a few sherds of Stafford-type ware (fabric 48, eg Fig 91 no 1). Quantities of Stafford-type ware found in Worcestershire have not been large and this ware has only been represented by a single sherd from Gloucester (Vince 1985, 63). The present evidence, therefore, may be taken to indicate that this pottery type was most typical of the central Marches area and north-west Midlands, rather than further south (see below).

Dating

The pottery of phase 4 could be divided into two groups on the basis of the stratigraphic evidence. That associated with phase 4i was all residual Roman wares, except for St Neots-type ware. At Oxford, assemblages with a major component of St Neots-type ware have been regarded as pre-Conquest, since, on the basis of the coin evidence, this ware was dominant in the Oxford region *c* AD 1015–1066 (Mellor 1980, 19). The Cotswolds ware began to replace it in popularity just before *c* 1070 (Mellor 1980, 21). The Droitwich range of St Neots-type ware forms was the same as that most commonly represented at Oxford, such as jars, bowls and dishes. The possibility of pre-Conquest 10th to early 11th century activity also seemed to be supported by the possible Kilmurray fabric D sherd associated with phase 4ii (see above), while the low incidence of Stafford-type ware may have been a local phenomenon connected with the presence of St Neots-type ware. Unfortunately it is not yet possible to compare the 10th and 11th century distributions of Stafford-type ware in the area. The possibility remains that the supply of Stafford ware may not have been as vigorous in the 11th century as earlier, which may have made the St Neots-type ware penetration of the Droitwich market more likely.

The phase 4ii pottery was characterized by the introduction of glazed Stamford ware, and Cotswolds and Worcester-type ware cooking pot. As no other internal dating evidence for this phase was available, dates derived from elsewhere for similar pottery types were used to suggest dating for this assemblage. At Hereford, Cotswolds ware (Hereford fabric D2) first appeared in the late 11th century (Vince 1985, 55), and Worcester-type ware cooking pots (Hereford fabric C1) were introduced at a similar date (Vince 1985, 53). As at Droitwich, Stamford ware also seemed to be commoner in the later 11th to 12th centuries than at other periods at Hereford (Vince 1985, 58). At Gloucester, Cotswolds ware (Gloucester TF41b) use was well established by the late 11th century (Vince 1984b, 254) and at Oxford, too, Cotswolds ware supplanted St Neots-type ware in the second half of the 11th century (Maureen Mellor pers comm). The range of Stamford wares present, including both the coarser Kilmurray A and finer B and G fabrics, was datable to *c* 1025–1125 (Howard Leech pers comm). The absence of any vessel forms made closer dating of the Stamford ware very difficult. Phase 4 therefore may be 11th century in date, with 4i most likely to be pre-Conquest and 4ii belonging to the mid to late 11th or early 12th century.

The non-local sources of pottery used at Droitwich throughout the late Saxon to early medieval period corresponded closely to the pattern of salt routes from Droitwich for this period, derived from documentary and place-name evidence. Thus Hooke (1985, 125, fig 31 and reproduced as Fig 92) has shown that the principal pre-Conquest saltways were orientated towards the south and south-east of Droitwich, into Gloucestershire and Oxfordshire respectively. Later medieval saltways are recorded to the east and west, but salt transportation was not attested by any saltways beyond the northern boundary of Worcestershire. The impact of the salt trade on pottery supplies may therefore have affected sources of supply, especially during the late Saxon and early medieval periods, and the result of this may have been (to some extent at least) to limit the

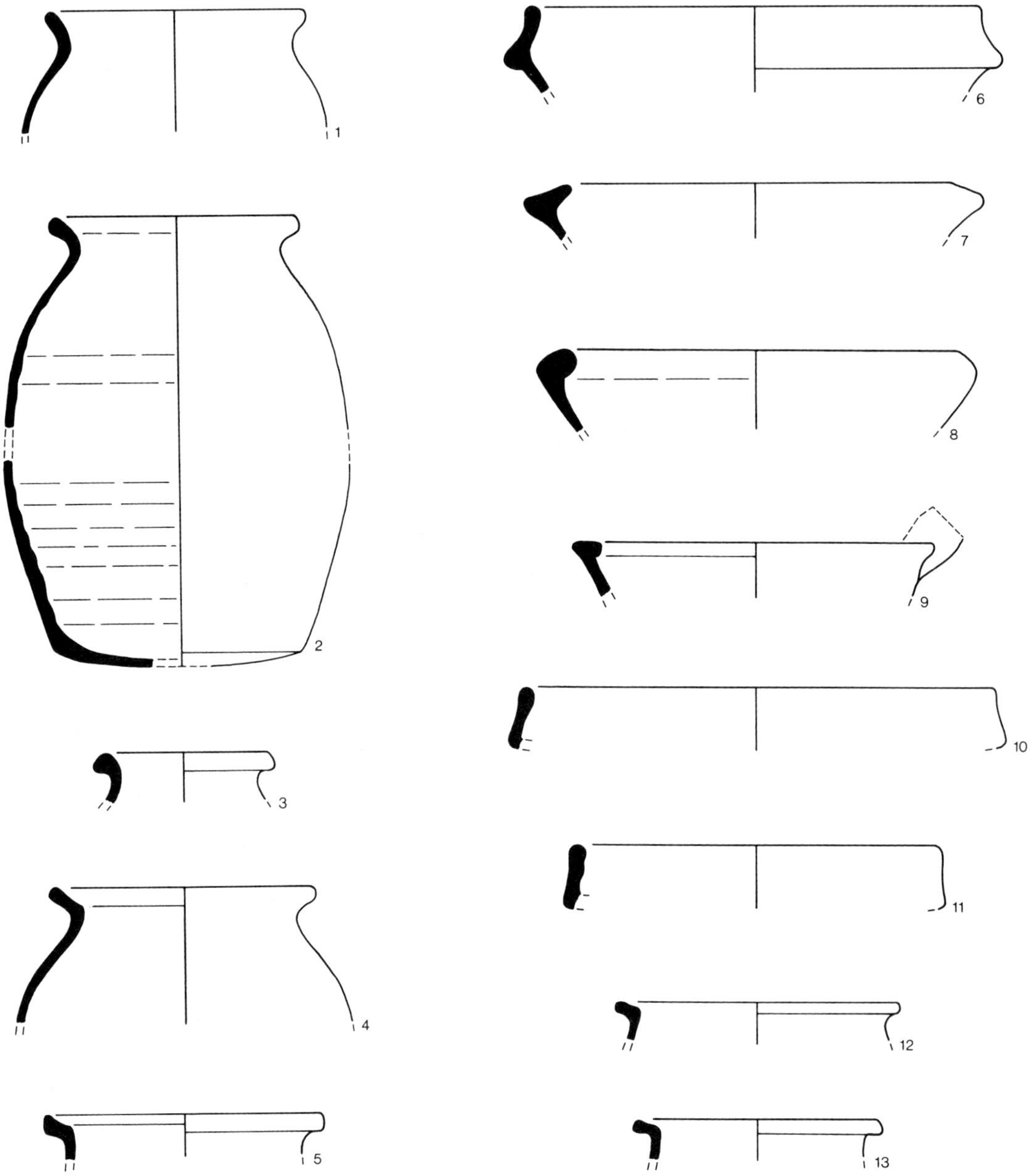

Figure 91 Saxo-Norman pottery: fabrics 48 (1 and 9), 46 (12, 13) and 49 (2–8, 10, 11). Scale 1:4

market penetration of pottery production centres such as Stafford to the north. Importantly, the absence of a more successful local pottery industry would have contributed to this by making ceramics from other production centres, for instance in Lincolnshire and the Cotswolds area, a profitable trade commodity in Droitwich at this period.

Illustrated pottery

Saxo-Norman pottery (Fig 91)

Stafford-type ware (fabric 48; nos 1 and 9), Stamford ware (fabric 46; nos 12–13) and St Neots-type ware (fabric 49; nos 2–8 and 10–11)

1 Jar with sooting on rim exterior (552, S70, P4)
2 Jar (1016, S19, P4)
3 Narrow necked jar with external sooting (476, S35, P5)
4 Jar with slight external sooting (281, S43, P7)
5 Lid seated jar (1032, S38, P5)
6 Inturned rim bowl (971, S35, P5)
7 Inturned rim bowl (545, S32, P4–5)
8 Inturned rim bowl with soot impregnated exterior (548, S32, P4–5)
9 Spouted bowl with sooted exterior (226, S108, P6–7)

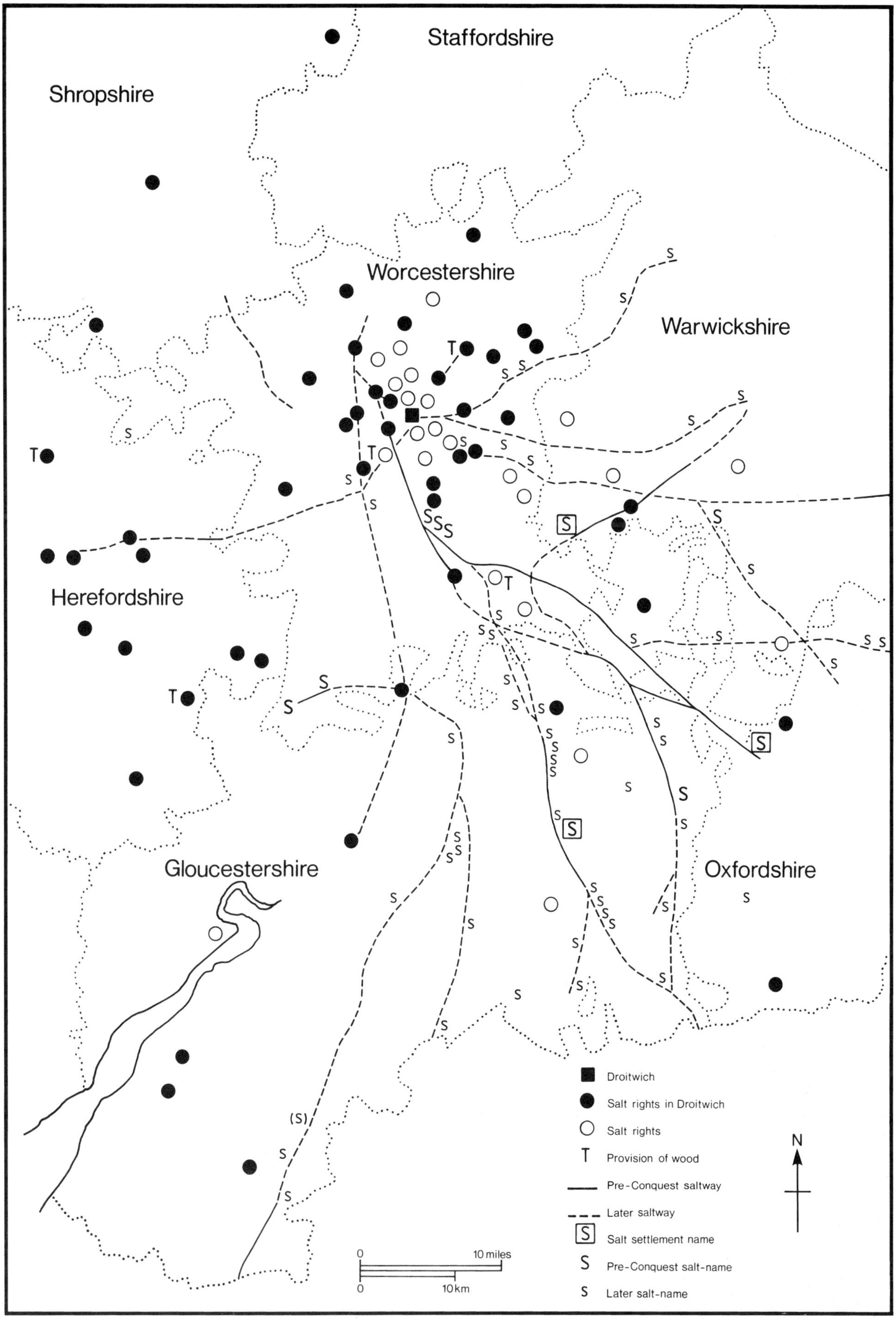

Figure 92 Pre-Conquest saltways (from Hooke 1985, 125, fig 31)

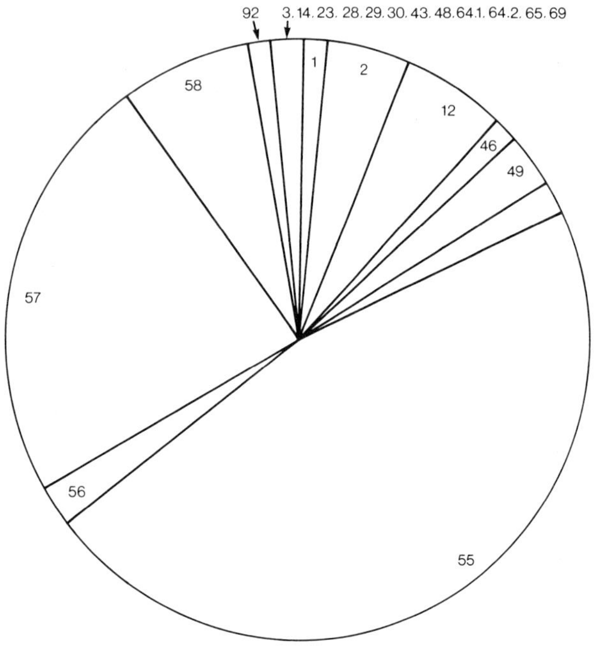

Figure 93 Fabric proportions by weight in phase 5

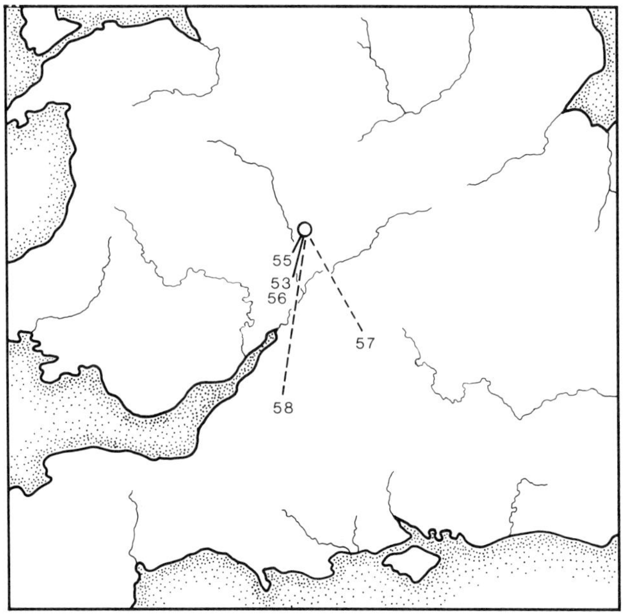

Figure 94 Principal pottery sources in phase 5

10 Dish (984, P5)
11 Dish (552, S70, P4)
12 Possible spouted pitcher (Kilmurry form 5) (423, S98, P6)
13 Possible spouted pitcher (Kilmurry form 5) (522, S34, P4ii)

Phase 5: 12th to early 13th century (Figs 93 and 94)

Fabrics present: 1, 2, 3, 12, 12R, 12.2, 12.4, 14, 23, 28, 29, 30, 43, 46, 48, 49, 53, 55, 56, 57, 58, 64.1, 64.2, 65, 69 and 92

Much of the phase 5 pottery came from the southern part of the excavation where the contemporary stratigraphy was deepest. The fills of two pit groups (S34 and S35) and boundary ditches (S36) accounted for the greatest quantities of pottery. There was also a greater quantity of residual Roman pottery than in phase 4, indicating an increase in site activity disturbing earlier deposits in the vicinity.

The principal fabric/form type was the Worcester-type sandy cooking pot, which comprised just under half the assemblage (47%). A variety of different rim forms was represented, as noted for phase 4ii, though in phase 5 the everted rim (eg Fig 97 no 1) was the commonest type. Other rim shapes were clubbed with an external thickening (cf Fig 96 no 11) or T-shaped (cf Fig 96 no 9) in section. Both these types differed in general body shape from the everted rim cooking pot, in that the narrowest point of the vessel was at rim level. A lid could have been positioned on these flat topped rims, but would not have been as securely held in place as on the everted rim types. Similar rim shapes and vessel profiles were used for both the Worcester-type (fabric 55) and Cotswolds (fabric 57) cooking pots of this period. Occasionally decoration was applied to the Worcester-type fabric, either using a diamond pattern roller-stamp (Fig 96 no 15) or stab marks in a neat horizontal row. This former vessel was also unusual for displaying an impressed wavy line along the top of the rim. A similar, though smaller, cooking pot in the same fabric was noted at Sidbury, Worcester (Morris 1980, fig 69).

The two other most prominent fabrics of this phase were both calcareous. Cotswolds ware cooking pots were present in quantity, as in phase 4ii, and a similar range of rims was noted. An additional form, however, seemed to be a globular bodied vessel with an almost upright rather than everted rim (Fig 95 no 7). One Cotswolds ware vessel, possibly a pitcher, was decorated with rosette stamps and the strap handle with incised lattice and a grid-pattern stamp (Fig 95 no 5). Stamp decoration of this type may have enjoyed a 12th to early 13th century vogue in parts of the south-west, as it was a feature of Bath and Bristol assemblages of this date. Bath fabric B (Vince 1979a, 28), for instance, displayed examples of grid-pattern and rosette stamping and was also, like fabric 57, tempered with oolitic limestone, though not to the same degree. Fabric 58 was not as homogeneous as fabric 57 and may consequently be capable of subdivision into several fabric groups. It first appeared in phase 4 and was most common in phase 5. Stamp decoration occurred rarely, for example a rosette stamped body sherd (Fig 95 no 6). This fabric closely resembled Bath fabric A, as described by Vince (1979a, 27), which also sometimes exhibited stamp decoration. Vince considered that this fabric was probably a local product and so it may serve to emphasise further the southerly aspect of Droitwich trade already noted in the previous phase.

Other cooking pots comprised a small amount (2%) in Malvernian fabric associated with structures 36, 38, 74 and 76, though none of these contexts was sealed by other phase 5 components and so the possibility remains that they were intrusive in this phase. However no other definite indications of contamination were detected, and so fabric 56 may indeed have been characteristic of the later phase 5 assemblage and thus probably an early 13th century introduction.

The only glazed wares of this phase were Stamford wares (fabric 46), Cotswolds oolitic tempered ware (fabric 65), sandy glazed ware (fabric 64.1) and early Malvernian pitcher (fabric 53). The first type was very fragmentary and the only form that seemed to be represented was the spouted pitcher (cf Fig 91 no 13). Some sherds could have been residual and none could be dated more closely than to the 11th to 12th centuries (Howard Leech pers comm). Early glazed Malvernian ware included a large tripod pitcher (Fig 101 no 1) which conformed to the later type of this vessel, as classified by Vince (1977, 269) and dated to the early 13th century. This vessel was associated with sherds of Malvernian cooking pot. Another context also produced sherds of both early Malvernian glazed pitcher and Malvernian cooking pot, indicating that both vessel types may have become available in Droitwich at about the same period.

The few glazed oolitic limestone tempered sherds (fabric 65) were probably from tripod pitchers, which may have been present throughout phase 5. One sherd displayed the combed decoration typical of this fabric type. A similar fabric was in evidence at Sidbury, Worcester (Morris 1980, 231, TV161). The sandy pitcher type (fabric 64.1) was also represented by a sherd with a worn foot.

Dating

The contexts of phase 5, as of phase 4, could be divided on the grounds of stratigraphic evidence into an earlier (5i) and a later (5ii) phase.

Generally, a 12th to early 13th century date was suggested for phase 5. Though Cotswolds ware was still a major component of the assemblage, it was proportionally less important in phase 5 than the probably more locally produced Worcester-type ware (fabric 55). At Hereford, the former declined in importance throughout the 12th century (Vince 1985, 55), and the Worcester-type ware had become commoner than the Cotswolds ware by the end of that century. However, the 12th century assemblage from Sidbury, Worcester (Morris 1980, 225) showed a much higher proportion of the Worcester-type fabric (82%) and a lesser proportion of Cotswolds ware. The amount of fabric 58 in absolute terms was at its highest during this phase, which also indicated a 12th to 13th century date, as this type was commonest at Bath during this period (Vince 1979a, 27–8). The small amount of associated glazed wares, together with the forms and fabrics represented, also generally supported a 12th to mid 13th century date for phase 5.

Comparison between phases 5i and 5ii pottery demonstrated a progressive decline in the amount of Cotswolds ware from 31% to 12%. The decline in Cotswolds ware was, therefore, similar to that evidenced at Hereford in the 12th century. A later 12th to early 13th century date for phase 5ii was also indicated by the later type of Malvernian tripod pitcher from context 407 (Fig 101 no 1). This was contemporary with the appearance of Malvernian cooking pot indicating that this type may have been a relatively late introduction at Droitwich compared with other sites in the region such as Gloucester, where it occurred in early 12th century assemblages (Vince 1984b, 255). The Friar Street ceramic sequence suggested that Cotswolds glazed ware (fabric 65), though confined to the 13th century at Sidbury, Worcester (Morris 1980, 231), was introduced at Droitwich during the 12th century.

Illustrated pottery

Medieval pottery (Fig 95)

Limestone tempered wares: fabric 57 (Cotswolds ware; nos 3, 5, 7, 9–10 and 12–17), fabric 58 (nos 1–2, 4, 6 and 11) and fabric 65 (no 8)

1 Cooking pot (473, S35, P5)
2 Possible pitcher with crudely stamped rosette decoration (182, S137)
3 Possible pitcher with stamped rosette decoration (226, S108, P6–7)
4 Rosette stamped sherd with lug handle (921, S120, P9)
5 Handle with grid pattern stamp (226, S108, P6–7)
6 Rosette stamped sherd (444, S33, P5)
7 Globular cooking pot, heavily tempered and with cracked exterior surface (473, S35, P5)
8 Green glazed pitcher decorated with combing and applied strips (412, S96, P6)
9 Handmade cooking pot (984, P5)
10 Handmade cooking pot with everted rim (473, S35, P5)
11 Possible handmade pitcher (576, S34, P4ii)
12 Handmade cooking pot with extensive external sooting (671, S37, P4ii)
13 Handmade cooking pot with occasional thumbing on top of rim and external sooting (585, S34, P4ii)
14 Handmade cooking pot with thumbed rim and some sooting (300, S76, P5)

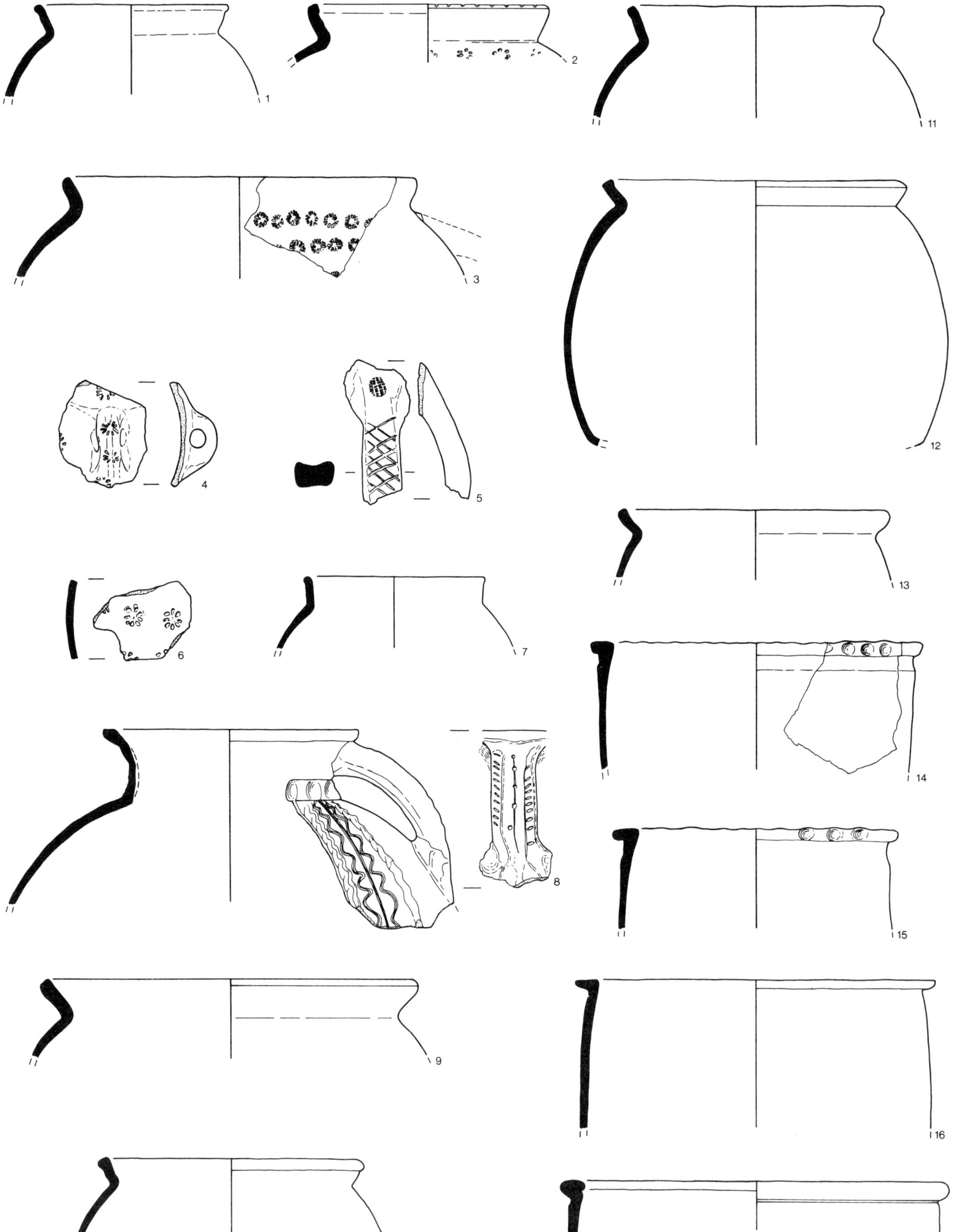

Figure 95 Medieval pottery: limestone tempered wares, fabrics 57 (3, 5, 7, 9, 10, 12–17), 58 (1, 2, 4, 6, 11) and 65 (8). Scale 1:4

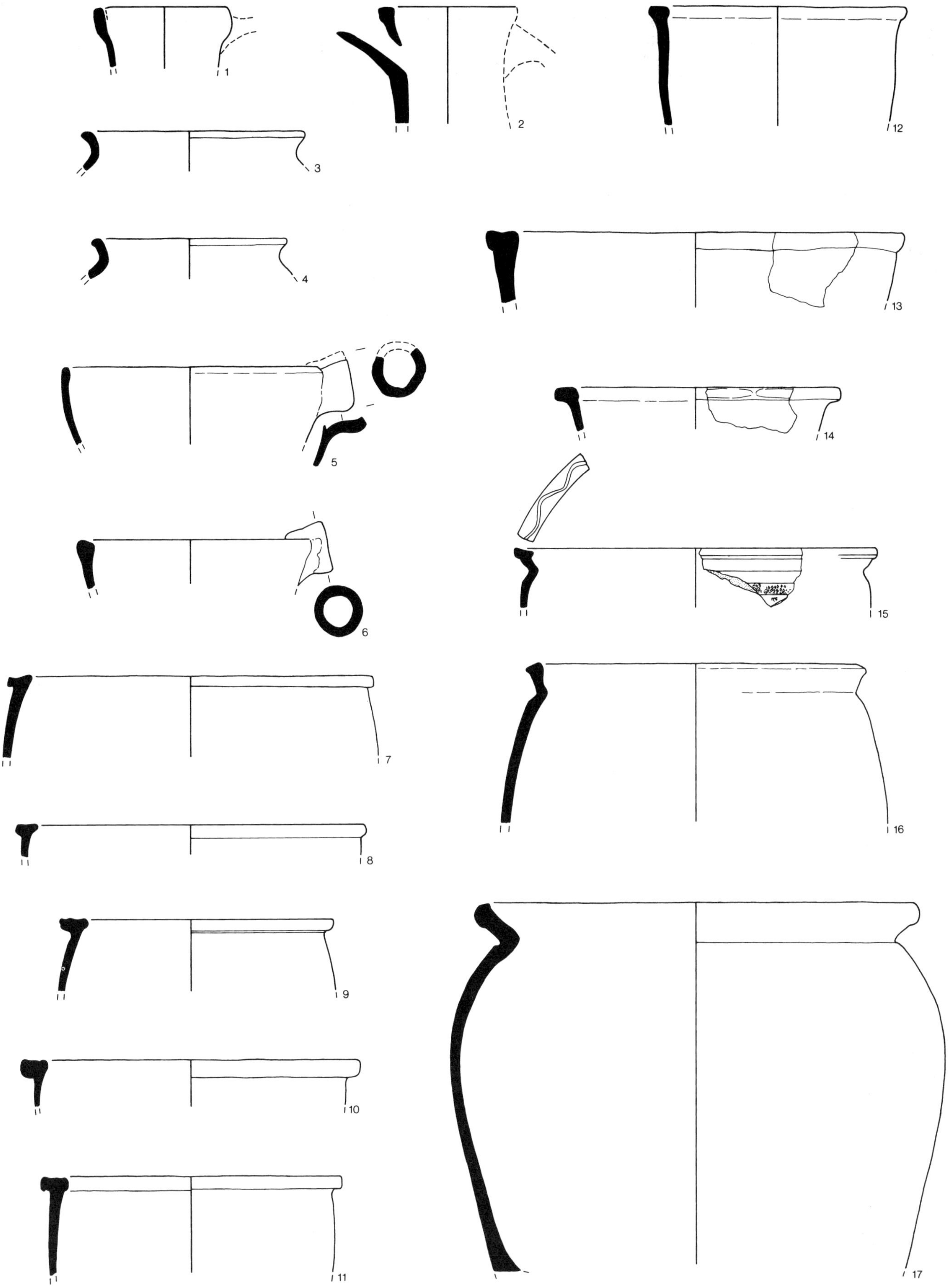

Figure 96 Medieval pottery: Worcester-type wares, fabrics 55 (3, 4, 6–17), 64.1 (1, 2) and 92 (5). Scale 1:4

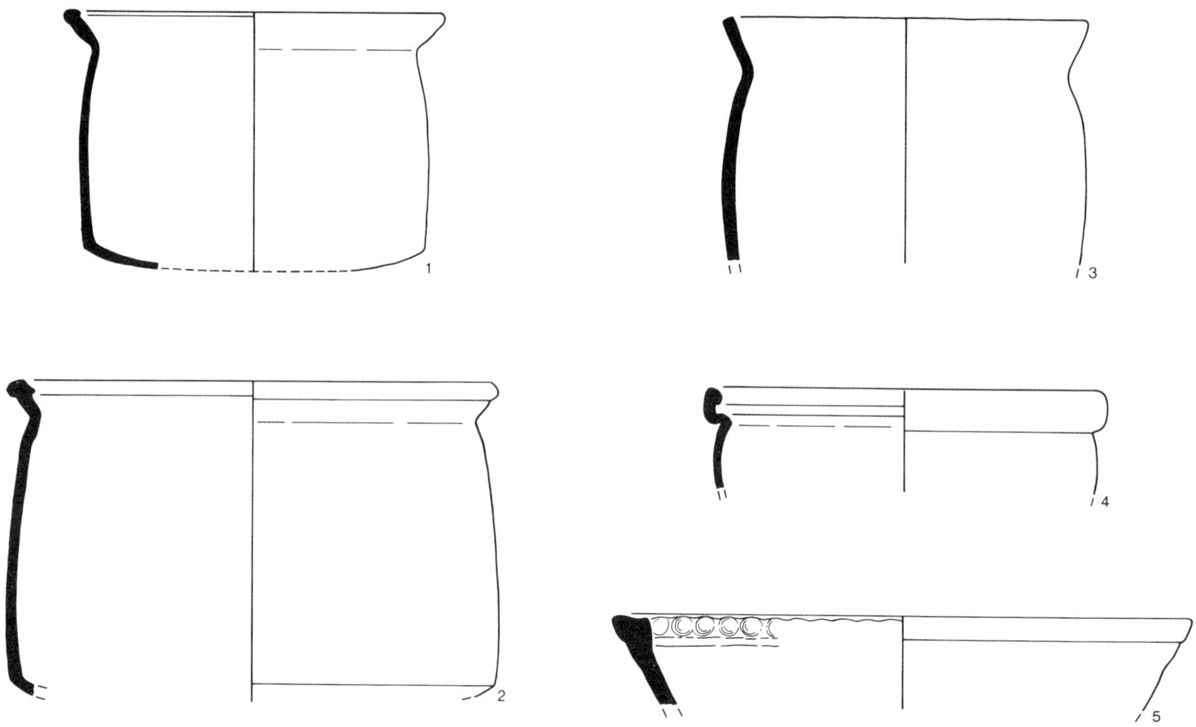

Figure 97 Medieval pottery: Worcester-type wares, fabrics 55 (1–3, 5) and 64.1 (4). Scale 1:4

15 Handmade cooking pot with thumbed decoration (545, S32, P4–5)
16 Handmade straight-sided cooking pot (444, S33, P5)
17 Handmade cooking pot (300, S76, P5)

Medieval pottery (Fig 96)

Worcester-type wares: fabric 55 (nos 3–4 and 6–17), fabric 64.1 (nos 1–2) and fabric 92 (no 5)

1 Green glazed jug with pulled lip (328, S42, P6–7)
2 Bridge-spouted jug with lustrous external green glaze except on spout (338, S42, P6–7)
3 Jar with sooted exterior (297, S42, P6)
4 Jar (398, S96, P6)
5 Spouted bowl (182, S137)
6 Spouted bowl (421, S94, P6)
7 Handmade cooking pot with slight external sooting (518, S31, P4)
8 Handmade bowl (984, P5)
9 Handmade cooking pot with sooted exterior (571, S34, P4ii)
10 Handmade bowl (984, P5)
11 Handmade bowl (654, S34, P4ii)
12 Handmade bowl (182, S137)
13 Large handmade cooking pot (333, S42, P6)
14 Handmade bowl (333, S42, P6)
15 Bowl with lozenge pattern roller stamping, and impressed wavy line decoration (308, S74, P5)
16 Handmade bowl with occasional surface spalling (585, S34, P4ii)
17 Large handmade cooking pot (576, S34, P4ii)

Medieval pottery (Fig 97)

Worcester-type ware: fabric 55 (nos 1–3 and 5) and fabric 64.1 (no 4)

1 Handmade cooking pot with external sooting (555, S36, P5)
2 Handmade cooking pot with some sooting (996, S39, P5i)
3 Handmade cooking pot with heavily sooted exterior (671, S37, P4ii)
4 Bowl with oxidized surfaces patchily reduced (952, S114, P6)
5 Large bowl with thumbing on inside of rim (670, S37, P4ii)

Phase 6: 13th to 14th century (Figs 98 and 99)

Fabrics present: 1, 2, 3, 12, 12.4, 14, 15, (?)16, 17, 19, 29, 30, (?)33, 43, 46, (?)48, 49, 53, 55, 56, 57, 58, 63, 64.1, 64.2, 64.3, 65, 69, 70 and 92

This phase marked a fundamental change of site use, as domestic structures were superimposed on earlier industrial features. Much of the pottery was from the make-up and surfaces of a road (S42) and adjacent deposits (S108 and S116).

The assemblage of this phase contained a very high percentage of sandy fabrics (60%), which continued a tendency already noted in phase 5. There was also a higher proportion of glazed wares

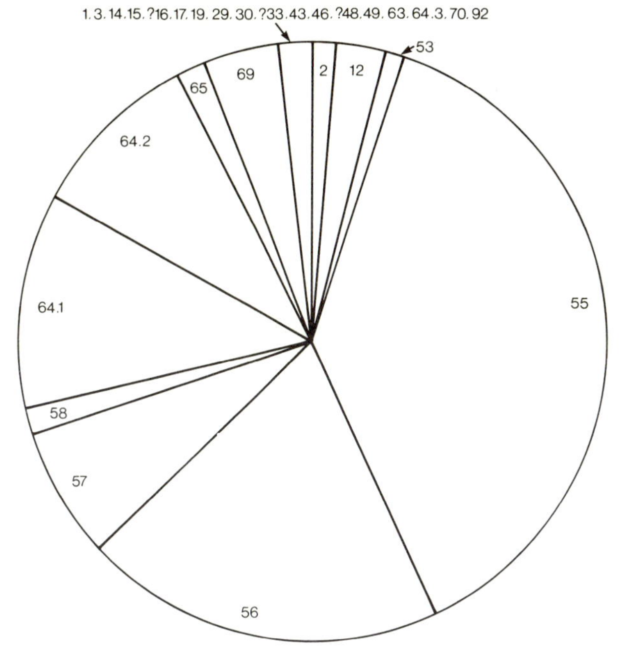

Figure 98 Fabric proportions by weight in phase 6

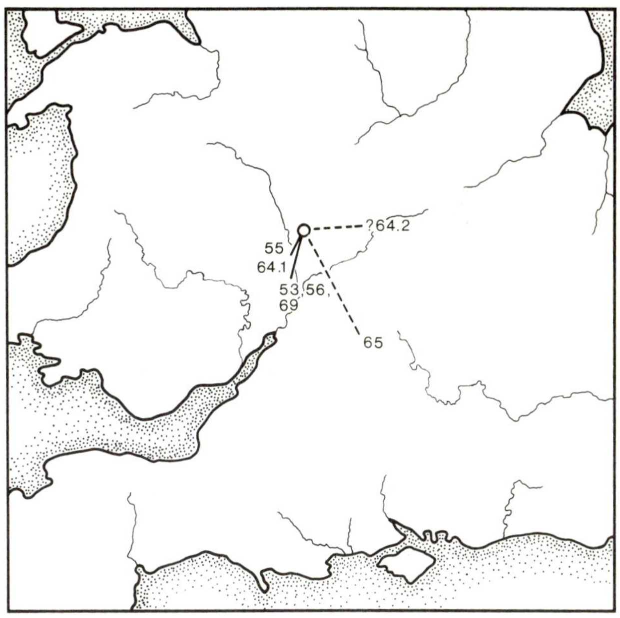

Figure 99 Principal pottery sources in phase 6

(28%) compared with only 3% for the phase 5 assemblage, and this coincided with the widespread introduction of jugs to replace pitchers.

The commonest fabric was, as in the previous phase, the sandy Worcester-type cooking pot. However this exhibited much less variability in rim form than in phases 4 and 5, and the everted rim type was found almost exclusively. Both the sandy (fabric 55) and Malvernian (fabric 56) cooking pots shared the same everted rim vessel style but, despite this similarity, rim forms in either fabric were normally quite distinctive. The Malvernian rims had a more rounded and rolled appearance (cf Fig 101 no 9) and the sandy cooking pot rims were more angular (cf Fig 97 no 2). The decoration of cooking pots was not at all common, but wavy impressed line decoration was sometimes employed along the top of the rim, or on the body. Other examples of decoration were ribbed patterning along the top of a rim and diamond roller-stamping. There were some exceptions to this general uniformity of shape for cooking pots in this phase (eg Fig 96 nos 13–14) but these may have been residual. In general the cooking pots seem to have been fired on their own, and accidental glazing was only noted on one sherd.

There was a small group of sandy unglazed bowls and jars (eg Fig 96 nos 3–4) with sooty external surfaces, which bore some resemblance to Stafford-type ware in both fabric and form. Since any Stafford-type ware would have been residual by this date, the fresh condition of these sherds and absence of other associated Saxo-Norman fabrics suggested strongly that they were medieval rather than late Saxon in date.

The Malvernian industry supplied Droitwich with a much higher proportion of its pottery in phase 6 than earlier. The cooking pot outnumbered glazed forms in a proportion of about 4:1. The cooking pots were consistently shaped, and not decorated except for two examples, where thumbed bands of clay were applied. The Malvernian cooking pot seems to have displaced the last remaining non-local bulk supplier of kitchen ware (Cotswolds ware), while the Worcester-type ware cooking pot continued to be well represented.

At the beginning of phase 6 glazed tripod pitchers were available in either sandy (fabric 64.1), oolitic limestone (fabric 65), or Malvernian rock tempered (fabric 53) fabrics. During this period, however, the jug became the commonest glazed type, and these were typically well tempered with medium to coarse quartz sand. The uniformity in the choice of tempering agent led to fabric distinctions being drawn on the basis of iron content in the clay. The basic division was between a red firing Worcester-type glazed fabric which was rich in iron and iron-poor buff/white firing wares. The former was used for producing bridge-spouted jugs, often highly decorated with stamping or roller-stamping and covered in a lustrous dark green glaze (eg Fig 96 no 2). This glaze, on refiring in an oxidizing atmosphere, proved to be a reduced iron rather than copper coloured. This distinction between the Worcester-type glazed ware and other glazed fabrics, with the former being a reduced glaze coloured by iron, seemed to extend to the sandy glazed tile, which probably also had a Worcester origin (Chapter 23).

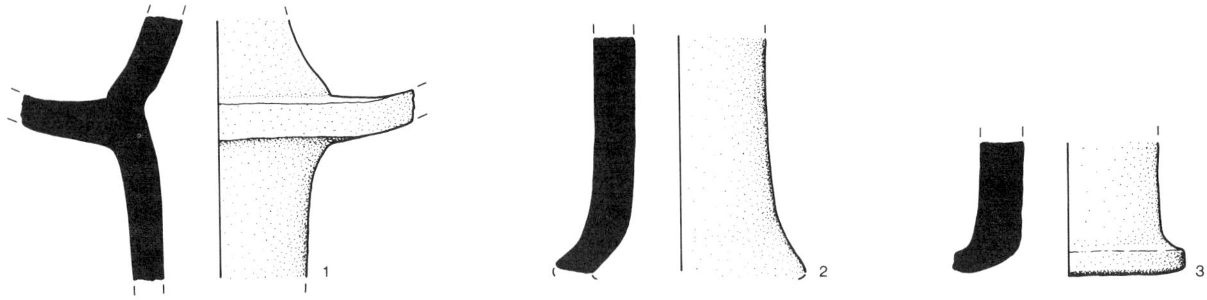

Figure 100 Possible candlesticks. Scale 1:2

Three unusual fragments in a sandy fabric may be candlesticks (Fig 100 nos 1–3). These were wheel-thrown, and the most complete (Fig 100 no 1) exhibited a wide flange, which was glazed dark green on one side.

There was a small quantity of oxidized glazed Malvernian jugs (fabric 69), though no forms could be illustrated. Other iron-rich fabric types present only in small quantities were Deritend-type ware (fabric 62) with lattice decoration in painted white slip and Brill/Boarstall type ware (fabric 63).

The buff firing fabric 64.2 was extensively available in Droitwich at this period, and was used for jugs, dishes, fish dishes, bowls and pipkins (Fig 102 nos 3, 6 and 9). The pipkins were hollow handled and the handle attached by pushing in the vessel wall opposite the handle void (eg Fig 102 no 6). A thin pink wash covered with copper flecked green glaze was used on most examples of this fabric type. Another fairly lightly coloured fabric resembled Nuneaton ware, but it was not present in any quantity. Such pale firing clays, probably from coal measure sources, were also used for fabric 64.3 which appeared to imitate the Tudor Green type ware of the later 14th to 15th centuries.

Dating

Phase 6 dating depended principally on the presence of glazed jugs instead of pitchers, and in particular the Worcester-type jugs, for which Morris (1980, 225) has suggested a mid 13th century introduction, while Vince (1983) has proposed a post-1220 date for this type. The continuation of this fabric type into the 14th century occurred at Worcester (Morris 1980, 225), when it was joined by oxidized glazed Malvernian ware. The earliest datable incidence of the latter fabric at Hereford was late 14th to early 15th century (Vince 1985, 52), while the absence of fabric 69 jar, pipkin and conical bowl forms in this phase suggested that it did not extend beyond the end of the 14th century and, therefore, was of post-*c* 1220 to 14th century duration.

Illustrated pottery

Medieval pottery (Fig 101)

Malvernian wares: fabric 53 (no 1), fabric 56 (nos 7-9) and fabric 69 (nos 2–6 and 10–14)

1 Pitcher with thin, patchy green glaze (407, S36, P5)
2 Green glazed jug (226, S108, P6–7)
3 Jug with slightly green speckled orange glaze (226, S108, P6–7)
4 Jug with pulled lip, and sparse orange glaze speckles (339, S167, P7)
5 Unglazed cistern, or large jar (220, S44, P6)
6 Jar glazed on inside of rim only (226, S108, P6–7)
7 Handmade cooking pot with slight external sooting (333, S42, P6)
8 Bowl (225, S108, P6–7)
9 Bowl, oxidized throughout except for patchy internal reduction (995, S39, P5)
10 Bowl with traces of orange glaze internally (226, S108, P6–7)
11 Bowl with internal green speckled orange glaze (226, S108, P6–7)
12 Bowl with sooted exterior and patchy internal glaze (220, S44, P6)
13 Bowl with internal green speckled orange glaze (226, S108, P6–7)
14 Pipkin with patchy internal orange glaze and sooted exterior (329, S131, P9)

Medieval pottery (Fig 102)

Buff sandy ware (fabric 64.2)

1 Jar with an exterior lime green glaze (227, S41, P8(7))
2 Green glazed tubular spouted jug with anthropomorphic decoration (232, S42, P7)
3 Bowl with yellow glaze speckles (333, S42, P6)
4 Unglazed bowl (996, S39, P5i)
5 Unglazed bowl (232, S42, P7)
6 Pipkin with body pierced at point of handle attachment (421, S94, P6)
7 Dish with internal green glaze and hollow handle (328, S42, P6–7)

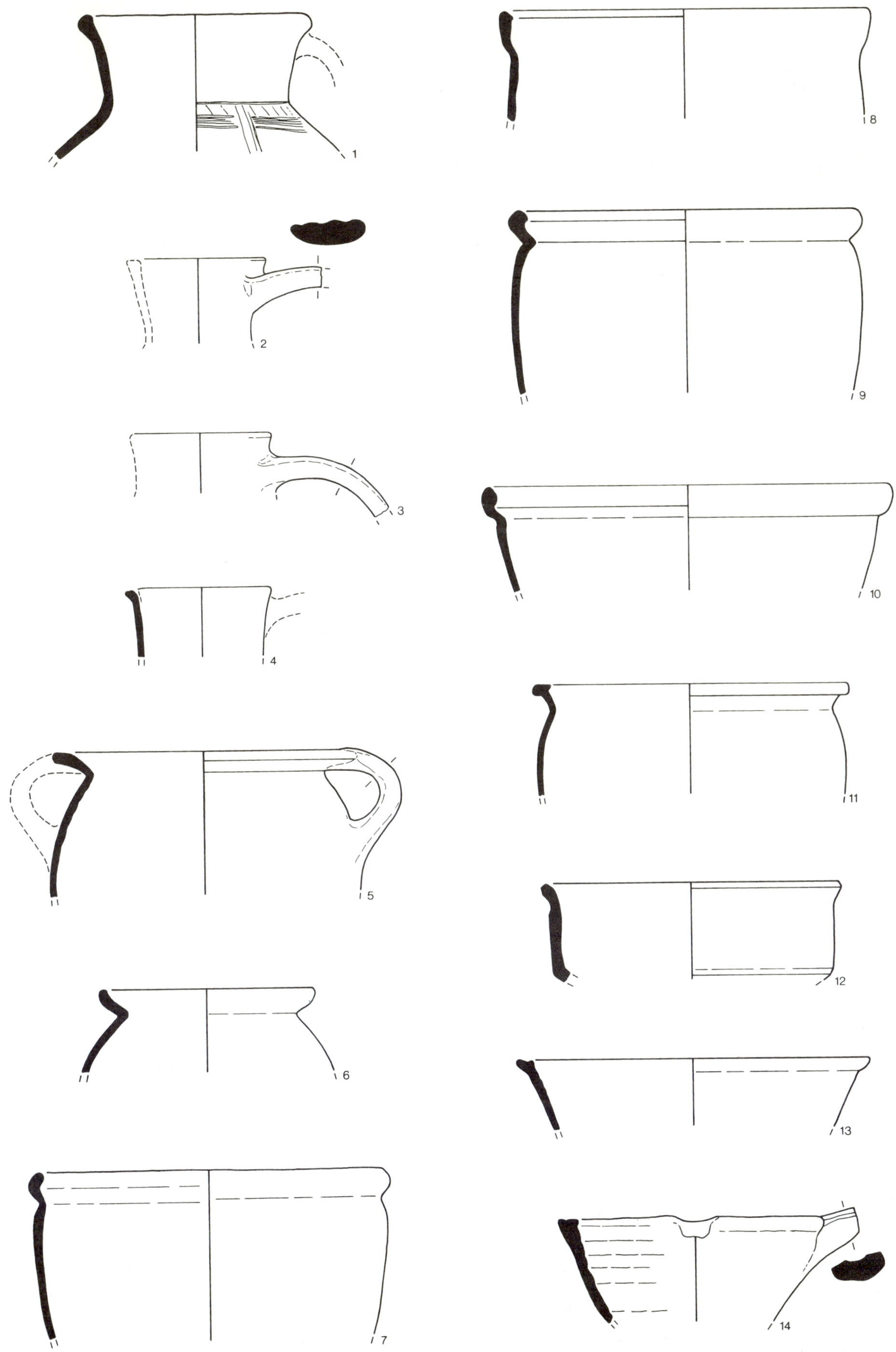

Figure 101 Medieval pottery: Malvernian wares, fabrics 53 (1), 56 (7–9) and 69 (2–6, 10–14). Scale 1:4

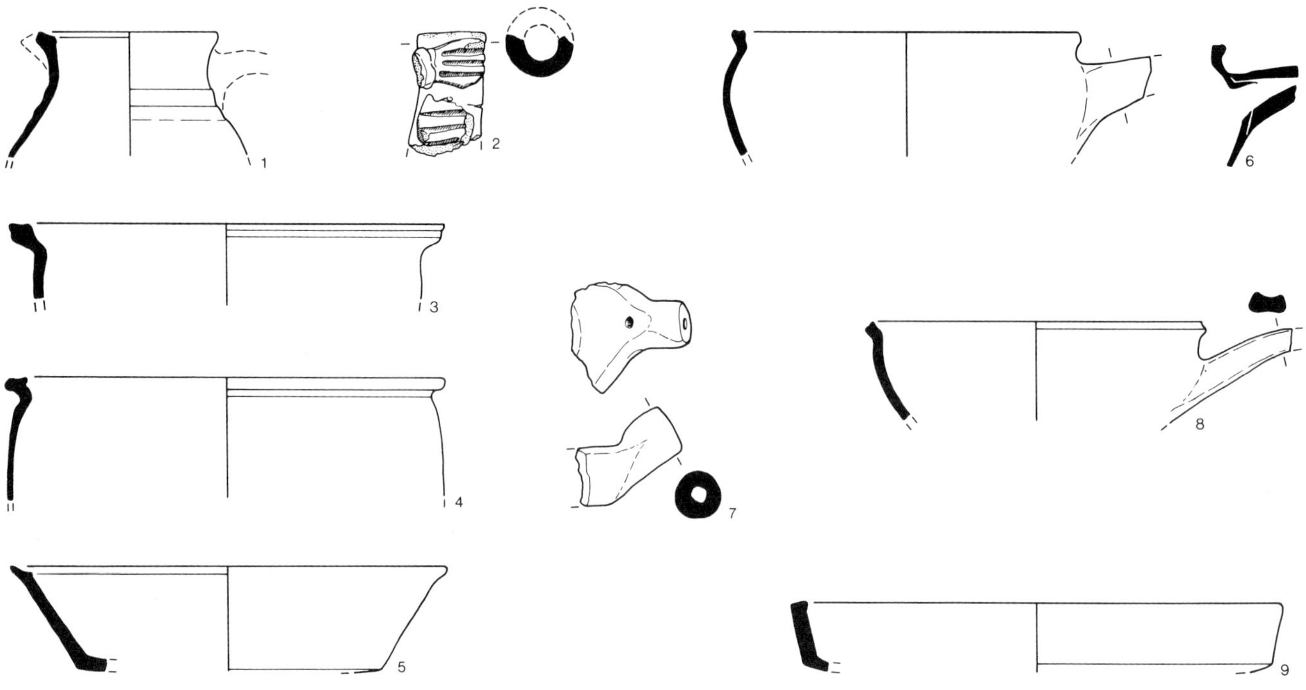

Figure 102 Medieval pottery: buff sandy ware, fabric 64.2. Scale 1:4

8 Pipkin with patchy glaze similar to no 6 (334, S42, P6–7)

9 Dish with internal dark green glaze (333, S42, P6)

Phases 7 and 8: 15th to mid 17th century (Figs 103 and 104)

Fabrics present: 1, 2, 12, 12R, 12.4, 19, 22, 43, 46, (?)48, 49, 55, 56, 57, 58, 64.1, 64.2, 65, 69, 70, 71, 72, 78.1, 81.8, 82 and 100

Domestic occupation on the street frontage established in phase 6 continued in this phase. Much of the pottery was associated either with the area between two properties (S108 and S135), and probably resulted from a general accumulation associated with the use of these buildings, or with the surface of the street (S42).

Malvernian wares, both glazed and unglazed, accounted for 51% of the assemblage and so the medieval sandy wares no longer dominated the Droitwich market. The proportion of glazed wares also continued to rise to 50% and this included a higher proportion of oxidized glazed Malvernian ware (48%) than in phase 6. The expansion in supply was accompanied by a greater range of forms in this fabric, and large jars or cisterns (cf Fig 101 no 5), smaller jars (eg Fig 101 no 6) and new conical bowl types (eg Fig 101 no 13) were now available. An oval dish form probably used for baking fish, or as a dripping tray, was also represented. The proportion of unglazed Malvernian cooking pot also increased in this phase.

The other main glazed ware fabrics of this phase were generally similar to those that occurred during phase 6, and may have included a residual element, though this was difficult to quantify. The iron-rich Worcester-type glazed ware, which consisted mainly of jugs, included an aquamanile

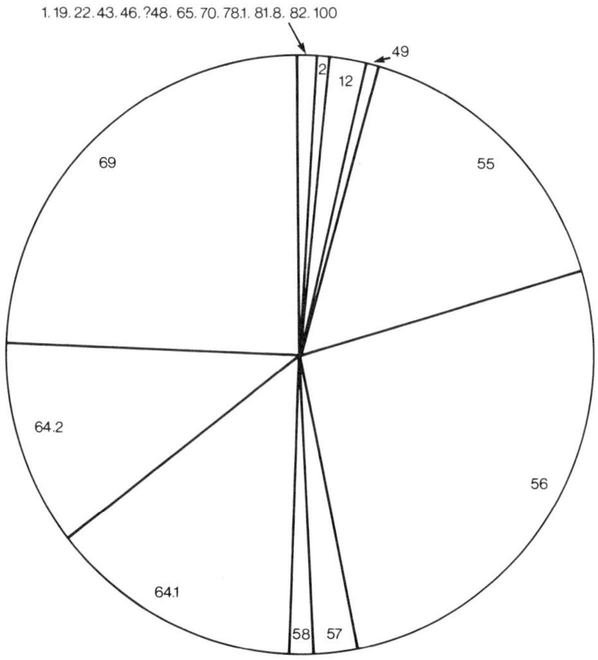

Figure 103 Fabric proportions by weight in phases 7–8

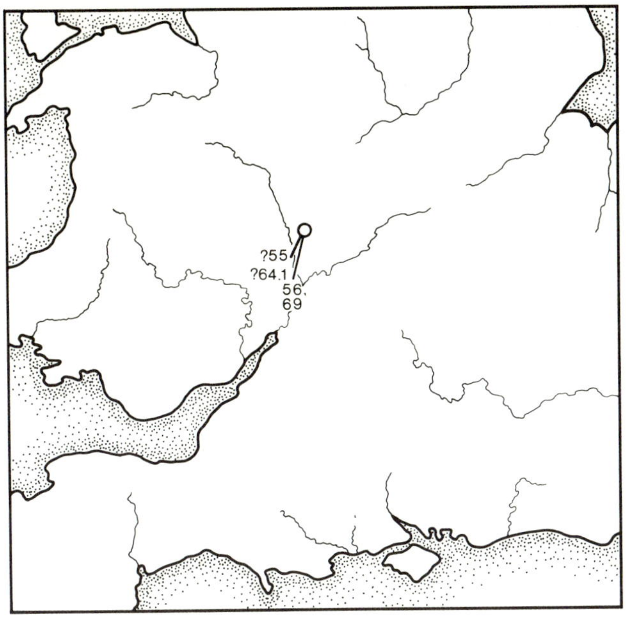

Figure 104 Principal pottery sources in phases 7–8

(Fig 105), possibly in the form of a boar. Buff/white glazed wares were still fairly common, and an unusual form was a tubular spouted jug with anthropomorphic decoration (Fig 102 no 2). A few glazed jug sherds suggested sources to the west and north for some pottery during this period. Thus there were a few sherds of fine micaceous ware (fabric 71), which resembled fabrics produced in the Hereford area (eg Vince 1985 fabric A7b) and, as in phase 6, white painted orange ware of 'Deritend' type (fabric 62).

Several new fabrics were present for the first time in this phase and they were mainly used for drinking vessels. There were southern white wares (fabric 70), brown glazed ware (fabric 72, Fig 106 no 8) and black iron-glazed Cistercian-type ware (in fabric 78). The only imports were a German drinking mug of Raeren type (Fig 106 no 5) and the base of a South Netherlands Maiolica drug jar (J G Hurst pers comm, Fig 106 no 3).

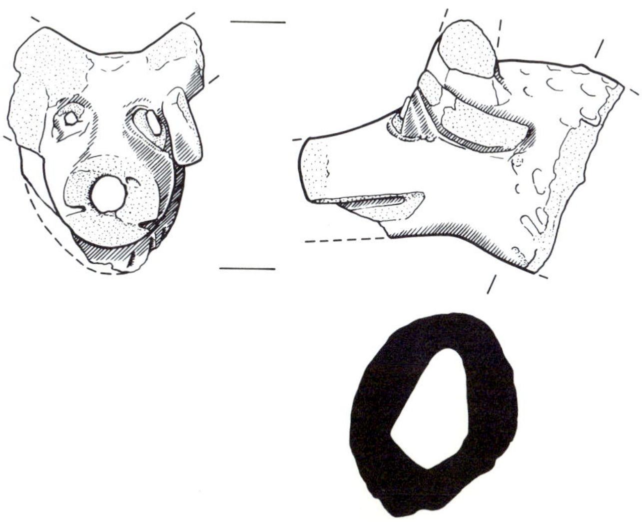

Figure 105 Green glazed aquamanile possibly representing a boar (fabric 64.1). Scale 1:2

Dating

Several forms in the oxidized glazed Malvernian fabric could be dated to the 15th to early 17th centuries. These were the jar with thumbed decoration below the rim, large jar or cistern (probably double handled), pipkin, conical bowl and oval fish dish. The overall proportion of this fabric also indicated a similar date as it was most common at both Hereford (Vince 1985, 52) and Worcester (Morris 1980, 226–7) during the late medieval and early post-medieval periods.

The introduction of ceramic drinking vessels also occurred in this phase (see above) and this has generally been dated elsewhere to the later medieval period. Thus the introduction of southern white ware has been established as early 15th century in London (Mathews and Green 1969, 6–7), while the more locally produced brown glazed jugs (fabric 72) were regarded as a 16th century introduction at Worcester (Morris 1980, 226). Black iron-glazed drinking vessels were also present, but were not as common as at Hereford or Gloucester (Vince 1985, 64–5). The Raeren drinking jug was dated to the late 15th or 16th century.

Illustrated pottery

Post-medieval pottery (Fig 106)

1 Jug (fabric 81.7) with roller-stamped animal motifs (103, S147, P11(10))
2 Blackware (fabric 78.1) jar (183, S43, P11)
3 South Netherlands *albarello* (fabric 82) with faded blue banding (292, S46, P8–9)
4 Blackware (fabric 78.1) tankard (183, S43, P11)
5 Raeren (fabric 81.8) drinking jug (222, S108, P6(7))
6 Ointment pot (fabric 78.3) with silver lustre glaze (55, S169, P10)
7 Buff ware (fabric 91) small bowl (102/103, S147, P11(10))
8 Tyg (fabric 72) glazed externally and inside rim (931, S41, P7–8)

Phases 9 and 10: Mid 17th to 18th century (Figs 107 and 108)

Fabrics present: 3, 12, 12.4, (?)13, 19, 22, 29, (?)46, 49, 55, 56, 57, 58, 64.1, 64.2, 69, 75, 78.1, 78.2, 78.3, 81.2, 81.3, 81.4, 81.5, 81.7, 82, 83, 83.1, 84.1, 84.2, 84.3, 85, 91, 92 and 99

Much of the pottery in phases 9 to 10 was associated either with an area of cultivated ground (S119) adjacent to one of the properties established

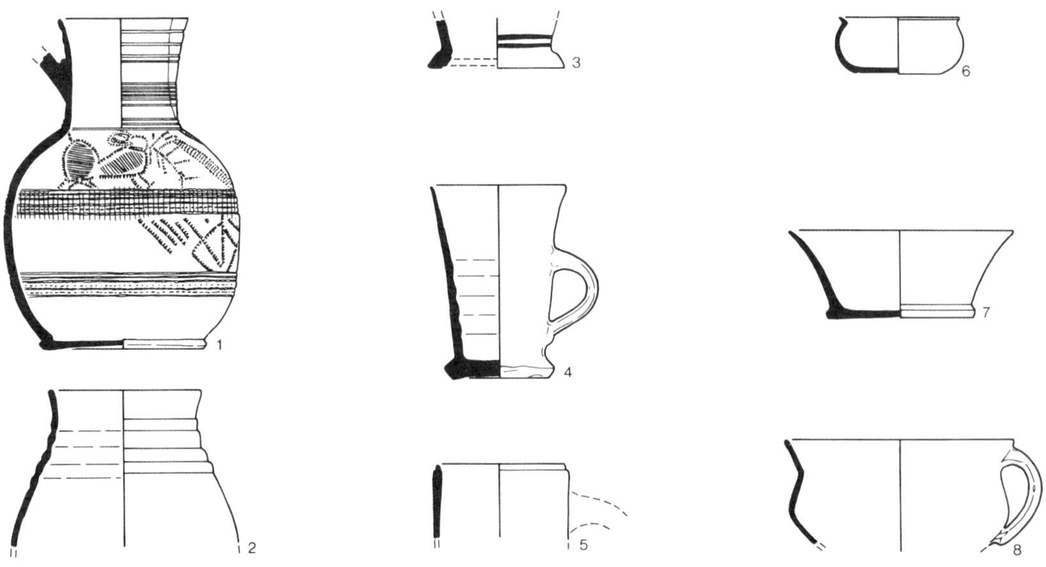

Figure 106 Post-medieval pottery: fabrics 81.7 (1), 78.1 (2, 4), 82 (3), 81.8 (5), 78.3 (6), 91 (7) and 72 (8). Scale 1:4

in phase 7, or with the demolition horizon (S150 and S152) for these properties (Chapter 21). There was a high proportion of residual (and especially Roman) pottery and this may have occurred because of extensive disturbance of earlier deposits during demolition. The pottery assemblage of this phase was characterised by post-medieval red and buff earthenwares (fabrics 77, 78 and 91) and a range of stonewares (fabric 81). The red and buff wares were decorated with slip trailing or combing under a pale yellow lead glaze, or finished with a black iron glaze over a dark red slip. Typical forms were flanged dishes, press moulded baking dishes, deep conical bowls and tankards (cf Fig 106 no 4). In contrast both North Devon gravel tempered ware (fabric 75) and Midlands yellow wares (fabric 77) were but poorly represented in phases 8 to 10.

The stonewares included examples from London, Nottingham/Chesterfield, Staffordshire and possibly south Shropshire. The latter area seems to have been involved in the manufacture of white stoneware (Malam 1981, 45) and possibly some slipwares as well. There was also a small amount of imported Westerwald stoneware (fabric 81.2).

Creamwares (fabric 84) were present, and some had 'tortoiseshell' decoration. There was also a small amount of porcelain (fabric 83), including some originating from Worcester.

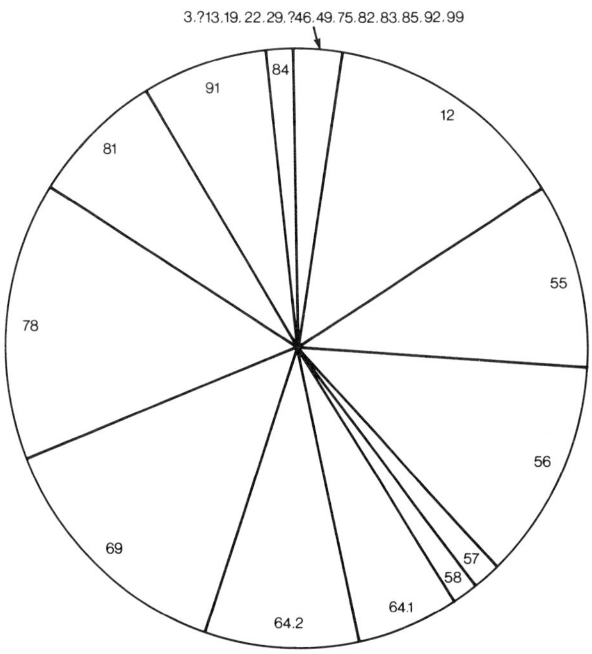

Figure 107 Fabric proportions by weight in phases 9–10

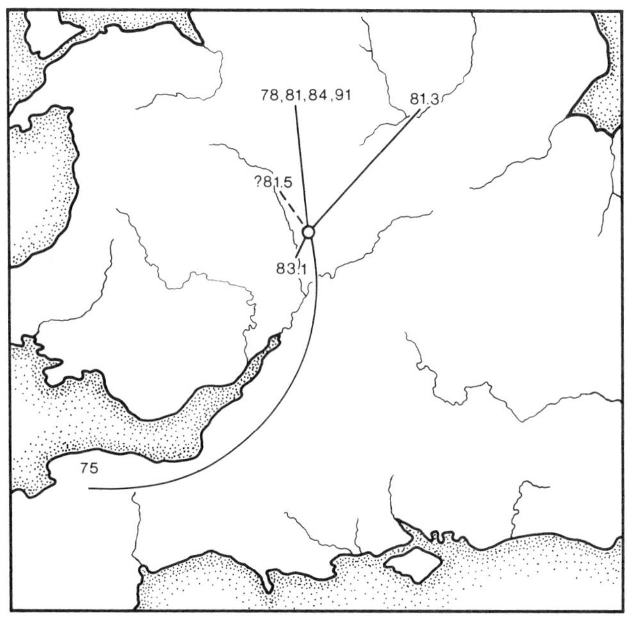

Figure 108 Principal pottery sources in phases 9–10

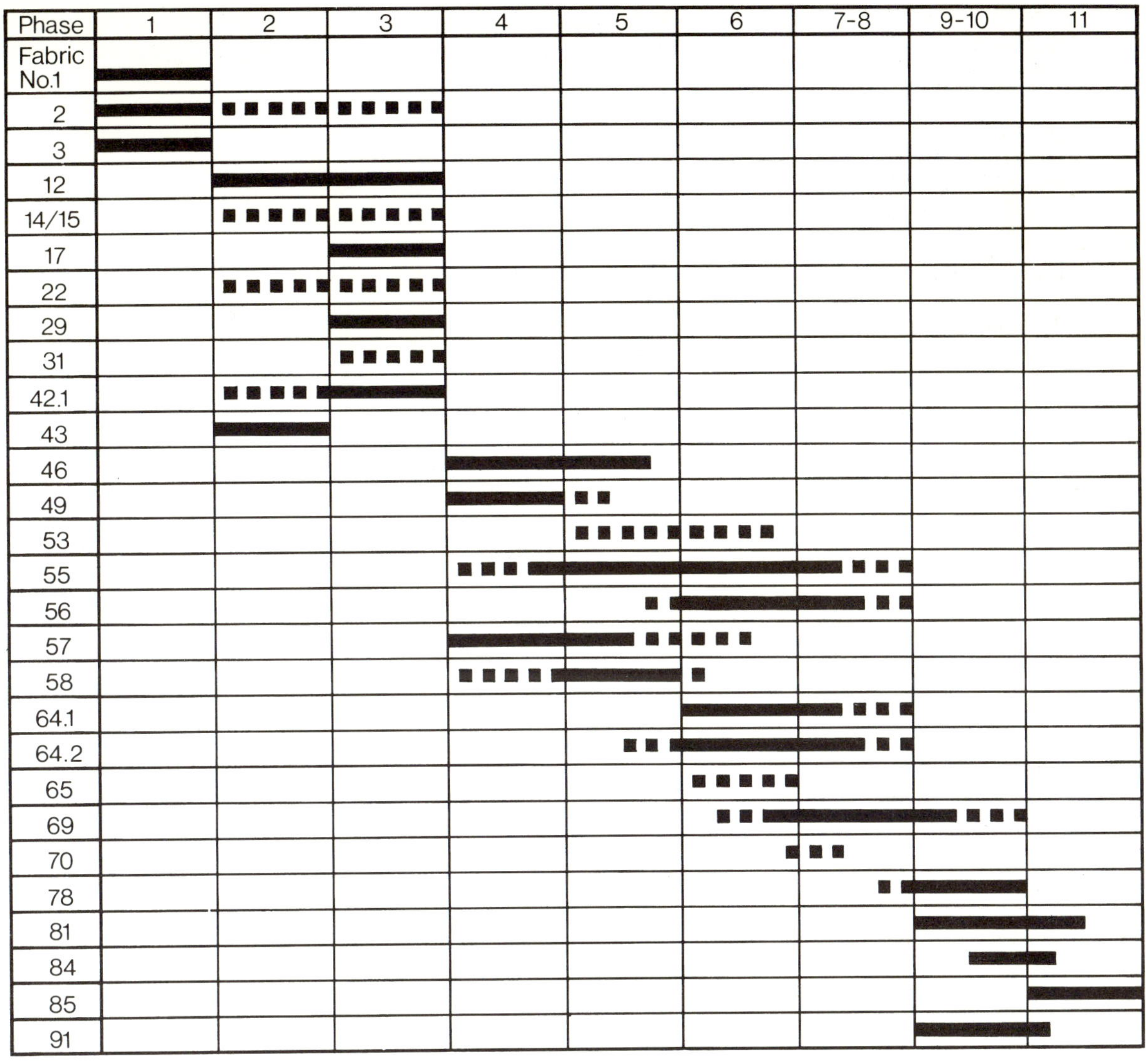

Figure 109 Sequence of principal pottery fabrics

Dating

Both 17th and 18th century pottery types were present. The former included thrown slip trailed redware dishes and possibly some of the less well-made black wares. The 18th century was represented by white stonewares and brown Nottingham-type stonewares, as well as the manganese mottled and combed slip decorated wares and fine black glazed wares. During this period new techniques of manufacture were introduced, including slip casting, press moulding, lathe turning and the more careful preparation of the clay body. Creamwares and porcelain were developed during the second half of the 18th century.

Conclusions

The site produced a well stratified ceramic sequence (Fig 109) from the Iron Age onwards, with the Saxo-Norman and early medieval periods being particularly well represented. The importance of Droitwich as an industrial and trading centre was well demonstrated in the Iron Age by the availability of non-local pottery types brought into the town. The Roman pottery assemblage was not particularly large and, though it contained examples of imports from Spain, the Rhineland, and Gaul, this was not exceptional for the period. Subsequently, however, the special importance of the salt trade for the development of the late Saxon and medieval town was demonstrated by the close correlation of pottery supply and the saltways during these periods.

The pattern of pottery supply, however, rarely remained stable for very long. It could be affected by many disparate factors, and progressive fluctuations in supply could clearly be observed in the Friar Street assemblage. For instance, there was a well marked tendency towards supply by more local production centres from the 12th century, after a period when more distant sources had been dominant suppliers. However, the trend reverted to a network incorporating less local sources in the later 17th to 18th centuries, after improvements in transportation and the growth of large-scale factory production.

A notable characteristic of the medieval assemblage was the number of vessels only represented by single rim sherds. As the Friar Street excavation was at the front of burgage plots, the general scarcity of conjoining sherds perhaps suggests considerable organisation of rubbish disposal at Droitwich, as is known for several other medieval towns from documentary evidence (McCarthy 1979, 225). This could involve rubbish being taken out of the town or being buried inside, especially at the rear of burgage plots (Platt 1976, 56).

There were few medieval foreign imports. Generally such pottery has been found to be absent in the West Midlands region, except at sites of the highest status, for example Stafford Castle and Dudley Castle (Deborah Ford pers comm), or at important ecclesiastical centres such as Worcester (Dunning 1968). The evidence from Droitwich, where (despite the economic significance of the town) such pottery was not found, has therefore tended to confirm this pattern of distribution.

The assemblage has thus provided an insight into the main trends of pottery usage, especially in post-Roman periods. Though larger and better preserved pottery groups (especially for the medieval period) may result from future excavation, the ceramic sequence of fabrics and forms established for the Friar Street and Old Bowling Green assemblages may be taken as substantially representative of pottery types available in the area from the Iron Age onwards. Further petrological study elsewhere, notably perhaps at Worcester, and future discoveries of kiln sites, will indubitably bring greater precision to our understanding of pottery production and exchange in the Droitwich region.

Samian, by Brenda Dickinson

The small collection of samian ranges from the Neronian period to the late 2nd or early 3rd century, with no obvious gaps. There are several pre-flavian pieces, including four decorated bowls, one of which was made at Lezoux (Fig 110 no 1). Any 1st century samian from this source reached Britain in small quantities from *c* AD 50 to 75, but was even rarer thereafter. There is a different bowl, probably by the same potter, from the nearby Old Bowling Green excavation (Chapter 3).

In the 2nd century, both Lezoux and the other large Central Gaulish factory at Les Martres-de-Veyre are represented. The three sherds from Les Martres-de-Veyre are all Trajanic, while the Lezoux ware ranges from the Hadrianic or early Antonine period to the latter part of the 2nd century. The single East Gaulish sherd (232) is probably from Rheinzabern (Fig 110 no 2). It is not closely datable, but could be contemporary with the latest Lezoux ware, rather than 3rd century. The full catalogue is in fiche (3:B11–12).

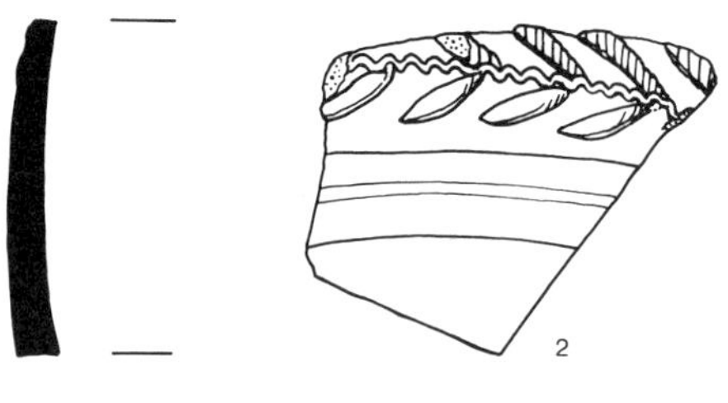

Figure 110 Decorated samian (fabric 43). Scale 1:1

23 Ceramic building material *Derek Hurst*

There was an assemblage of 94.063kg of ceramic building materials associated with the Roman and later periods (phases 2–11). This comprised mainly roofing tile until phases 7–8, when the use of brick became widespread.

Fabric descriptions

The definition of terms used to describe fabric inclusions and hardness follows Orton (1977, 28–30). Full descriptions are available in archive.

Fabric 1 Hard modern type

Colour	Usually red, but sometimes distinctly purplish
Inclusions	Abundant, coarse, angular inclusions usually visible
Form	Tiles may be double nibbed, and are usually quite thin ie <10mm
Hardness	Very hard

Fabric 2a Common sandy type

Colour	Usually red or orange throughout
Inclusions	Moderate to abundant, variably sorted quartz (<1mm)
Form	Both brick and tile
Hardness	Usually hard

Fabric 2b

Fabric description as for 2a, except that coarser inclusions may be present (ie >1mm)

Fabric 2c

Fabric description as for 2a except for moderate clay pellet/grog inclusions

Fabric 2d

Colour	Red throughout with pale buff banding usually evident
Inclusions	Sparse to moderate, medium (0.1–0.5mm) quartz
Form	Used for Roman roofing tile
Hardness	Hard dense fabric

Fabric 2e

Colour	Pale buff, sometimes with off-white laminations
Inclusions	Abundant, well-sorted, medium (0.1–0.5mm) quartz
Form	Used for Roman roofing tile
Hardness	Soft

Fabric 2f

Colour	Red, may have grey margins
Inclusions	Moderate, medium (0.1–0.5mm) quartz, and sparse shelly limestone (<2mm)
Form	Used for flat roof tile
Hardness	Very hard

Fabric 2g

Colour	Light grey iron-free fabric
Inclusions	Sparse, medium (0.1–0.5mm) quartz
Form	Possibly box flue tile
Hardness	Very hard

Fabric 2h

Colour	Oxidized red throughout
Inclusions	Rare, medium (0.1–0.5mm) quartz
Form	Box flue tile
Hardness	Very hard

Fabric 2i Delftware

Colour	Pale buff throughout
Inclusions	Moderate, medium (0.1–0.5mm) quartz
Form	Wall tile. Tin glazed with transfer decoration
Hardness	Soft

Fabric 2j

Colour	Oxidized to red or orange throughout
Inclusions	Moderate, well-sorted medium (0.1–0.5mm) quartz
Form	Used for Roman roofing tile
Hardness	Soft

Fabric 3 Malvernian

Colour	Usually oxidized orange/buff throughout; patchy copper green glaze
Inclusions	Sparse Malvernian rock inclusions (usually <5mm); moderate medium (0.1–0.5mm) quartz
Form	Ridge tile and flat roof tile; thinner than examples in fabric 2a
Hardness	Hard

Phase 2: Early to mid Roman

There was a small quantity of Roman tile from this phase and this was mostly associated with the infill of the well (S6). Box flue tile, *tegula* and *imbrex* were all represented, and the two commonest fabrics were 2d and 2j. These were similar, except for the variegated appearance of the clay used for the former.

Phase 3: 3rd century to Anglo-Saxon

The same range of tile types was represented as in phase 2, though in increased quantity, and the same fabrics were most common. This material was evenly distributed across the site and, since no standing structures incorporating tile were known on the site during this phase, the assemblage probably constitutes rubbish disposal from elsewhere. The abraded condition of tile from contexts 694, 793, 870, 1018 and 1103 was notable, tending to corroborate an Anglo-Saxon date for some of the deposits in this phase.

Phase 4: Saxo-Norman

All the tile in this phase was Roman (fabrics 2d and 2j) and therefore residual. The quantity was small and most was rather abraded.

Phase 5: 12th to early 13th century

During this phase most contexts containing tile included residual Roman tile. However, the principal tile type represented was the glazed ridge tile in sandy fabric 2a. This generally displayed a green glaze coloured by reduced iron, for when fired in an oxidising atmosphere the glaze turned brown. This characteristic was shared by the Worcester-type glazed wares (fabric 64.1) and this general similarity of fabric type suggested that both the tile and pottery could be derived from the same source. Sandy ridge tiles were noted at Sidbury, Worcester (Morris 1980, 228), where their introduction was dated to the 14th century. At Hereford a mid 13th century date has been assigned to the earliest datable ridge tile (Vince 1985, 68), while they have been recovered from late 12th and early 13th century contexts in Gloucester (Vince 1985, 69). This earlier dating from Gloucester for the introduction of ridge tiles bears comparison with that inferred from associated pottery for the Droitwich glazed roof tile assemblage.

Phase 6: 13th to 14th century

Ceramic roofing tile was much commoner in phase 6 than earlier, and it comprised mainly a new type, flat roofing tile. This was present in the same sandy fabric as ridge tiles in the previous phase. It was generally 15–18mm thick and usually nibbed for hanging, though sometimes perforated for attachment using nails. Occasionally both methods of fastening were available on one tile. The use of such flat roofing tiles was well developed during phase 6, whereas Vince (1984c, 39) considered that this type of tile was not used in Hereford and Worcester, or Gloucestershire, before the late 16th century. However, both archaeological and documentary evidence indicate otherwise. Thus at Sidbury, Worcester, Morris (1980, 225) recorded flat lugged tiles from the early 14th century onwards, while excavation at Blackfriars, Worcester (HWCM 378) in 1985–6 has shown that flat roof tile was in use throughout the lifetime of the friary, which was founded in the mid 14th century (Charles Mundy pers comm). The documentary evidence has also indicated a pre-16th century commencement for the tile industry in Worcester, though the exact nature of its production is less clear and no documentary evidence has yet confirmed tile production contemporary with phase 6 (13th to 14th centuries). However, a series of local regulations in Worcester in 1467 contained references to tile working in the city (VCH ii, 275; iv, 387) and some of these obviously promoted the industry by stipulating that wooden chimneys should be rebuilt in brick or stone and that thatching be replaced by tiles. They also ordered that tile makers should stamp their wares, and since the only stamped tile found in Worcester (Carver 1980, 213) or Droitwich (Chapter 4) has been flat roofing tile, there is a strong possibility that at least some of this is of 15th century date.

The use of glazed ridge tiles continued during phase 6. These were usually in the same sandy fabric as in phase 5, but there were also some examples in a Malvernian fabric (274, 537). The latter were more common in the next phase and so their pattern of introduction and use conformed closely to that established for Hereford (Vince 1985, 69) and Worcester (Morris 1980, 228).

Phases 7 and 8: 15th to mid 17th century

There was a greater range of ceramic building materials in evidence during this period than earlier. Sandy flat roof tiles were abundantly available, and were occasionally accompanied by flat roof tile in a Malvernian fabric. Glazed ridge tile was also represented in both these fabrics. Floor tiles occurred, though the largest group had been reused as rubble in a foundation (375). These included rectangular and triangular (*c* 120 x 90 x 90mm) types, which were brown glazed without any slip decoration. The use of brick was also well attested on the site during this period.

As observed at Worcester (Morris 1980, 226) and Droitwich (Chapter 22), production of sandy Worcester-type pottery wares declined in the 15th to 16th centuries as the products of the Malvernian industry became more popular. An increased demand for mainly sandy, probably local, ceramic building materials, for which there is some documentary evidence (see above) may, therefore have counteracted this decline in the local pottery production.

Phases 9 to 11: Mid 17th to 20th century

During this period pantiles were introduced, and there was a tendency towards more highly fired tile, especially in fabric 1, which became common in the 19th and 20th centuries. Some of the bricks were remarkably similar to examples found recently at White Ladies, Worcester, where brick and tile kilns were operating from soon after the dissolution, and it is possible that these products were widely traded by canal to the surrounding area (Ian Walker pers comm).

24 Other ceramic objects *Derek Hurst*

Fig 111

1 Spindle whorl made from a Severn Valley ware (fabric 12) sherd; context 460, structure 159, phase 5–6.

2 Slab-built fragment in a fabric tempered with Malvernian rock inclusions; context 419, structure 79, phase 5i (not illustrated).

Fragments of this type were first observed by Peacock (1965–7, 24, fig 4, nos 80–2). It also occurs at the Old Bowling Green excavation (Chapter 5), and at other sites in the region, for example Beckford (HWCM 359 and 497) and Wick (HWCM 5568).

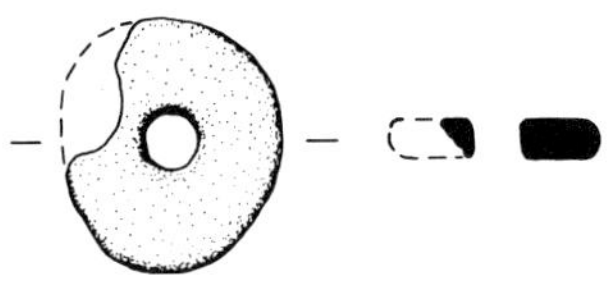

Figure 111 Spindle whorl. Scale 1:2

25 Clay pipe *Derek Hurst and Allan Peacey*

Pipe smoking became quite widespread during the 17th century and most of the Droitwich clay pipes (Table 8) date from *c* 1650 onwards. They were mainly of local origin, though there were also several pipes from the Broseley area of south Shropshire. Local sources in Worcester and Stourbridge (Oswald 1975, 199) probably account for many of the unmarked and unburnished pipes (David Higgins pers comm).

Table 8 Clay pipe marks

Context	Mark	Pipe type	Date	Reference
Unstrat	JOHN LEGG 1696	Broseley 5	1696	Atkinson 1975
79	RH	Broseley 4	?early 18th century	–
85	?T	Broseley 4	?early 18th century	–
88	WD	Broseley 4	?early 18th century	Atkinson 1975
88	SAM RO*DEN	?	*c* 1720–1750	Atkinson 1975
88	JOHN RODEN BROSELEY	?	*c* 1820–1840	Atkinson 1975
102	RICH SMITH	?	–	–
901	IW	?Bristol	?	–
906	.......SON	?	–	–

26 Glass *Louise Monk*

The glass assemblage weighed 5.68kg and included 0.248kg of Roman glass. The remainder was post-medieval bottle glass.

Roman (Fig 112)

1 Large mould-blown hexagonal bottle of bluish-green metal with pin-head bubbles. A crude graffito in the form of a plant was incised on the vessel prior to breakage and the multi-reeded handle was drawn into 27 sharp claws to grip the bottle. Context 1115, structure 55, phase 2.

2 *Unquentarium*. Context 102, structure 147, phase 11(10?). Another fragment of the same vessel type (not illustrated) was noted from context 553 (S70, P4?). Both may be dated to the 2nd or 3rd century.

Other contexts of Roman date (1037, S68, P3; 1103, S64, P3; and 1117, S16, P3) also produced bottle glass and all of these vessels were datable to the mid 1st to early 3rd centuries (Charlesworth 1966).

Post-Roman

The more common use of glass did not resume until phases 7 and 8 and it was not extensively represented until phase 9.

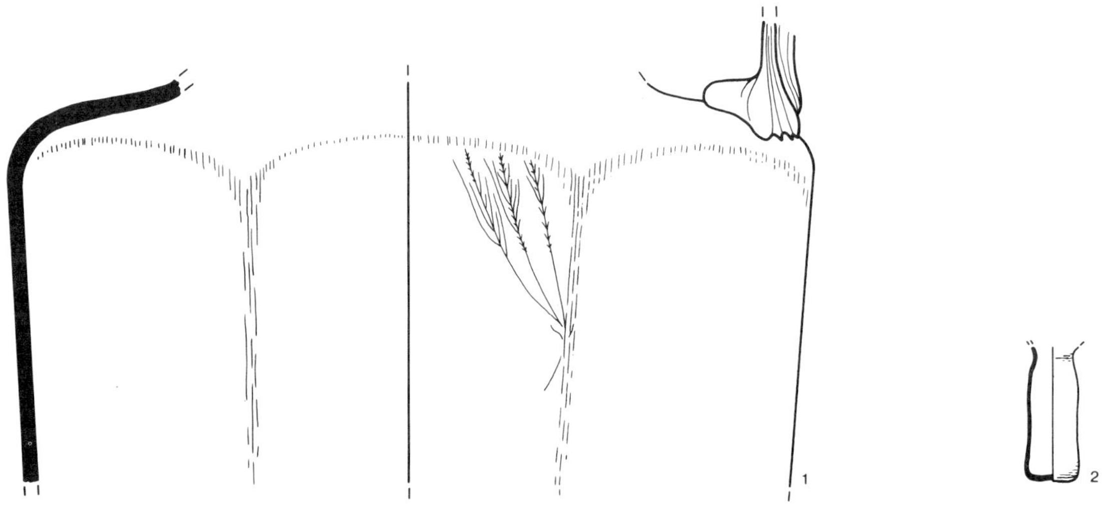

Figure 112 Roman glass. Scale 1:2

27 Worked stone *Derek Hurst, with stone identification by Fiona Roe, flint by Alan Saville and discussion of the Ruding memorial slab by Jerome Bertram*

There was a small quantity of worked stone artefacts. Other worked stone, however, had been sampled during the excavation, and was no longer available for study. This included many examples of split sandstone which may have been used for roofing, and which were mainly from phase 3.

Flintwork (Fig 113)

1 Mesolithic/early Neolithic core; context 419, structure 79, phase 5.
2 Mesolithic bladelet core; context 1037, structure 68, phase 3.
3 Mesolithic edge-trimmed blade fragment; unstratified (not illustrated).
4 Possible fragment of edge-trimmed blade; context 1068, structure 11, phase 2 (not illustrated).
5 Gun flint; context 183, structure 43, phase 10 (not illustrated).

Whetstones (Fig 113)

Mica schist

6 Context 240, structure 47, phase 9 (not illustrated).

This whetstone was made from a distinctive grey micaceous schist, and a thin section confirmed that it was imported from the Telemark area of Norway. The material is known variously as Eidsborg schist or Norwegian ragstone, and is a quartz-muscovite-biotite-chlorite-ore schist, described by Ellis (1969, 136–43). The sample from Friar Street additionally contains grains of epidote, a variation noted amongst hones from medieval sites (Ellis 1969, 141).

7 Context 297, structure 42, phase 6 (not illustrated). Probably belongs within the same lithic range as no 6.
8 Context 232, structure 42, phase 7 (not illustrated). Probably belongs within the same lithic range as no 6.

Phyllite or siltstone

9 Context 297, structure 42, phase 6.
10 Context 412, structure 96, phase 6.

The materials for these two whetstones are also likely to have been imported to the site. The whetstone from context 412 was thin sectioned to test whether it was made of blue phyllite, which is known to occur in association with Eidsborg schist at medieval sites (Moore 1983, 285). The thin section however showed a different rock, a siltstone consisting of tiny grains of iron ore and clay minerals too fine to identify under the microscope.

Building stone (Figs 113 and 114)

Oolitic limestone

11 Window tracery fragment; context 32, structure 172, phase 11.

Sandstone

12 Red sandstone window tracery fragment; context 374, structure 41, phase 7? (not illustrated).
13 Pink, micaceous sandstone roofing tile; context 222, structure 108, phase 7.

Other sandstone roofing tile of this type was retrieved from contexts 226 (S108, P6–7), 294 (S125, P6), 366 (S96, P6), 553 (S70, P4), and 1116 (S16, P3) (none illustrated).

14 Pink, micaceous sandstone *tessera*; context 256, structure 108, phase 6–7 (not illustrated).
15 Grey/buff, micaceous sandstone roofing tile; context 541, structure 30?, phase 4ii.

Other sandstone roofing tile of this type was retrieved from contexts 552 (S70, P4) and 697 (S28, P3) (none illustrated).

16 Pale sandstone moulded block; context 374, S41, phase 7?

The pink and grey varieties of flaggy sandstone both range over several periods and are not necessarily all from the same source. The most likely provenances lie amongst the Lower Old Red Sandstone of the Welsh borders and in particular amongst the Downton Castle 'tilestone'.

Chalk

17 Two *tesserae*; context 223, structure 108, phase 7 (not illustrated).
18 *Tessera*; context 421, structure 94, phase 6 (not illustrated).

Blue Lias mudstone

19 Possible roofing tile; context 297, structure 42, phase 6 (not illustrated).

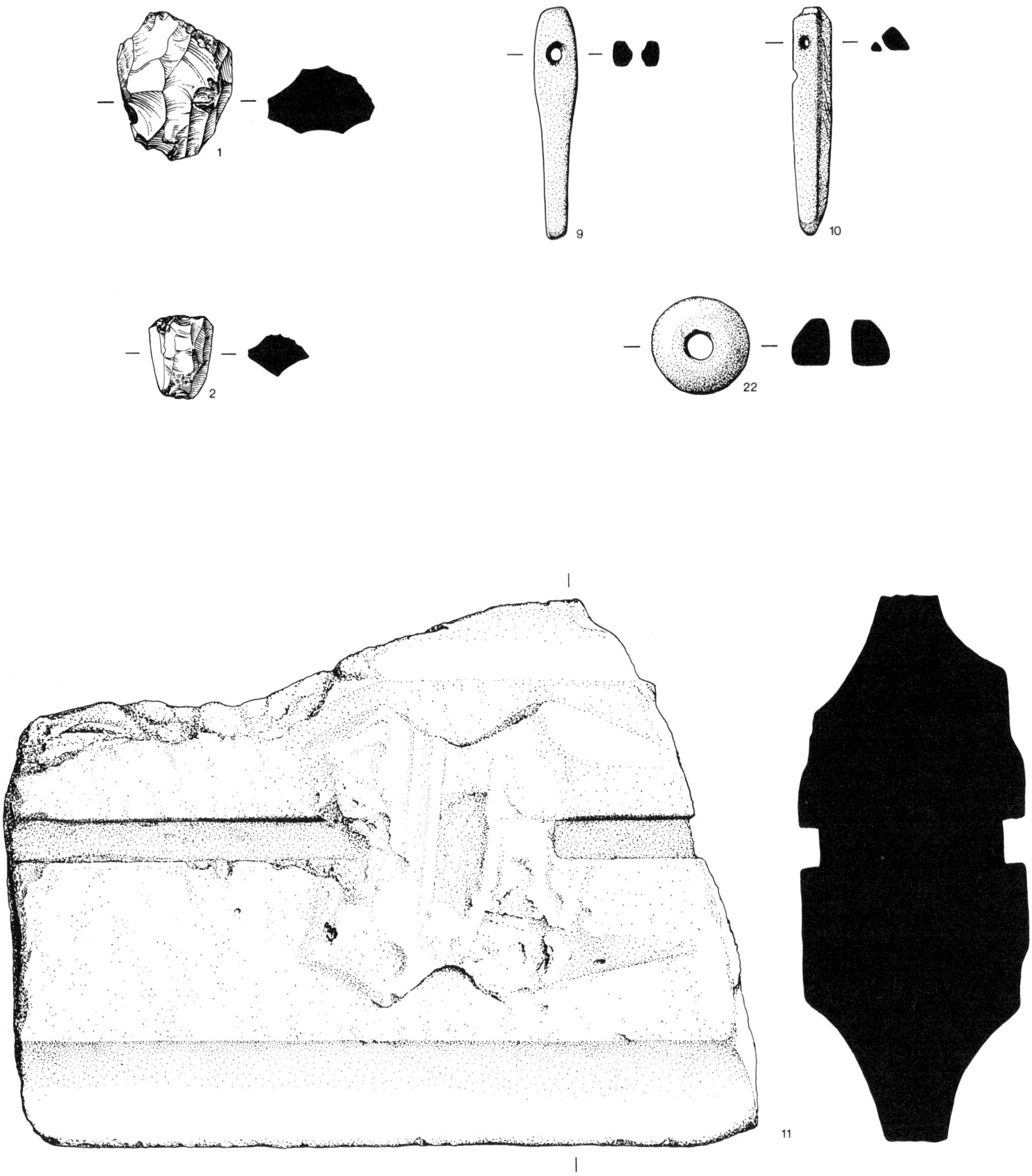

Figure 113 *Worked stone: flintwork (1,2), whetstones (9,10), building stone (11) and spindle whorl (22). Scale 1:2*

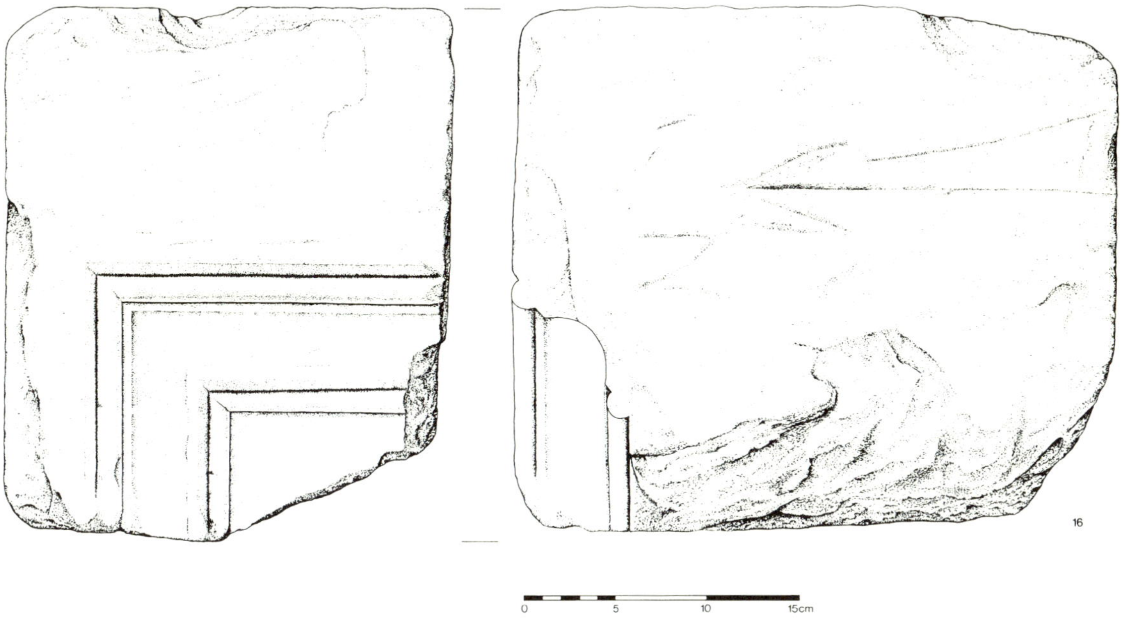

Figure 114 Worked stone: building stone (16)

Green Porphyry

20 Wall veneer fragment; context 444, structure 3, phase 5 (not illustrated).

The green porphyry is a distinctive material, consisting of pale green phenocrysts set in a dark green groundmass. It derives from the Sparta area in mainland Greece, and is known from Roman sites in London, Colchester, and Fishbourne (Frances Pritchard pers comm).

Ruding memorial slab (Fig 115)

21 Decorated memorial slab of Triassic sandstone; context 375, structure 41, phase 7.

This is an apparently unique survival from a small local workshop, cut in local sandstone from beds in the Elmley Lovett area and the design is unlike any other incised slab yet recorded. Other incised figure slabs in Worcestershire are located at Cofton Hackett (1514), Rock (*c* 1540), and Queenhill (1584). The surviving fragment is 770mm wide and 180mm to 200mm thick. The design was cut mainly using deep V–shaped incisions, though the wider areas in the shields of arms were flat-bottomed. The heraldry, and probably the remainder, were originally filled with black pitch.

The slab depicts a man in armour and probably his wife. The armour, which is very stylised, seems to date from the end of the medieval period when armour had become simpler and closer fitting. The helmet above the man's head is probably an extremely poor representation of a sallet with a pointed skull, similar to examples in St Mary's Hall, Coventry, and in the Victoria and Albert Museum, London. This type of helmet may be dated to the late 15th century. The female effigy is badly damaged, though the shoulder straps of a sideless surcoat are clearly visible. The head-dress is the most distinctive item of costume and is of the 'forked' variety of the popular 'horned' head-dress. This style may be dated 1440 to 1473, the latter date being that of a slab in St Mary's, Shrewsbury, which offers the nearest parallel to the costume on this slab. Both the man and woman rest their heads on stylised cushions.

At the top of the slab are two shields of arms depicted hanging from their straps. Both bear the arms of Ruding. The sacred monogram *ihu*, also inlaid with pitch, is placed between the shields and adorned with tendrils ending in roundels above and below. In the top corners of the slab are small fleurs-de-lys, which are probably purely decorative. There are faint traces of a border line around the slab. Any inscription would probably have been at the feet of the figures. It is likely that the shields and fleurs-de-lys were repeated at the bottom of the slab.

The arms were readily identifiable as those of Ruding, a prolific family in Worcestershire and Leicestershire. However, the pedigree of Ruding, given by Nichol (1811, 568), does not go back beyond the early 16th century. Before this date the only information on the Ruding family is derived from scattered references, though there is some

Figure 115 The Ruding memorial slab

evidence to link the Rudings with Droitwich in the description of the Chapel on the Bridge by Nash (1781–2, 329), which mentioned that the arms of Beauchamp, le Despenser, Ruding, and others were displayed in the chapel windows. This memorial to members of the Ruding family was found upside down, built into the base of a chimney stack (S41), which was erected during phase 7. As the nearby church of St Nicholas (HWCM 255) was in a state of disrepair, possibly following the Dissolution as it belonged to a local nunnery at Westwood (HWCM 2574), there was some possibility that it had been transferred at this time from an original position within this church. The memorial was probably erected *c* 1470 according to the dating suggested for the female head-dress.

Other worked stone (Fig 113)

22 Limestone (Lias mudstone) spindle whorl; context 102, structure 147, phase 11 (10)?

23 Purbeck marble mortar fragment; context 256, structure 108, phase 6 (not illustrated).

24 May Hill Sandstone quern or rubber fragment; context 488, structure 3, phase 1 (not illustrated).

This is a pinkish gritstone which was recorded in some quantity at Beckford (HWCM 359 and HWCM 497) where it was used for querns and rubbers. At this site it occurred mainly in Iron Age contexts (Roe pers comm).

Brief discussion of worked stone

It is notable that there is very little local stone in this assemblage, only the two pieces of Jurassic mudstone and the Triassic sandstone used for the Ruding stone. In contrast, the stone brought in from overseas is worthy of comment, both the mica schist from Norway and the green porphyry from Greece, while the possibility cannot be excluded that the phyllite or siltstone is also of non-British origin.

The British materials are wide ranging, with Devonian sandstones from west of Droitwich, Carboniferous Millstone Grit from northwards, Jurassic oolite and Purbeck Marble from a southerly direction, as also Cretaceous chalk, and perhaps even Cretaceous flints.

28 Copper alloy *Derek Hurst*

Fig 116

Studs

1 Plain head; context 301, structure 128, phase 6.
2 Scallop shell head; context 223, structure 108, phase 7.

Other studs were retrieved from contexts 226 (S108, P6–7) and 290 (S46, P7–8).

Pins

3 White metal plating on head and shaft; context 274, structure 127, phase 6.
4 Spiralled wire head; context 222, structure 108, phase 7.
5 Head formed with spiralled wire; context 222, structure 108, phase 7.
6 Looped head; context 299, structure 126, phase 7.
7 Fragment with spherical head; context 218, structure 133, phase 8.

Other pins were retrieved from contexts 299 (S126, P7), 226 (S108, P6–7) and 218 (S133, P8).

Thimbles

8 Fragment; context 264, structure 108, phase 6.
9 Large thimble; context 235, structure 136, phase 8.

Other thimbles were retrieved from contexts 264 (S108, P6), 182 (S137, P7) and 183 (S43, P10?).

Buckles

10 Possible loop fragment from buckle; context 366, structure 96, phase 6.
11 Fragment with iron rivet; context 222, structure 108, phase 7.
12 Decorated buckle plate; context 230, structure 135, phase 7.
13 Buckle loop; context 315, structure 123, phase 9–10.

Rings

14 Fragment with round section; context 959, structure 36, phase 5ii.
15 Cast in one piece with oval section; context 341, structure 41, phase 7?

Other objects

16 Fragment of decorative mount; context 552, structure 70, phase 4ii.
17 Tapering implement with perforated head (?large needle); context 552, structure 70, phase 4ii.
18 Perforated object; context 406, structure 36, phase 5ii.
19 Button with illegible inscription on one side; context 257, structure 173, phase 6.
20 Jetton of English type with arms of Evreux on obverse dated to *c* AD 1355–70 (Barnard 1916, 105, no 57); context 423, structure 98, phase 6.
21 Mount with linear decoration; context 226, structure 108, phase 6–7.
22 Double thickness of riveted sheet; context 292, structure 46, phase 8.
23 Lace-end; context 223, structure 108, phase 7 (not illustrated).
24 Lace-end; context 218, structure 133, phase 8 (not illustrated).

Discussion

No identifiable copper alloy objects were recovered from phases 1 to 3. A similar assemblage of scraps and lumps was retrieved from each of phases 4 to 5, except that there were three objects from contexts 552 and 406. However, from phase 6 onwards there was a wider range of artefact types. Thimbles and pins were relatively well represented in phases 6 to 8, probably reflecting the domestic nature of the site during this period. The jetton from phase 6 indicated a period before the 16th century, when reckoning in Roman numerals was achieved using a board and counters.

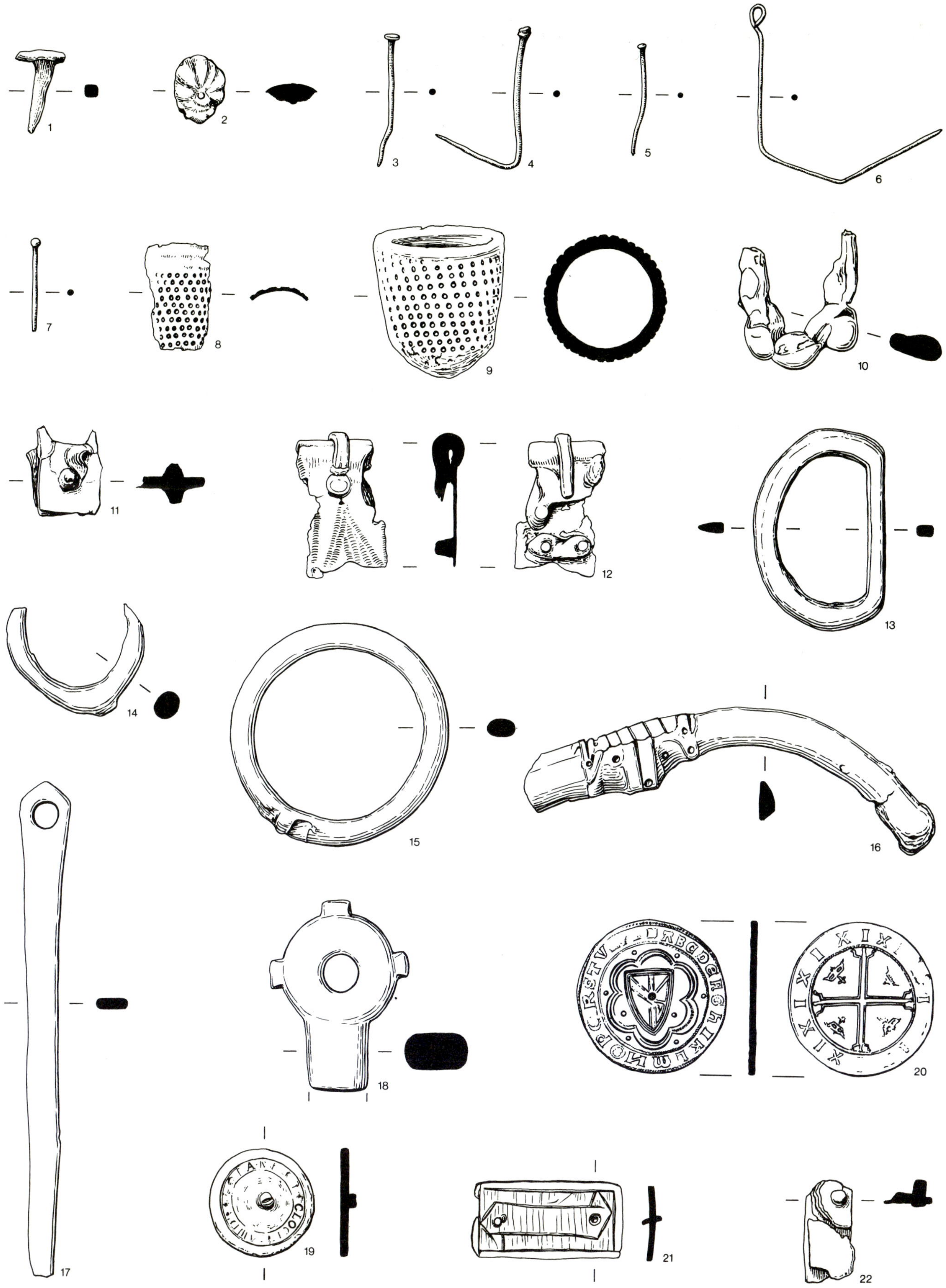

Figure 116 Copper alloy objects. Scale 1:1

29 Ironwork *Derek Hurst*

Fig 117

Tools

1 Possible point fragment; context 552, structure 70, phase 4ii.
2 Padlock key cf LMMC fig 45, no 3 possibly of late 12th century date; context 962, structure 36?, phase 5.
3 Ferrule; context 981, structure 81, phase 5.
4 Knife fragment; context 995, structure 39?, phase 5.
5 Possible knife; context 334, structure 42, phase 6 (not illustrated).
6 Half of shears of specialised late medieval type (cf LMMC pl 32, no 11) with maker's mark evident on radiograph; context 226, structure 108, phase 6–7.
7 Point; context 178, structure 150, phase 11 (not illustrated).

Of the tools, the shears were notable for displaying a maker's mark in the form of a small rosette. The implement type to which the ferrule belonged was unknown.

Weapons

8 Dagger chape; context 1103, structure 64, phase 3.

Personal ornaments

9 Possible buckle fragment; context 226, structure 108, phase 6–7.

Structural fittings

10 Lock plate with traces of plating in the engraved decoration around the keyhole; context 226, structure 108, phase 6–7.
11 Hinge fragment; context 226, structure 108, phase 6–7.
12 Rotary key (cf LMMC pl 31, no 52 where type dated to 15th century); context 182, structure 137, phase 7.
13 Wall hook; context 230, structure 135, phase 7.

Horseshoes

14 Horseshoe with wavy outline characteristic of the earlier type; context 1016, structure 19, phase 4.

Other horseshoe fragments from contexts 182 (S137, P7) and 222 (S108, P7).

Nails

Nails were the most commonly represented type of iron artefact throughout the site development (Table M31, 3:C1), except during phase 4. They appear to have become very common from phase 6 onwards in the Friar Street region of the town. A small cluster of six nails was associated with phase 3 (1103), and these had corroded together as if they had originally been tied.

Other objects

15 Plate fragment with traces of plating in fretting along one edge; context 366, structure 96, phase 6.

Other plate fragments from phase 6 (955, S116), phase 8 (218, S133), and phase 11 (32, S172; 85, S43; 89, S151; 174, S150 and 179, S51).

Discussion

The ironwork assemblage was generally too small to permit valid conclusions to be drawn. It was noticeable, however, that iron structural fittings were not represented until phase 6, when the domestic nature of the site had become well established. Nails also became more common during this phase, probably as a result of nearby buildings being erected or repaired (Chapter 21). Better quality ironwork with decorative plated finishes was also attested from phases 6 to 7.

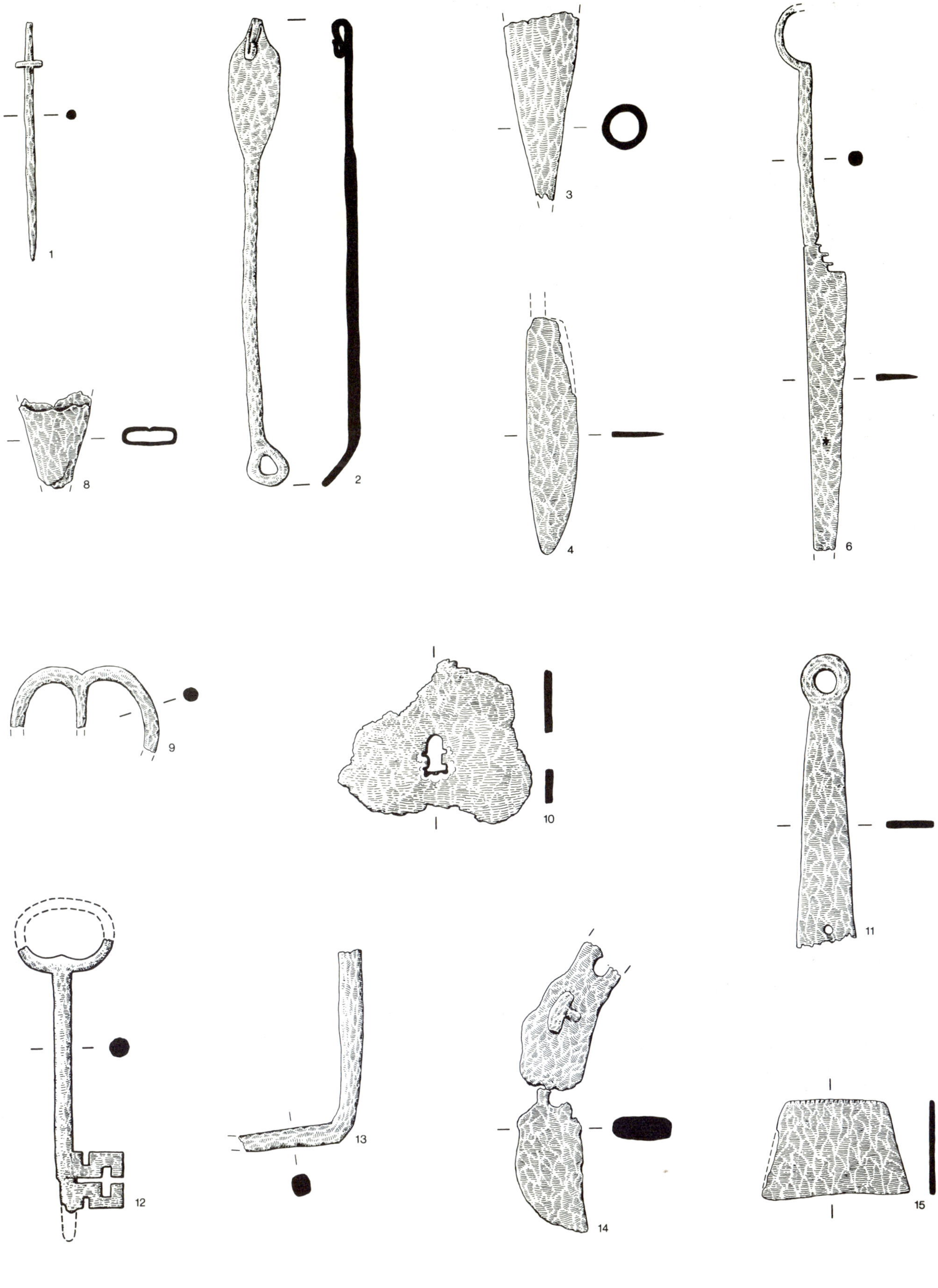

Figure 117 Ironwork. Scale 1:2

30 Lead *Derek Hurst*

Fig 118

1 Spoon with leaf-shaped bowl decorated with ribbed border and cross motif on the underside. The ribbing continues along the upper side of the handle, which is twisted along its length and has a knob terminal. This spoon is of early medieval type and similar examples are known from Beverley and Carlisle (Brian Spencer pers comm); context 1032, structure 38, phase 5ii.

2 Bar with hooked end; context 301, structure 128, phase 6.

Miscellaneous fragments

A number of small lead fragments (Table M32, 3:C2) occurred, especially in phases 5 to 7, and these were mainly pieces of lead sheet or wire, the latter probably being trimmings from larger sheets. They were most likely to be remnants related to the extensive use of lead by the Droitwich salt industry, though when this use commenced is, as yet, uncertain. However, evidence from other sites for the extensive production and utilisation of lead in the Roman period suggests that any association between the salt industry and the use of lead would probably date from this period.

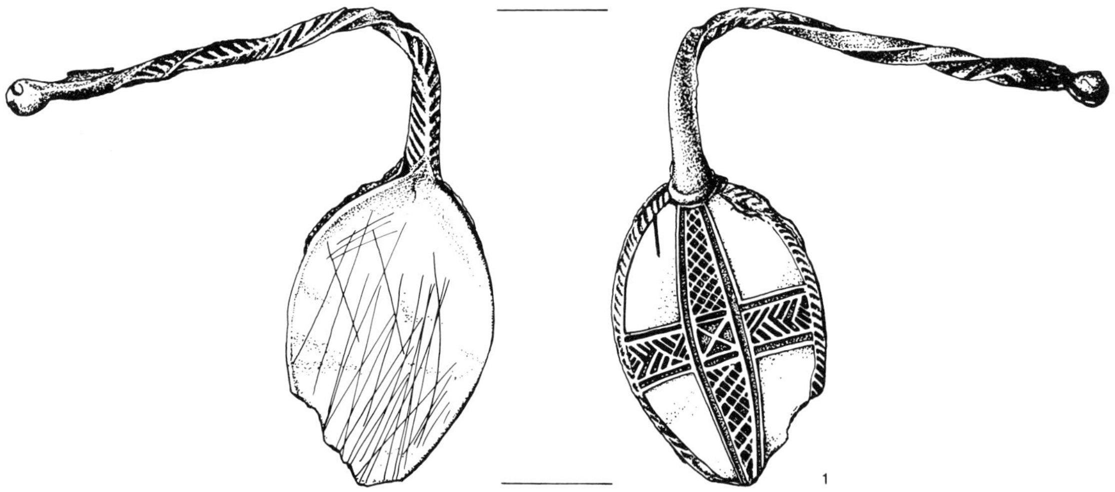

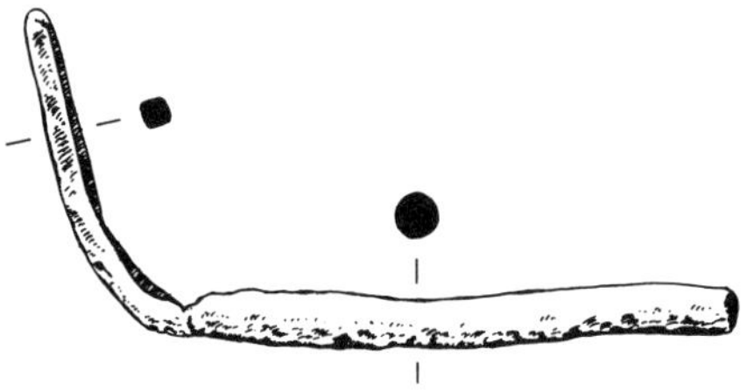

Figure 118 Lead objects. Scale 1:1 (1) and 1:2 (2)

31 Pyrotechnological note *Derek Hurst and Gerry McDonnell*

A scatter of residues from high temperature processes, mainly fuel ash, was recovered. However, the quantities were too low for there to be any suggestion that industrial activities were being carried out on, or in the immediate vicinity of the site. There were four examples of smithing slag, including one complete hearth bottom (226, S108, P6–7), from medieval levels. These seem to have been isolated finds redeposited from elsewhere, though in the case of the example from pit 995 (S39?, P5i) there was an association with ash and charcoal, suggesting that a larger scale dumping of waste had taken place.

Table 9 Quantification of smithing slag

Phase	Structure	Context	Weight (kg)
5–6	39?	995	0.15
6	42	333	0.504
6	96	366	0.438
6–7	108	226	0.812
		Total	1.904

Table 10 Quantification of fuel ash

Phase	Structure	Context	Weight (kg)
3	8	898	0.010
4	70	552	0.066
5	38	960	0.002
6	137	257	0.004
6	125	294	0.016
7	42	232	0.022
7	135	251	0.006
7	59?	353	0.064
11	149	88	0.010
		Total	0.200

32 Animal bone *Alison Locker*

Summary

Approximately 12,000 animal bones were identified (Table 11). The earliest phase was associated with brine extraction during the Iron Age, while later Roman occupation was of a domestic character. The bone samples from these early phases were small but statistically similar to the butchered bone debris from later occupation. Subsequent medieval activity included tanning and amongst the domestic debris a relatively small number of ox, sheep and goat horn cores were found. In the succeeding medieval and later settlement small numbers of horn cores were also found, but the great proportion of animal bone was domestic debris in which ox and large ungulate fragments were most numerous. Ovicaprid and small ungulates were the next most frequently occurring groups, followed by pig. Goat has only been positively identified from the horn cores. Small numbers of horse, dog, cat and less frequently red and roe deer, rabbit and hare were also found. Birds were largely represented by domestic fowl and goose. The final section compares the animal bone assemblages from Friar Street and the Old Bowling Green excavations.

Introduction

Approximately 12,000 bones were recovered from contexts dating from the Iron Age to the 17th century. Some later deposits also contained bone but these have not been recorded. Moreover, no sieving was carried out, with a consequent bias in favour of larger fragments.

The following species were identified: ox (*Bos* sp domestic), sheep (*Ovis* sp domestic), goat (*Capra* sp domestic), pig (*Sus* sp domestic), horse (*Equus* sp domestic), dog (*Canis* sp domestic), cat (*Felis* sp domestic), red deer (*Cervus elaphus*), roe deer (*Capreolus capreolus*), rabbit (*Oryctolagus cuniculus*), hare (*Lepus* sp), fowl (*Gallus* sp domestic), goose (*Anser* sp domestic), duck (*Anas* sp), wood pigeon (*Columba palumbus*), and crow (*Corvus corone*).

In addition some indeterminate groups were used, primarily large and small ungulate, which are most likely to be ox and ovicaprid respectively. The bone was recorded using the method of Jones *et al* 1981.

Within each phase a number of structures were recognised. Detailed tables indicate the number of bones for each species/group in the contexts comprising each structure (Table M33, 3:C3–D14). The animal bone from phases 9 to 11 was recorded but because of the relatively recent dates for these levels, analysis was deemed inappropriate.

In the following sections the occurrence and butchery of the three main economic species ie ox, ovicaprid and pig, are discussed. This is followed by the metrical and ageing data, and the occurrence of other species.

Phase 1: Iron Age

Animal bone was recovered from three of the five structures associated with brine extraction. In this small sample of 56 bones there did not appear to be any predominance of species or anatomy. Ox and ovicaprids were equally represented. Clear signs of butchery were also scarce; knife cuts were observed on the midshaft of an ox humerus, the acetabulum of an ovicaprid pelvis and midshaft of an ovicaprid tibia, and the midshaft of a pig radius. The cuts were probably the result of filleting. Two ox bones showed canid gnawing.

All the bone from this phase, with the possible exception of five bones of horse which were fragmented but did not show definite indications of butchery, was probably food debris. As activity on the site at this period was of an industrial nature, deposition of large quantities of animal bone refuse was unlikely.

Phase 2: Early to mid Roman

Extraction of salt continued into the 1st century, but the main activity was associated with early Roman occupation. Structure 6, a well shaft, contained the largest sample of bone. Here two skeletons of fox were found. Both mature animals, the corpses may have been conveniently disposed of in the well shaft, or may have had a ritual purpose. A number of complete skeletons were found at the base of a Roman well in Staines (Chapman pers comm) and the skeleton of a lap dog and another larger dog together with sheep skeletons were found in a 1st century shaft in Keston, Kent (Locker unpublished). There does seem to be some significance in the placing of dog carcasses (though not specifically fox) in well or shaft fills during this period. Other bones from this same context (1142), were eighteen fragments of ox, ovicaprid and pig, as well as six fish fin rays.

Ox and large ungulate fragments dominated the assemblage. An ox scapula was chopped across the neck and a tibia was chopped across the distal joint surface. Knifecuts were observed across the lateral side of the diastema of a mandible, and on the midshaft and distal surface of a metacarpal. On

Table 11 Animal bone summary

Phase	1		2		3i		3ii		4i		4ii		5i		5ii		6		7		8	
Ox	8	14%	26	7%	43	11%	2	2%	4	4%	140	10%	257	13%	90	12%	554	11%	58	10%	13	7%
Sheep	–		–		–		–		–		9	1%	13	1%	3	<1%	10	<1%	1	<1%	–	
Goat	–		–		–		–		–		–		–		2	<1%	2	<1%	–		–	
Ovicaprid	9	16%	11	3%	9	2%	7	8%	10	10%	94	6%	146	8%	59	8%	355	7%	59	10%	8	4%
Pig	2	3%	5	1%	11	3%	8	9%	6	6%	93	6%	110	6%	48	6%	292	6%	38	6%	8	4%
Horse	5	9%	2	<1%	6	2%	2	2%	1	1%	5	<1%	–		27	4%	120	2%	4	<1%	–	
Red deer	–		–		–		–		–		1	<1%	–		12	2%	1	<1%	1	<1%	–	
Roe deer	–		–		–		–		–		1	<1%	–		1	<1%	5	<1%	–		–	
Red/fallow	–		–		–		–		–		–		–		–		2	<1%	–		–	
Dog	–		1	<1%	4	1%	1	1%	1	1%	9	1%	19	1%	4	<1%	24	<1%	4	<1%	–	
Fox	–		156	45%	–		–		–		1	<1%	7	<1%	–		–		–		–	
Cat	–		–		–		–		–		5	<1%	7	<1%	2	<1%	11	<1%	–		–	
Rabbit	–		–		–		–		–		–		–		–		2	<1%	1	<1%	1	<1%
Hare	–		7	2%	–		–		–		–		–		–		1	<1%	–		–	
L Ungulate	10	18%	72	21%	145	37%	36	41%	31	30%	547	37%	614	31%	261	35%	181	37%	238	39%	75	40%
S Ungulate	11	20%	41	12%	48	12%	17	19%	27	26%	323	22%	467	24%	155	21%	693	14%	110	18%	46	25%
Fowl	–		–		–		1	1%	2	2%	28	2%	11	<1%	–		17	<1%	1	<1%	1	<1%
Goose	–		–		–		–		–		4	<1%	1	<1%	1	<1%	4	<1%	–		–	
Duck	–		–		–		–		–		–		–		–		–		–		1	<1%
Wood pigeon	–		–		–		–		–		–		1	1%	–		–		–		–	
Crow	–		–		–		–		–		1	<1%	–		–		–		–		–	
Bird	–		–		1	<1%	1	2%	1	1%	12	1%	17	1%	11	1%	13	<1%	2	<1%	5	3%
Fish	–		6	2%	–		–		–		9	1%	–		–		–		–		–	
Frog/toad	–		6	2%	–		–		–		–		–		–		–		–		–	
Unidentified	11	20%	15	4%	123	31%	12	14%	20	19%	156	11%	248	12%	79	10%	975	20%	91	15%	27	15%
Total	56		348		290		88		103		1458		1937		755		4898		608		185	

In phase 2 fox is represented by 2 skeletons

two ovicaprid pelves the acetabulum was chopped and knifecuts were also observed in this area.

Apart from the fox skeletons all the bone was domestic food debris, although none of the features appears to have been used for intensive bone waste disposal.

Phase 3i: 3rd to 4th century

In the 3rd and 4th centuries small scale domestic settlement was evident. Less than 400 bones were recovered from nine structures, although the greater concentrations of bone in three particular contexts (1116, 1103 and 1037) was a characteristic not apparent in earlier phases. They produced 282 bones, 85% in the large and small ungulate and unidentifiable mammal categories, which may indicate a high level of fragmentation. Again the ox and large ungulate categories were dominant. These larger samples of bone in particular features were more typical of settlement debris than in earlier phases.

Butchery marks were largely observed on ox; knifecuts were seen below the alveoli on the lateral side of two ox jaws. Two ox scapulae were chopped laterally across the neck. This area and the glenoid cavity are commonly chopped on ox scapulae, and is often complemented by chopping through the midshaft and or distal end of the humerus (the latter was observed in later phases). A calcaneum was cut around the proximal surface, possibly as a result of tendon cutting, and cuts were also observed around the midshaft of a metacarpal, providing evidence of skinning. An ovicaprid humerus had been cut around the midshaft. No chopping or knifecuts were seen on pig bones.

Table 12 Animal bone totals for each species

Structure	30	32	34	37	70
Total	149	91	541	44	633
Total for ox	21	10	55	10	44
Ox skull frag	3	2	3	5	2
Ox mandibles	2	2	4	1	1
Ox maxillae	3	–	1	1	1
Ox teeth	–	5	2	–	4
Ox horn cores	3	1	11	2	10
Ox metacarpal/tarsal	1	–	10	–	7
Sheep skull/horn cores	3	–	5	–	2
Goat horn cores	–	–	18	–	19
Total for ovicaprids	13	3	22	3	53
Ovicaprid skull frag	–	1	2	–	3
Ovicaprid mandibles	4	1	6	1	11
Ovicaprid maxillae	3	–	1	–	2
Ovicaprid teeth	2	–	–	–	9
Ovicaprid horn cores	–	–	3	–	–
Ovicaprid metacarpal/tarsal	–	–	1	1	6

Phase 3ii: Sub-Roman and Anglo-Saxon

Tables in fiche indicate the small quantity of bone from this period. Linear features thought to be associated with field boundaries or soil cultivation were defined and might not be expected to produce significant quantities of bone. Only 88 bones were recovered from the eight structures. Of the entire sample 76% came from a pit (S28).

Few examples of butchery were evident; a large ungulate scapula was chopped across the neck, a large ungulate humerus had been split and a pig radius showed knifecuts across the shaft.

A high degree of fragmentation was evident (even with only hand picked bones and no sieved samples) with 74% of all bone attributed to large or small ungulates or unidentifiable mammals. This intensive fragmentation could reflect abrasion during soil cultivation.

Phase 4i: Earlier Saxo-Norman

The first part of this phase was associated with a series of pits and ditches. The numbers of bones from structures 22, 31/158, and 170 were insignificant (three, thirteen, and two respectively). Eighty-five bones were recovered from structure 19. No butchery was observed, although some bones showed evidence of gnawing (S19 and S22) on ox, ovicaprid, pig, large and small ungulates.

Phase 4ii: Later Saxo-Norman

During this phase a series of features thought to be associated with tanning was identified. Structures 30, 32, 34 and 37 were pits and structure 70 was composed of black clay and loam layers. Of the 1,458 bones, 79% were recovered from structures 34 and 70. In structures 30, 34 and 70, ox and large ungulates dominated numerically. However, in structure 32 ovicaprid and small ungulates were more numerous.

Apart from domestic food refuse, skull fragments and horn cores from ox, sheep and goat were found. These were probably debris from tanning where the untanned hide may be acquired with the horns still attached. Partial skulls and horn cores of sheep and goat were only found in structures 34 and 70, whereas ox occurred in all five structures. Table 12 compares the number of skull fragments and associated pieces with the total number of bones for ox, sheep, goat and ovicaprid.

It is evident that although there were relatively high instances of horn cores, particularly for goat and ox in structure 34, they may not be significant within the overall bone sample. Similarly there were not high numbers of metacarpals and metatarsals, which are often associated with tanning waste.

Part of an ox skull from structure 30 may have been poleaxed, judging from the fracture of the frontals, while other ox horn cores had been chopped at the base or showed cut marks there. A pair of goat horn cores had been chopped away from the neurocranium, and others had been chopped or cut at the base. Similarly sheep horn cores showed evidence of chopping and cut marks at the base. The horn from these cores may also have been utilised. One sheep skull was naturally polled.

The skulls and horn cores only totalled 70, less than 5% of the bone sample from this phase. The remainder was largely food refuse except for a few bones of horse, roe deer, dog, cat, and a fragment of a red deer antler. One of the changes in butchery practice seen in this phase and onwards was the more frequent occurrence of axially splitting pig skulls and long bones. A high incidence of butchery was seen on ox and large ungulate bones, and to a lesser degree on ovicaprid and pig (552, S70). Ox scapulae were split, a metacarpal was split proximally and a humerus chopped across the midshaft. Gnaw marks were also quite common on ox, ovicaprid and pig.

The identification of goat in this and later phases was solely from horn cores. Boessneck's (1969) indices for separating the metacarpals and metatarsals of sheep and goat suggested that these were all from sheep except for one metacarpal which was just within the range for a goat. So it would appear that goat was only present in the context of tanning.

Although the archaeological evidence indicated that tanning took place in this phase the small proportion of horn core and associated remains compared with the domestic food debris is echoed in later phases. The majority of hides were possibly brought without horn cores attached or, if present, they were disposed of elsewhere. As there were only small numbers of metapodials it may be that the horn cores indicate horn working waste.

Phase 5i: 12th century

The main activity of this phase was associated with drainage and rubbish pits. Structures 29, 33, and 35 yielded the largest sample of bones. The remaining structures were layers, except for a pit (S83) which only contained seven bones.

Forty-five ox horn cores and four partial skulls were found. These were in a variety of contexts and showed no particular concentrations. Sheep and goat were less frequent, ie one horn core and one partial ovicaprid skull. One horn core and eight skulls were positively attributed to sheep and four to goat. These were not restricted in their distribution but formed only 4% of the total bone sample. However, there was only a 1% difference between the proportion of partial skulls and horn cores in this phase and phase 4ii. As the features associated with tanning comprised all the activity within phase 4ii the animal bone alone would not appear to be a reliable indicator of tanning.

An almost complete ox skull was found (971, S35) in which all teeth were in wear and the horn cores were of a 'short horned' type. From the same structure, two sheep skulls had the horn cores chopped off. Some of the ox horn cores had been chopped at the base.

Butchery marks were most clearly observed on ox. Knife cuts were seen on the diastema and below the alveoli on the mandible. Chop marks were also seen at the latter point. The scapula was chopped across the glenoid cavity and across the neck. The mid point of the shaft of the humerus was also chopped and cut, along with the distal articulation of the bone. The radii were occasionally split axially and knife cuts were apparent on the proximal surface, as were chop marks across some of the midshafts. The ulna was frequently chopped on the joint surface. Metacarpals and metatarsals were infrequently butchered, sometimes on the proximal surface and across the midshafts. Phalanges were largely complete. On the hind limb the pelvis was chopped across the top of the ilium, across the acetabulum and across the ischium. Femora were generally fragmentary, but knifecuts were observed on the proximal and midshaft areas. Tibiae were chopped about the midshaft or split. Ribs were chopped into sections or had cut marks on them and vertebrae were trimmed laterally. This butchery pattern for ox is typical for the post-Roman periods, largely reflecting the preparation of the carcase into joints.

Within the assemblage of ovicaprids, one scapula was chopped on the glenoid cavity, knifecuts occurred around the midshaft of the humerus, with chop marks through the distal end and two radii were split. On the hind limb a pelvis was chopped across the ilium, and a femur had been chopped across the midshaft. Limb extremities were generally complete.

A pig humerus was chopped proximally and across the midshaft, and cut marks were seen across the midshaft of a radius, and one radius was proximally burnt. All three species were subject to gnawing.

Phase 5ii: Early 13th century

A total of 692 bones was recovered from the property boundary ditches (S36), including two ox, three sheep and two goat horn cores. The remainder were associated with domestic food debris. Ox and large ungulate fragments dominated the assemblage. The butchery mirrored that seen in phase 5i.

A humerus of roe deer, rarely identified from this site, was found in the larger of the two enclosure ditches. Twelve red deer antler fragments were recovered from the smaller north-south enclosure, one of which had been chopped at the base. However, this species was not represented by any post-cranial fragments.

Phase 6: 13th to 14th century

A total of 4,894 bones was recovered from eighteen structures, 65% of which were found in structures 42 and 108. A number of partial skulls and horn cores from ox, sheep and goat were recovered. A particular concentration of ox horn cores and skulls (23 out of a total of 45) were found in structure 42, along with one horn core of a sheep and one of a goat.

In some contexts there were relatively high concentrations of ox metapodials, for example in structure 42. Of the 58 bones positively attributed to ox, ten were metacarpals and thirteen were metatarsals. Similarly, of the 112 ox bones present in structure 42, eighteen were metacarpals and seventeen were metatarsals. Although these bones are easy to identify, and are easy to attribute to ox, rather than to large ungulates, they are also less fragmentary than other long bones bearing little meat. Specific disposal of metapodials seems to have taken place, rather than whole feet, since the metapodials were not matched by two first, two second and two third phalanges for each bone.

Ox and large ungulates were dominant in the phase, particularly in the structures containing larger quantities of bone. For example, in structures 42, 96 and 108, ungulates respectively formed 57%, 55% and 38% of the total. The butchery of ox, ovicaprid and pig was generally similar to that noted in phase 5ii, but included chop marks around the gonion and alveoli of ox mandibles and showed indications of having been gnawed by dogs.

Phase 7: 15th to 16th century

Only 608 bones were recovered from twelve structures; a reflection of the purely urban residential use of the site in the 16th century.

Ox and large ungulates dominated the assemblage. Only ox and one sheep horn core were recovered. A fragment of red deer antler was found in context 230 (S35). The butchery practices were not notably different from the two preceding phases.

Phase 8: 16th to mid 17th century

Only 185 bones were recovered, largely from structure 46, with twenty or less bones from structures 118, 129 and 166. Ox and large ungulates were numerically dominant. A few bones were butchered and gnawed. The bone largely represented food debris, with the exception of some loose teeth and limb extremities.

Ageing of ox, sheep and pig

Mandibles were aged using Grant (1975) and the values calculated shown in Table 13. Where a phase is absent from the table no ageable mandibles were found in it. The high proportion of estimated values is indicative of the number of incomplete mandibles. For a mandible to have an estimated value at least two permanent teeth were present and then compared with the range of tooth rows shown by Grant.

Until phase 5i the sample size for ox is very small. If the third permanent molar erupts at around value 30, after two years, it would appear that there is a higher proportion of animals slaughtered at less than two years in phases 5 and 6 than in the preceding phases. Although the absence of younger animals in phases 2, 3 and 4 could be a feature of small sample size, it might indicate a use for milk, traction, etc, for a longer period before slaughter. In the later periods there appears to be a greater surplus and therefore young animals can be killed for meat, although the small size of the sample makes any interpretation tentative.

Sheep (assumed from the scarcity of goats indicated by the metapodial index) represented a larger sample in phase 4ii than for ox. If between one and a half and two years is within the range of 20 to 25, no specific patterning is discernible, except that where the sample size is large enough (ie from P4ii onwards) a certain proportion of mandibles are slaughtered at less than eighteen months. The overall low proportion of mandibles of less than value 25 could support an expected emphasis on wool production, but the samples are too small to make any further comment.

Similarly the pig mandibles, where 20 to 25 indicates the end of the second year, from phase 4ii onwards (the sample size is too small in preceding phases) a small proportion of animals are slaughtered at less than two years.

Metrical data

Full details of all the metrical data are available in archive. In Table 14 the withers heights for ox and sheep are displayed. Some of the smallest oxen seem to occur in phase 4ii and 5, which could accord with the reduction of size in the medieval period.

The horn cores of ox were measured along the length of the outer curve when complete and across the maximum and minimum diameter at the base. The lengths indicated that the 'short, medium and long horn' types identified by Armitage and Clutton-Brock (1976) were present. This may suggest that the metapodial size differences are more related to breed.

For sheep withers, heights are greater in phase 5ii than in 5i or 6, and may be a reflection of a

Table 13 Ageing of ox, ovicaprid and pig mandibles (using Grant 1975)

Phase 2	**3**	**4ii**	**5i**	**5ii**	**6**
Ox					
49–50e	25e	11–12e	12–21e x2	20–32e	4e
	31	30	13–25e	27e	8–25e
	37–49e	32e	20–32e x2	31–39e	20–32e
	42–49e	39–40e	21–26e	33–34e	22e
	45–48e	46–50e	23–26e	35–38e	23–26e
			25	41–46e	25e
			26 x2	41–50e	26e x3
			27 x2	44e	27e
			30		29e
			32		32
			34 x2		34
			38		37–39e
			39 x2		37–46e
			40 x2		38e
			40–59e		42–46e
			41–46e		44
			42		46+ex2
			42–47e		47
			46–47e		49+e
			47		5+e0
					51+e

1	**4i**	**4ii**	**5i**	**5ii**	**6**	**7**
Ovicaprid						
32–34e	29–36e	3–6e	5–22e	8e	7e	8e
		4–12e	11e	26e	24e	8–10e
		7–8e x2	21e	36+e	28e	30e
		12–20e	25–26e	41	31e	36–38e
		23–25e	30e		32e	
		30–32e	31		32–37e	
		30–36e	32		33 x2	
		32	33e		34	
		33	36		35e	
		36e	38 x2		36 x3	
		37	39 x3		39–41e	
		38	41e		41e	
			41+e			
			47+e			
			48e			

e = estimated value or range due to incomplete jaw

Table 13 (continued)

Phase 2	3i	4ii	5i	5ii	6	7
Pig						
26–30e	41+e	13–14e	10–15e	24–31e x3	13+e	10+e
	3ii	22e	17e	27e	20–26e	27–31e
	36	27e	19e	29e	21e	29e
		30e x2	28e	32e	23–25e	33e
		31	29		24–26e	
		31e	30e		26e x2	
		34–38e	31e		26–33e	
			37e		28e	
			40e		29e	
					31e x2	
					31–33e	
					34e	
					35e	
					35–38e	
					38e	
					42+e x2	

change in site function from 5i to 5ii, although activity was clearly domestic in phases 5ii and 6.

For pig, one of the more common measurements is displayed, the width of the proximal joint surface of the radius indicating a particularly large and possibly wild specimen in phase 4i.

Occurrence of other species

Horse

Horse bones were recovered from contexts in all phases. These were in small quantities and the bones relatively complete. In Table 14 it is clear that estimated withers heights of between 13 and 14 hands indicate ponies rather than horses. These were largely mature except for a mandible in which the third incisor had not erupted and was aged less than four years (Thompson 1949, 240). Knifecuts were observed on the ilium of a pelvis from context 453 and from the midshaft of femora from contexts 475 and 476 (all in P5i), and from the midshaft of a tibia from 412 and of a metacarpal from 366 (both in P6). The proximal end of a metatarsal from 233 (P6) was chopped and indeterminate metapodials were split from 338 (P6) and 552 (P4ii). There is no real evidence that horse was eaten; these marks may just be associated with skinning and or dismemberment of the carcass. Exostoses were observed on the proximal joint surfaces of two metatarsals from 460 (P5i). These may be a pathological reaction to stress.

Dog

Dog bones were frequently found as isolated occurrences in many features. They were all from adult animals except for one proximally unfused humerus from context 460 (P5i). The shoulder heights were calculated using Harcourt (1974) and are shown in Table 14. None were extreme in size at either end of the scale. The fox skeleton from 1142 (P2) has already been described.

Other mammals

The occurrence of red deer, only represented by antler, has already been noted in the phase descriptions. Roe deer was represented by two post cranial bones, a mandible and a radius. Cat occurred in a number of contexts as disarticulated bones. All have been measured and the data is available in the metrical archive. None of the cat bones had any cut marks that might indicate skinning. Rabbit was identified from a few medieval and later contexts, as was hare. Both were probably eaten.

Birds

Domestic fowl was the most commonly identified species. Some bones were immature and porous and others had been butchered. A fowl tibiotarsus had been chopped distally and a tarsometatarsus had knifecuts across the midshaft (585, P4ii). From context 959 (P5ii) a fowl tibiotarsus had been cut distally and a goose tibiotarsus cut across the

Table 14 Withers heights in cm for ox, ovicaprid, pig, horse and dog (using the factors of Fock, Teichert, Kiesewalter and Harcourt respectively)

Phase 1	**4ii**	**5i**	**5ii**	**6**	**7**
Ox					
mc 118	mc 102	mc 104	rad 113	mc 116	mc 107
Phase 2	mc 110	mc 105	mc 119	mc 106	**Phase 8**
tib 114	mc 99	mc 104	mc 115	mc 94	mt 109
mt 113	mt 102	mc 104	mt 106	mc 108	
mt 112	mt 107	mc 109	mt 97	mc 116	
Phase 3i		mc 119	mt 108	mc 110	
mc 127		mc 115	mt 105	mc 121	
mc 103		mc 98		mc 103	
		mc 106		mt 103	
		mt 113		mt 114	
		mt 114		mt 102	
		mt 110		mt 108	
		mt 112		mt 111	
		mt 108		mt 105	
				mt 105	
				mt 119	

4ii	**5i**	**5ii**	**6**
Ovicaprid (Boessneck's index for separating metapodials suggests these are all from sheep rather than goats)			
mt 56	rad 57	rad 64	hum 56
	mc 57	mc 61	rad 52
	mc 59	mc 67	mc 55
	mt 52	cal 61	mc 51
	mt 53	mt 59	mc 53
	mt 59		cal 54
			cal 52
			mt 51
			mt 60
			mt 58
			mt 53

3ii	**4i**	**5i**	**5ii**	**6**
Pig (The proximal width of the radius in mm is shown since shoulder heights cannot be estimated)				
30	37	31	29	27
	25	27		28
		25		24
		28		23
		27		26

Table 14 (continued)

1	4ii	5i	5ii	6
Horse				
rad 136	mt 134	rad 143	mc 141	rad 142
			mc 133	mc 140
			mt 135	mc 137
				mc 144

4ii	5i	6	7
Dog			
tib 46	hum 56	fem 30	hum 48
	hum 36		
	hum 36		
	rad 39		
	tib 48		
Dog/fox skeleton from well fill 1142 approx 39 cm at the shoulder			

Key:
hum = humerus rad = radius mc = metacarpal fem = femur tib = tibia cal = calcaneum mt = metatarsal

midshaft. A goose tarsometatarsus from the same phase was cut across the proximal midshaft area. All these cuts are likely to have been made in preparation for the table. As well as fowl and a few goose bones, pigeon and wood pigeon were identified from phase 5i. Two crow bones from phases 4ii and 6 may be incidental.

Comparison between the animal bone assemblages from Friar Street and the Old Bowling Green

The occupation pattern of the two sites is very similar. From the Iron Age (and in the case of the Bowling Green possibly the Late Bronze Age) into the Roman period, salt production was the main activity. The animal bone forms part of the disuse fill of the associated features. After relatively little evidence from the Anglo-Saxon period, occupation continued from the medieval period (including buildings, pits, ditches and depressions).

Recovery methods at both sites were similar: no sieving was carried out and hand collection will have favoured recovery of the larger fragments and species.

At both sites the animal bone samples are dominated by ox and large ungulate fragments. Of secondary occurrence numerically are ovicaprid and small ungulates. The large and small ungulate categories refer to bones which probably belong to ox and ovicaprids respectively but, because of heavy fragmentation, poor preservation etc, could not be specifically identified. Some goats were identified from the horn cores but, judging from the separation index used for metapodials, other parts of the skeleton seem to be mostly sheep.

Small concentrations of horn cores from ox, sheep, and goat were associated with evidence of tanning from Friar Street, phase 4ii of late 11th to early 12th century date. However, skulls and horn cores were less than 5% of the bone from the phase. Similarly there were small quantities of horn cores from other phases where there was no evidence of tanning activities, but goat was not identified earlier than the medieval period. Horn cores from ox, sheep, and goat were also found in small numbers in deposits of Roman date and later at the Old Bowling Green, mixed in with other domestic butchered debris.

Specific concentrations of certain anatomies were difficult to identify. This may be influenced in part by the small size of the samples when divided by phase and structure. However, at Friar Street in phase 6, of 13th to 14th century date, some contexts contained relatively high concentrations of ox metapodials. This can sometimes be related to tanning waste, however in this phase there is no other archaeological evidence for it. In terms of meat quality these limb extremities are poor, although they do contain some marrow.

At the Old Bowling Green in phase 7, comprising a late Roman stone building and access track, the relatively high concentration of ox, loose teeth and skull fragments is thought to reflect a high level of fragmentation. In phase 6 (late Roman) the disuse fills of barrels contained a high proportion of ox skull fragments and mandibles. In phase 9, of similar date but a different location, a linear cut contained high proportions of ox skull, mandible, and scapula fragments. There is therefore some evidence to suggest the specific disposal of ox heads, which have been butchered extensively to remove the horns, brain, tongue and marrow (from the mandible), over all three areas occupied in the late Roman period. For ovicaprids an increased proportion of mandibles were found in phase 4 and possibly 5, compared with other parts of the carcass. However, apart from ox and large ungulate, other species/groups were too sparse to indicate any particular distributions.

The butchery practices on the two sites are also similar. Any superficial differences are largely attributable to the sites being recorded by two people. Although the same recording system was used, personal preference for the level and order of recording influences the data.

In conclusion, the animal bone from these two closely related sites suggests that, from the Iron Age until the post-medieval period, ox and large ungulate remains are generally the most numerous of all species/groups, this trend increasing through time. In some instances there is evidence for specific disposal of certain parts of the carcase. Ovicaprid (both sheep and goat) and small ungulate are the second most commonly occurring species/group, with pig occurring in small numbers at both sites. The bones generally come from disuse fills and therefore would not be expected to reflect any changes in site use (ie from brine extraction to settlement or agriculture), except for the association of horn cores with tanning at Friar Street. Remains of horse, red deer, domestic fowl and goose were consistently found but in small numbers. Dog was represented both by isolated bones and partial skeletons at both sites.

33 Worked bone *Derek Hurst, with bone identification by Alison Locker*

Fig 119

1 Toggle or bobbin with central perforation, made from pig metapodial; context 308, structure 74, phase 5.

2 Perforated ovicaprid scapula; context 997, structure 39?, phase 6.

3 Tip of point with polished surface; context 223, structure 108, phase 7 (not illustrated).

4 Musical pipe made from horse or possibly ox metatarsal; context 232, structure 42, phase 7.

5 Handle notched at butt end; context 251, structure 135, phase 7.

6 Plain handle; context 240, structure 47, phase 9 (not illustrated).

7 Button. Flat disc with central perforation and stained copper green; context 240, structure 47, phase 9 (not illustrated).

8 Button. Dished with four symmetrically placed perforations, and also stained green; context 948, structure 41, phase 9 (not illustrated).

9 Double-sided comb fragment; context 7, structure 149, phase 10.

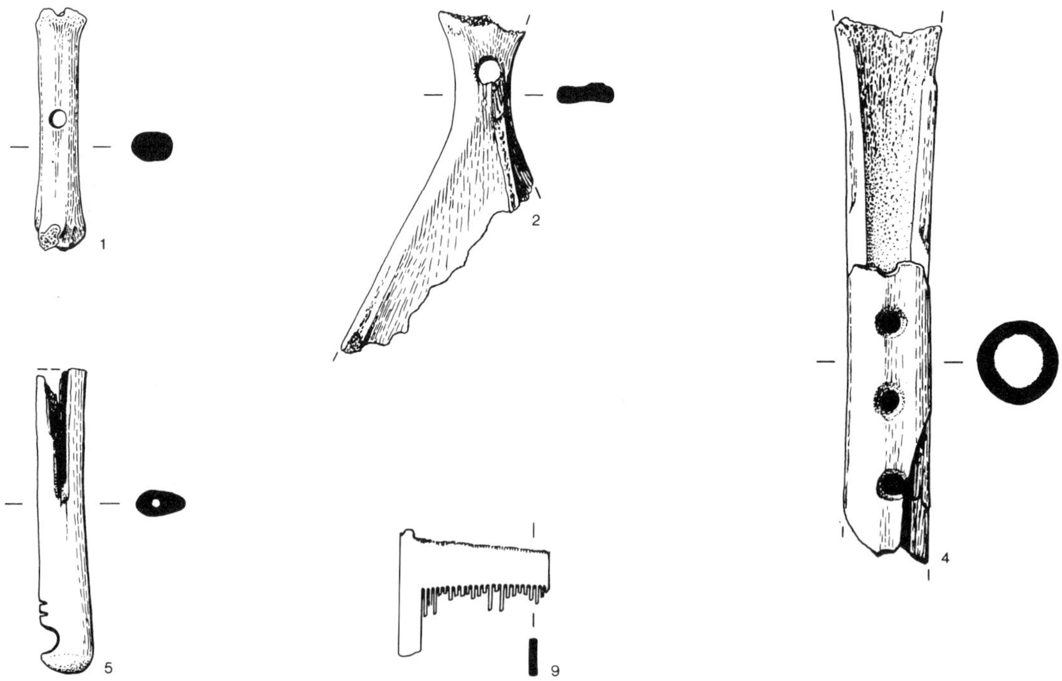

Figure 119 Worked bone. Scale 1:2

34 General discussion *Simon Woodiwiss*

This section aims to place information derived from the excavations into the context of contemporary research. It also aims to suggest new directions for future research. General background information is contained in the introduction to this volume, however, for the historic period discussion of the excavations is integrated with an outline of the documentary and topographical evidence. The excavations yielded especially useful information on two aspects, late Iron Age and early Roman salt production, and the development of the Saxon and medieval town.

Late Iron Age and early Roman salt production

Being soluble, salt leaves no trace in the archaeological record, except in very dry conditions not usually encountered in this country. When produced by the 'open pan' method, salt produces few waste or by-products which may easily be recognised in the archaeological record. Silt (from settling of sediments contaminating brine), water vapour (from boiling brine) and fuel waste (from hearths) are all produced, but are hardly unequivocal identifiers of salt production. Additionally, few of the structures or implements used in salt production are solely characteristic of that activity. There is however an exception to this in the Iron Age, as Morris (1985) has conclusively argued for the association of briquetage with salt production and export. However, the precise role of these vessels in the production process is not yet clear. The recording of large quantities of briquetage (especially at the Old Bowling Green) indicates that it formed part of the industrial process involving much wastage. Here an analogy may be suggested with the wastage of moulds often incurred in the casting of metals. Briquetage itself is closely associated with the hearths and tanks recorded at Friar Street and the Old Bowling Green, confirming their involvement in the salt production process. The association of the clay- and timber-lined tanks and the hearths with salt production, rests on their association with briquetage as well as the proximity of the raw material.

No later Roman equivalent of briquetage has been identified, and the structures of the Old Bowling Green's phase 8 (especially the lines of barrels set into the ground) cannot be directly related to salt production in the same way as the tanks and hearths of earlier phases. The evidence for the function of the barrels of phase 8 is inconclusive, however, secondary industries (such as preservation of meat and processing of leather) are to be expected in an area of primary salt production. The failure to identify the processing of animal products in the bone assemblage of phase 8, as an alternative interpretation of the barrels, may also be invoked to support their interpretation as part of the process of salt production. However this is hardly conclusive evidence for drawing an analogy between the barrels of phase 8 and the brine tanks of phases 2 and 3. It is possible that lead was used to construct pans for boiling and its presence from phase 7 onwards, albeit in very small quantities, may be significant. Additionally, the equivalent of briquetage used in phase 8 may have been made from a less durable material, and perhaps the use of wicker baskets for draining dates from the Roman period. However, in the absence of briquetage or an equivalent diagnostic feature of salt production, the function of the structures of phase 8 is suggested solely by the availability of brine. Given this absence of conclusive evidence, the discussion of salt production will consider only the Iron Age and early Roman periods. Consideration of the later salt industry must await further work.

The process of salt production

The general process of salt production by the open pan method (outlined in the introduction to this volume) may be used here, as no evidence was found to suggest that it may be invalid. As brine extraction did not occur in either of the two excavated areas, detailed discussion of the process must start with the brine tanks. Their presence, if used for settling, does indicate that brine was being drawn from sources which would have required removal of sediment prior to boiling (ie unlined or poorly lined pits or springs). The field record was not sufficiently detailed to identify the fine silts which may be expected to have existed if settlement had occurred and the tanks had not been cleaned. On the subject of post-medieval salt production, Rastel (1678, 1062) and Agricola (Hoover and Hoover 1912, 548) mention the close spatial relationship of stored brine to the boiling hearths. A similar relationship can be demonstrated for the late Iron Age and early Roman periods at Droitwich. The functioning of briquetage vessels for the drying of salt crystals, argued by Rees (1986, 51) from evidence from the Old Bowling Green, is still the most obvious. Further experimental work may help to identify a more precise function for these vessels in their industrial context. A number of wooden artefacts (forked and pointed stakes, the shovel, and perhaps the inscribed pottery) are likely to be associated with salt production but their precise function remains uncertain.

Though the evidence from coastal salt production sites is also incomplete, there is a general correlation with the inland process. Variation in details such as the presence of hearth furniture and boiling pans from coastal sites, and their absence from inland ones, do however occur.

Topography

The development of the town's non-industrial topographical aspects will be discussed in a later volume. In the late Iron Age two separate areas of salt production may be identified. The substantial area in the western half of the Old Bowling Green void of features associated with salt production separates the rows of brine tanks (probably aligned with the river) from those of the Friar Street excavation. There may also have been other areas and it would seem plausible that each had separate access to points of brine extraction. The similarity of dendrochronological dates further indicate that salt production in these two areas was contemporaneous.

The alignment of the tanks and furnaces with the river at the Old Bowling Green may reflect the importance of the river for transport of exported salt. As mentioned earlier, the river was navigable in the medieval period (Whitley 1923, 1).

The character of early Roman salt production appears to have changed little from that of the Iron Age, though for how long it continued and whether it ended abruptly or not is uncertain. There is less 1st century samian and coarse ware than might be expected, supporting the view that the areas covered by the Friar Street and Old Bowling Green excavations retained an industrial, rather than a domestic character. Droitwich was also at the junction of a number of Roman roads, providing another means of export. By the 2nd century the Friar Street area was no longer used for the production of salt, though this may have continued in some form in the area of the Old Bowling Green.

Organisation

Megaw and Simpson (1979, 421) have summarised the situation as '... by the end of the Iron Age there was in southern and eastern Britain a complex urban society, with a highly organised system of production and distribution, market centres where a small denomination coinage was used in exchange, and a political authority which minted coins and kept records.'

Whilst Droitwich does not form part of the area referred to, some aspects of the level of economic organisation are comparable. Extensive cropmarks in the Severn and Avon valleys (eg Webster and Hobley 1964), a proportion of which must be Iron Age in date, indicate a significant intensity of agricultural exploitation and settlement, at least in certain areas. Excavation of the mainly middle Iron Age settlement at Beckford has demonstrated intensive agricultural activity which continued into later periods (Wills pers comm). Coinage was also in circulation in the area of the Dobunni (Cunliffe 1978, 109). Though this does not include the small denomination silver minims and bronze coins alluded to by Megaw and Simpson there is a range of gold and silver denominations with some of the silver coins being quite base (Sellwood 1984, 196). The existence of an Iron Age trade in pottery from several centres in the Malvern area has been demonstrated by Peacock (1968) and its production continued into the Roman period (Peacock 1965–7). At Droitwich, large-scale salt production is indicated by the number and capacity of the brine tanks and hearths. A degree of organisation is also indicated by the definition of at least two different salt making areas, and in the methodical layout of the features within them, especially from the Old Bowling Green. The distribution of salt from Droitwich, suggested by the distribution of briquetage (Morris 1985, fig 6), is extensive. These indications of an organised and large-scale industry do not help in identifying the detailed processes and participants involved. This will be resolved only with the investigation of the non-industrial aspects of the settlement at Droitwich, when questions on seasonality of production, the presence or absence of specialised salt workers and how salt was distributed, may be addressed. There is at present little information that may help with this. For example the relationship of salt production at Droitwich with the surrounding hillforts cannot yet be discussed. Hanbury (Bassett and Dyer 1980, 88; HWCM 815), the closest (and then only tentatively identified as a hillfort), has not been excavated; the closest excavated hillfort being Midsummer Hill (Stanford 1981; HWCM 931).

Distribution

Before discussing the distribution of salt from Droitwich, the biases to be expected from non-cultural variables should first be considered. This report is not the place for detailed consideration of even such an important research concern but some comment is necessary. Figure 4 defines the areas lying closest to their respective known sources of salt. To these variables of location of raw materials and distance, may be added that of ease of production. For instance Droitwich brine, with its 25% solution of salt, would have greater economic potential than sea water (at 3%), expanding its potential area of influence at the expense of coastal sources. Ease of transport would also have favoured some sources above others and rivers may be assumed to have had as great an effect on the distribution of salt as they had on many other commodities. Physical impediments to transport are another consideration. For instance, the mountains of central Wales may be expected to have affected the distribution of salt from both the Welsh coast (if salt was produced there) and Droitwich. If these non-cultural variables were systematically assessed it seems probable, even from subjective observation, that a bias in the distribution of Droitwich salt may be expected in favour of the east, south and west.

The operation of cultural variables on the distribution of Droitwich salt is much more complex. The contemporaneity of source exploitation, and any resulting competition, is one obvious variable to be considered. Is, for instance, the restricted northward distribution of Droitwich briquetage (Morris 1985, fig 6) to be explained by the contemporaneous exploitation of the Staffordshire salt field (for which there is no evidence) or, as suggested by Morris (1985, 373), by a more successful distribution from a competing salt industry in Cheshire? Availability of the raw material alone does not necessarily imply that it was being exploited. Coastal exploitation appears to be limited to certain areas. However, these are areas where larger-scale production has made them more easily visible (eg Dorset, Hampshire, Essex, Sussex and Lincolnshire) and the production of salt on a smaller scale may not yet have been recognised.

Salt derived from brine is more likely to be produced at source due to the relative difficulty of transporting liquids, as opposed to the solid materials used in salt production (fuel etc). Thus the possibility of the transportation of brine and the production of salt at some distance from the source may be assumed to be unlikely.

If salt may not be considered a purely subsistence commodity, in so far as physiological needs may be easily met in the ordinary diet (Carter 1975, 13), its large-scale production and distribution indicates a greater economic and social role. Of greatest importance is perhaps its use in the preservation of meat (Cunliffe 1978, 300), tanning and other manufacturing activities. The identification of the salting of meat in the archaeological record is difficult, if not impossible. Bradley (1975, 22) has, however, argued for the likelihood of its occurrence through seasonal exploitation of coastal resources and the presence of cattle on coastal sites. This cannot be applied to Droitwich in quite the same way though salt production may well have been a seasonal activity, perhaps also related to seasonal culling. Bradley (1984, 146) has already suggested its role in the early establishment of links, later followed by other commodities on the south coast. The historically documented use of salt as a form of currency (cf Alexander 1975, 82) may also suggest a similar function in the Iron Age, though this would again be very difficult to identify archaeologically. The limited areas of large-scale salt exploitation in the Iron Age include the developing regions of Wessex and Essex, with their contacts with the Continent, as well as to the cultural areas of the Dobunni and Coritani.

Morris (1985, 373) has demonstrated that inland salt production and distribution occurred throughout the latter half of the 1st millennium BC and increased in intensification, a similar pattern to contemporary coastal sites. In the late Iron Age the distribution of briquetage (Morris 1985, fig 6) is rather restricted compared to that of Dobunnic coins (Sellwood 1984, fig 13.1) and conforms fairly well to that of Malvernian pottery (Cunliffe 1982, fig 4). That the distribution in the later Iron Age is not merely a product of the paucity of information is indicated by the number of sites where briquetage is not recorded (Morris 1985, fig 6).

Bearing in mind the non-cultural variables on the distribution of salt, the following different approaches to explain the distribution of briquetage, may be pursued. Droitwich is some distance from the developments of the southern and eastern seaboards, with their continental contacts. Droitwich and its salt industry however, may not have been isolated, and it is worth investigating further the possibility of a link between what seems (albeit from only two excavations) to be an increase in salt production in the late Iron Age and socio-economic developments elsewhere. The significance of salt production to these developments has already been suggested by Bradley (1984, 146), who has identified an eastern 'core area' actively exploiting resources from its periphery, including salt from Droitwich (Bradley 1984, fig 6.5). Such is the significance of salt that Bradley (1984, 156) considers it worth adding to the slaves, cattle, gold, silver, iron, corn and hunting dogs in Strabo's well-known list of British exports.

Alternatively, a more insular approach may be sought, perhaps tying the distribution of salt from Droitwich more closely to the territory of the Dobunni. Salt from Droitwich (Morris 1985) and pottery from the Malvern area (Cunliffe 1978, 109) are two well developed industries. The distribution of their products indicates that they primarily supplied the indigenous area. Rather than being on the periphery and responding to external developments, the Dobunni themselves may be seen as developing socially and commercially in much the same way as the communities of the south and east. This was apparently without substantial contact with the continent, however the potential of the Severn as an important route of contact to the south, east and even west has not yet been tested.

Distribution in the Roman period is not so easily visible. Although briquetage is recovered from early Roman deposits away from Droitwich, Morris (1985, 352) has suggested that this is probably residual and that its export ceased, though its use in production possibly continued. The large quantities of briquetage associated with early

Roman pottery from the Old Bowling Green, however, supports its continuity of use in production.

The town: ***c*** **400 –** ***c*** **1900**, by James Bond and Alan Hunt

Introduction

This section summarises the development of Droitwich during the medieval and post-medieval periods, drawing upon topographical, documentary, and excavated evidence. The later industrial development of the town is accorded a fairly brief treatment here, mainly because it will be dealt with more fully in a later volume. In our examination of the origins, development, and character of Droitwich it will become evident that many problems remain to be resolved. What follows, therefore, is essentially an interim statement, and should be judged as such. In it we have not hesitated to offer controlled speculations, to 'fly kites', on the basis of the summarised evidence, in the hope of stimulating further research.

Anglo-Saxon Droitwich: early development of the salt industry

By the 5th century AD Droitwich had a long history as an industrial settlement (or group of settlements?) specialising in the production of salt from the brine-springs rising from the local Mercian Mudstone (Keuper Marl). Its special function and economic significance was reflected in its Roman name. It is very probably to be identified with the *Salinae* of Claudius Ptolemy's *Geography*, complied *c* AD 140–150. Ptolemy's map actually locates Salinae near the Wash and his description places it within the territory of the 'Catyeauchlauni' (Catuvellauni), but in neither case do these directions accord with any known Roman salt-producing centre, and they must be regarded as errors. However, Ptolemy also describes *Salinae* as 120 Roman miles from London, only ten miles more than the actual distance between Droitwich and London. The *Ravenna Cosmography*, compiled soon after AD 700, records a place-name *Salinis*, which can unambiguously be identified with Droitwich. The name clearly refers to the saltworks which have dominated the economy of the place throughout most of its history (Rivet and Smith 1979, 120, 451). It has sometimes been suggested that an echo of the Roman name survives in that of the River Salwarpe and the nearby village named from it; however, here the preferred etymology seems to involve the Old English *salu*, meaning 'dark-coloured' or 'brownish-yellow'; and a root-word meaning 'sediment' or 'silt', related to the Old English verb *weorpan*, 'to throw or twist' (Mawer and Stenton 1927, 306–7; Ekwall 1960, 403), and the superficial resemblance to the Latin name is almost certainly fortuitous.

The archaeological record appears to point to a fundamental contrast in the Roman period between the high-status villa-type buildings at Bays Meadow (HWCM 678) on the north side of the river, enclosed within their own defences in the later part of the 3rd century, and the more extensive urban and industrial settlement to the south. The Bays Meadow complex was destroyed by fire at the end of the 3rd century and reoccupied on a reduced and less ostentatious scale; however, the coin sequence continues to the end of the 4th century. South of the Salwarpe some of the sites examined suggest a reduction or cessation of activity in the later Roman period, particularly on the higher ground, but industrial activity and settlement continued along the valley through the late 3rd and 4th centuries (Crickmore 1984a, 74, 102–4).

As usual, however, direct archaeological or documentary evidence of what happened in the 5th century is elusive. To what extent there was any degree of continuity between Roman *Salinae* and Anglo-Saxon *Wich* therefore remains a vexed question. Certainly on present evidence there is little sign of any direct topographical continuity in the town plan, with the single exception of the continuing use of the main Roman road from Worcester to Metchley in Birmingham. This remark must be qualified by pointing out how little is known of the internal street-plan of the Roman settlement south of the Salwarpe. However, the fact that at Vines Park and elsewhere in the valley late Roman occupation levels have been found sealed beneath considerable depths of alluvium or soil does not encourage optimism about the widespread continuity of Roman plan elements. Similarly it would be difficult to make any claim for the continuity of Roman institutions or civil administration. In the broader context of estate organisations and boundaries, the evidence is perhaps more open, and this question will be examined further below. The most likely area for continuity, or at least for early resumption of function, is in the salt industry. Salt would have been such a scarce and valuable resource in the midlands that it is unlikely that any convenient source of it would have been neglected for long. Archaeological evidence for early/middle Saxon salt-working has been identified at Upwich (HWCM 4575); this will be described at length in a later volume.

Charters provide plentiful evidence of the importance of the salt industry in the middle and later Saxon periods (Hooke 1981). Unfortunately a number of the earliest-dated Worcester charters are now lost and are known only from 18th century transcripts or summaries; moreover, some of those represented by early medieval copies are partly or wholly forged. The earliest documentary reference, though of uncertain authenticity, is said to have occurred in a lost charter by which King Wulfhere of Mercia (AD 657–674) granted to Abbot Colman a holding of 50 *manentes* at Hanbury, with meadows, woods and brine-pits belonging to it (Finberg 1972, no 195; Sawyer 1968, no 1822). By a second lost charter Wulfhere's successor Æthelred (674–704 ,is said to have granted to Bishop Oftfor of Worcester in AD 691 a shed and two furnaces belonging to the great brine-pit at *Wic* (Finberg 1972, no 197).

The first surviving document of Saxon date, which all authorities regard either as the genuine original or an accurate early copy, is one of *c* 716–7 by which King Æthelbald of Mercia (716–757) granted to the church of Worcester, in exchange for six furnaces in two salthouses on the north side of the Salwarpe, a piece of ground on the opposite side of the river at places called *Lootwic* and *Coolbeorg* for the construction of three salthouses and six furnaces (Finberg 1972, no 206; Sawyer 1968, no 102). Nominally of the same date is a charter from King Æthelbald granting to the church of Evesham a *mansio in Wico emptorio, salis quem nos Saltwic vocamus* free of all taxes (Finberg 1972, no 207; Sawyer 1968, no 97), but opinions on its authenticity are mixed; a third charter of the same date by which Evesham Abbey claimed lands in various places including *Hamtona juxta Wicca emptorium* and *Wittona* (Hampton Lovett and Witton by Droitwich) is largely or wholly spurious (Finberg 1972, no 205; Sawyer 1968, no 83). A lost charter of Æthelbald is said to have granted to the nun Eafe part of a building with two salt furnaces in *vico emptorio salis* (Finberg 1972, no 212; Sawyer 1968, no 1824). Another lost charter of King Coenwulf of Mercia (796–821) reputedly granted ten houses with saltpits in *Wich* to the church of Worcester (Finberg 1972, no 237).

Although the failure of some of the early charters to survive, together with the dubious authenticity of some of those which are extant, makes their value as historical evidence somewhat questionable, there is, nonetheless, independent confirmation of the importance of this inland source of salt around the beginning of the 9th century. Part of the *Historia Brittonum* attributed to the Welsh writer Nennius (fl 796–809), describes the wonders of Britain. 'The fourth marvel consists of wells of salt which are found in...[the region of the Hwicce]...from which wells salt is refined. Thence divers kinds of food are salted, and....[the wells]... are not near the sea, but spring from the earth.' (Wade-Evans 1938, 116–7; Morris 1980, 40).

Later charters, often of surer authenticity, also sometimes contain more information on the topography or the organisation of the Droitwich saltworks. In 836 a charter of King Wiglaf of Mercia refers to the saltpits and lead-furnaces belonging to the minster of Hanbury (Finberg 1972, no 247; Sawyer 1968, no 190). Two charters of the ealdorman Æhelred refer to tolls on the carriage of salt: in 884 Æthelwulf received five *manentes* in Himbleton with liberty to make six boilings of salt, paying toll on cartloads only to the lord of Himbleton and not to the ruler of Mercia or to any public officer (Finberg 1972, no 265; Sawyer 1968, no 219); and some time before 901 Æthelred and Æthelflaeda granted to Bishop Waerfrith and the cathedral church of St Peter various dues from Worcester, reserving to the king the toll of one shilling on cart-loads and one penny on pack-loads at Saltwic (Finberg 1972, no 268; Sawyer 1968, no 223).

Three saltpits are mentioned by name for the first time in the 10th century. In 962 Bishop Oswald of Worcester was leasing out lands at Bentley in Holt to which were attached four saltpans at *Upwich* and the amount of woodland in *Bradanlaege* (Bradley near Feckenham) required to provide fuel for salt-making (Finberg 1972, no 282; Sawyer 1986, no 1301). In 972 the great charter of Pershore Abbey, whereby King Edgar (959–975) granted or restored to the newly-revived abbey many lands and privileges, included 'a site for ten vats at *Middelwic* and eight at *Neodemestanwic* for salt-making, and the site of two furnaces at *Wictune*' (Finberg 1972, no 302; Sawyer 1986, no 786). The balance of opinion is that the Pershore charter is substantially authentic, but another charter of the same date to the monastery of Worcester, which repeats word for word details relating to the saltworks at Middlewich, Netherwich and Witton, is certainly spurious (Finberg 1972, no 303; Sawyer 1986, no 788).

Brine boiling consumed considerable quantities of fuel and the supply secured from the woods of Bradley in 962 for the Upwich pit is noted above. Other woodland properties in the surrounding area also had significant links with Droitwich. Five *cassati* at Phepson granted by king Eadwy to Worcester Priory in 956 included five salt-furnaces in Droitwich (Finberg 1972, no 280; Sawyer 1968, no 633). In 1017 Archbishop Wulfstan of York granted to his brother Ælfwig six *manentes* at Bentley in Holt, the appurtenances of which included two furnaces at *Sealtwic* (Finberg 1972, no 339; Sawyer 1968, no 1384); the same property with its two salt-ovens was leased out by Bishop Lyfing of Worcester in 1042 (Finberg 1972, no 348; Sawyer 1968, no 1395).

Charter evidence provides a *terminus ante quem* for the three main salt-producing sites. Ultimately, however, their origins and early development can only be established archaeologically and the results of the Upwich excavations (HWCM 4575) are very significant in this respect. As we have indicated above, salt-working seems to begin here in the early/middle Saxon periods, but the significance of this cannot be fully assessed until the results of this very important excavation are more widely available (summarised since this was written in Hurst 1991).

Many aspects of the early organisation of the salt industry first recorded in the pre-Conquest charters find confirmation in the Domesday survey (Round 1901; Monkhouse 1954). Here again the three principal *puteis* (brine-pits) are mentioned by name (*Upewic*, *Middelwic*, and *Helperic* together with five other unnamed brine-pits and more than 300 *salinae* (saltpans or salthouses). The King's own Droitwich property included 149 and a half saltpans and 21 further saltpans belonging to the royal manors of Bromsgrove, Tardebigge and Grafton. In all, fifteen different landowners on 40 separate vills in Worcestershire alone had saltpans in Droitwich; rights to salt from Droitwich were also held by eleven manors in Herefordshire, ten in Gloucestershire, six in Warwickshire, three in Shropshire, two in west Oxfordshire and, most distant of all, Princes Risborough in Buckinghamshire. The web of saltways radiating out to these various places has been reconstructed in outline by Whitley (1923) and Houghton (1929–30). In almost every case the value of the saltpans is recorded, sometimes in money, sometimes in terms of the amount of salt rendered in *mittae* (one mitt was probably about eight bushels or a horse load). Glimpses of the equipment and processes involved in salt-making are provided: two of the royal manors possessing salt-rights, Bromsgrove and Tardebigge, included respectively six and two *plumbi* (leaden pans or vats). The church of Worcester's manor at Northwick possessed in Droitwich a *fabrica plumbi* (presumably a leadworks for making or repairing such vats). Four *furni* (furnaces) in Droitwich in 1066 rendered 60s and 100 mitts of salt. Three *salinarii* (salt-workers) belonged to the royal manor of Bromsgrove. There are several references to *hocci*, variously translated as 'salt-pit' or 'mound or shed for drying salt', though the meaning remains obscure. There are also several mentions of fuel supply. The saltworkers belonging to Bromsgrove are said to have been given 300 cartloads of wood in the time of Edward the Confessor for the 300 mitts of salt they produced. Northwick manor had in Droitwich one saltpan rendering 100 mitts of salt for 100 cartloads of wood. The Bishop of Worcester had woodland in Fladbury supplying fuel for the saltworks. Finally, Westminster Abbey's manor of Hussingtree rendered annually 100 cartloads of wood.

Droitwich was not in 1086 a royal manor, but the royal holding in the saltworks was considerable, amounting to more than half the total number of saltpans recorded. Fifty-one and a half saltpans are said to have been held before the Conquest by Earl Edwin and these may represent a direct inheritance from the earlier Mercian and Hwiccan kings. Hooke (1985, 123) has pointed out that most of the Worcestershire estates possessing Domesday saltpans, and in particular all of those possessing a large number, such as Wychbold (twenty-six *salinae*), Bromsgrove (thirteen *salinae*), Alvechurch (eight *salinae*), Tardebigge (seven *salinae*), Hampton Lovett (seven *salinae*) and Salwarpe (six and five *salinae*), lay in the north-western part of the county. She suggests that this provides evidence for some early territorial link between these estates and Droitwich, very probably due to early royal or princely ownership within a common land unit. The identification of such a land unit and of the key focal and central places within it are an essential preliminary to the investigation of the emergence of Droitwich as a town in the pre-Conquest period.

Anglo-Saxon Droitwich: urban origins

In 1976 Martin Biddle, recognising that definitions of towns based on their legal status as boroughs were inapplicable before the Norman Conquest, sought clarification of what constituted a town in the Anglo-Saxon period. He adopted a pragmatic approach by listing a dozen criteria, the possession of one or more of which he felt might indicate urban status: defences, a planned street system, markets, a mint, a legal autonomy, a role as a central place, a relatively large and densely-concentrated population, a diversified economic base, plots and houses of urban type, social differentiation, complex religious organisation and a judicial centre (Biddle 1976). Some of Biddle's criteria are concerned with the practical functions of towns, some with their administrative status and some with their physical appearance on the ground. Not all of these criteria would have been present in all towns and not all of them were necessarily important at the same time. More significantly, not all of them need necessarily have operated from exactly the same place. Indeed, their needs were sometimes recognisably in conflict. The prime desideratum of the market function, for example, is maximum accessibility, leading to a preference for open sites where routes can converge unhindered from all directions. By contrast, the need of the defensive function is precisely the reverse, with access needing to be restricted and controlled, often by the choice of hilltop or promontory sites. In the pre-urban context the dispersal of central-place functions among the two or more different locations within a given territory is characteristic (cf several of the papers in Grant 1986). To some extent the successful emergence of a town depends upon the permanent coming-together of several of those functions within a single location.

Leaving aside those of Biddle's criteria which are concerned primarily with size and physical layout and concentrating upon those central-place functions which are likely to have some recognisable topographical expression, we can select six functions which are perhaps of particular importance:

i *Estate centres*: places from which a large area of land was exploited for its economic resources. The richer or the more extensive the estate, the higher the social status of its owner is likely to be, and the more likely the centre of the estate is to be represented by high-status buildings, such as a palace.

ii *Administrative and judicial centres*: places from which a large area of land could be administered and judgement of disputes pronounced. In the Anglo-Saxon period the hundred was the most significant territorial unit in this respect, with the hundred meeting-place as its focal point.

iii *Ecclesiastical centres*: places from which religious services and pastoral care could be exercised. At sub-diocesan level in the Anglo-Saxon period the key component was the minster *parochia* with the minster church at its centre.

iv *Military centres*: places exercising a strategic defensive role and serving as local refuges, represented in the Anglo-Saxon period by the network of *burhs*.

v *Industrial centres*: places where raw materials are exploited or processed by specialist producers for wider distribution. The salt industry of Droitwich is a somewhat singular example of this function. The minting of coinage, though of much more widespread occurrence, was strictly controlled by the crown, and by the late Saxon period the distribution of mints was closely linked with that of the emergent boroughs.

vi *Commercial centres*: any of the above fields of operation will tend to bring together numbers of military or religious personnel, bureaucrats and artisans, people who are not themselves primary producers; so a need developed to bring in foodstuffs and raw materials and to establish a mechanism for the exchange of goods through markets or fairs.

Before these areas of activity can be examined individually, we need to establish the territorial framework within which they operated. It will immediately be apparent that different central-place functions may serve somewhat different hinterlands, where these can be defined. However, in the case of Droitwich, a peculiar significance seems to be attached to the territory which constituted its rural deanery in the Middle Ages. Although the bounds of the Worcestershire Deaneries cannot be defined precisely before 1291, there is some evidence for their existence as units of ecclesiastical administration in the early 11th century, and the general tendency of the church towards conservatism raises the possibility that the deaneries may be of much earlier origin, with their bounds remaining substantially stable over a long period (Bond 1988, 128–30). Della Hooke has underlined the probable antiquity of the Droitwich deanery as a territorial unit, pointing to its fairly compact outline and the fact that its bounds mostly follow clear topographical features, such as the Hadley Brook in the west, the Dean Brook and Seeley Brook in the south, the River Cole in the east and the River Rea in the north (Hooke 1985, 66). Moreover, the deanery bounds coincide to a remarkable degree with those of a group of Domesday hundreds, comprising the entire hundred of Came, the whole of the detached southern portion of Clent hundred, the Hanbury portion of Esch hundred and a couple of exclaves of Pershore hundred (this close relationship was destroyed by the reorganisation which resulted in the creation of Halfshire hundred around the mid 12th century). Even more significantly, the pattern of later manorial attachments and ecclesiastical dependences emphasises the number of interlinked estates which had an overall cohesion within the deanery boundary. It is tempting to see the territory fossilised by the later deanery as an Anglo-Saxon folk-area, with its heartland centred on the richer agricultural soils of the Salwarpe Valley, complemented by a less developed northern area spanning the higher plateaux beyond Bromsgrove (Hooke 1986, 88–9).

Within a large multiple estate of this character there will inevitably be a multiplicity of settlements, at least some of which developed tendencies towards specialization from an early period. We can now return to the various central-place functions identified above and see how they operate within the territory represented by the later deanery.

i *Estate centre*: Round's map of Domesday ownership shows that the land around the headstreams of the Salwarpe and the watershed to the north represented the biggest concentration of vills in royal ownership in the county (Round 1901, 253). The most important of these in 1086 was Bromsgrove with its eighteen berewicks and its other dependencies of Grafton, Cooksey, Chadwick and Willingwick, all held before the Conquest by Earl Edwin, as were the royal holdings in the Droitwich saltworks already noted. The other royal vill, Tardebigge, had belonged to Edward the Confessor before the Conquest. These properties seem to represent the remnants of a large royal estate, fragmented by 1086. The site of the palace associated with this estate appears to have been at Wychbold, four kilometres north-east of Droitwich. A charter of AD 692 by which ten *manentes* at *Wicbold* were granted to the church of Worcester (Finberg 1972, no 199; Sawyer 1968, no 75) is regarded by some authorities as substantially genuine (John 1964, 74–6), but by the 9th century this property was back in royal hands. Charters were signed here by King Coenwulf of Mercia in 815 and by Wiglaf of Mercia in 831 and were said to have been written in 'the royal vill'. The '-bold' element often translated merely as 'building' may imply a building or hall of a rather superior kind, in fact a palace (Campbell-Curtis 1913, 59; Hooke 1986, 89). The site of the Wychbold palace has not yet been located archaeologically and this is a major target for future research.

ii *Administrative and judicial centre*: of the Domesday hundred of Came, the whole of which lies within the territory under discussion, little is known other than its extent. The origin of the name and the site of its meeting-place are both lost. The same can be said for the hundred of Esch. Droitwich itself lay within the southern portion of Clent hundred in 1086, but Clent, the site of the hundred meeting-place, lay in the northern portion outside the Droitwich deanery boundary. The two detached portions of Pershore hundred had both belonged to Pershore Abbey since the 10th century. When Halfshire hundred was created in the 12th century it belonged to the king, and for a time in the 13th century was known as the Hundred of Wych, but there is nothing to suggest that any administrative or judicial function was carried out from or near Droitwich in the Anglo-Saxon period.

iii *Ecclesiastical centre*: in the Middle Ages there were two churches in Droitwich itself in addition to two in Witton and the church of Dodderhill on the fringes of the town. Multiple church provision, related to fragmented ownership, is a characteristic of pre-Conquest towns. In a recent general survey of Norman churches in Worcestershire a tentative claim was made for St Andrew's (HWCM 607) as a minster church on the basis of two Domesday priests on the Abbot of Westminster's demesne holding one hide which had never paid geld (Bond 1988, 123), but the evidence for this is far from conclusive. The dominant topo-graphical position of St Augustine's church (HWCM 606) in Dodderhill on the bluff immediately north of the town, its location within the south-east corner of the Roman fort on the extreme margins of a large parish and its cruciform plan in the Norman period provide more evocative hints of an early ecclesiastical centre. Here, however, the earliest surviving fabric accords well with the date of the first recorded institution in 1178; and while the likelihood that this represents a reconstruction of an older building cannot be discarded, there is nothing in the later records to suggest that Dodderhill had any special significance in the pre-Conquest period. Unfortunately excavations in the eastern part of Dodderhill churchyard in 1977 shed little if any light on the origins of this incongruously- placed church (Hurst pers comm; HWCM 603). On present evidence the case for Droitwich or Dodderhill as the centre of an early minster *parochia* is not a strong one. There can be little doubt that the principal minster in the southern part of the territory stood six kilometres to the east at Hanbury (HWCM 815), where the present church stands in a prominent position within an Iron Age hillfort, and where a minster church is unambiguously documented in Wiglaf's charter of 836 (Finberg 1972, no 247; Sawyer 1968, no 190). The predilection of early minsters for reoccupying prehistoric or Roman sites seems not infrequently to override the convenience of proximity to the royal vills with which they were linked, and the distance between Hanbury and Wychbold finds parallels elsewhere (Blair 1988, 40–6). The church of Bromsgrove (HWCM 1365), rebuilt on a cruciform plan in the 12th century, and the centre of a large medieval parish with five dependent chapels, can perhaps be seen as a complementary minster serving the northern part of the territory.

iv *Military centre*: two defensive strong-points have already been mentioned, the hillfort at Hanbury (HWCM 815) and the Roman fort at Dodderhill (HWCM 603), and the presence of churches within them both hints at some sort of reoccupation in the post-Roman period. Both churches are now comparatively isolated and unrelated to the present nucleations of settlement, and any settlement which did take place around them has not been sustained, either because of their comparative inaccess-ibility or because they were superseded in their defensive role by another centre elsewhere in the territory. There is a possible candidate six kilometres to the north of Hanbury. In 910 Æthelflaeda is recorded as building a *burh* at a place called *Bremesburh* immediately after the Danish defeat at Tettenhall (Whitelock 1955, 192). *Bremesburh* is not securely located, but Slater (1982, 180) has suggested an identification with Bromsgrove, already a royal estate, only 30 kilometres distant from Tettenhall, containing the same place-name prefix, and occupying an important strategic position on the Roman road from English Mercia into Danish Mercia. No trace of any 10th century fortification has yet been detected in Bromsgrove, but the absence of any later references to the *burh* suggest that it did not retain its military importance for very long. The site was, however, better suited to commercial growth, and a market was established in Bromsgrove in 1200, followed soon after by a new borough foundation.

v *Industrial centre*: the presence of the brine springs gave Droitwich a peculiar industrial function which transcended merely local needs, though the concentration of salt privileges within the territory under discussion has already been mentioned. An additional special function is represented by the presence of a mint in the time of Edward the Confessor (Stewart and Blunt 1978).

vi *Commercial centre*: the exceptional industrial component within Droitwich's population can only have been supported by some system of exchange or marketing. The description of Droitwich as *emptorium* (place of sale) in three of the charters claimed to be of 8th century date is, therefore, of great interest. One of these charters is lost, another is certainly spurious, but the third (Finberg 1972, no 207; Sawyer 1968, no 97) is possibly authentic, at least in part. Even the forged charter implies that Droitwich was regarded as a trading-place at the time when the forgery was made, though it did not possess a chartered market until as late as 1215. The very place-name may carry a similar implication. In the earliest sources Droitwich is either *Saltwic*, or, more frequently, merely *Wic*. The prefix *'Droit'* is not recorded before the 14th century and this is most likely to be the middle English word meaning 'dirty', reflecting either the low-lying and muddy situation of the town or the smoke from the saltworks, or both. The later forms of the name probably represent conscious attempts to disguise these unpleasant associations (Mawer and Stenton 1927, 285–7). The element *wic* remained in use over a long period of time, and like many other place-name elements it tended to change its meaning according to the period and context of its use. In some instances it seems to mean little more than 'settlement', in other cases more specifically 'dairy farm'; but one of its early meanings seems to be that of a place fulfilling a special economic function as a trading-centre. Droitwich may not have rivalled other great *wics* or trading emporia of the 7th and 8th centuries such as *Eoferwic* (York), *Hamwic* (Southampton), Ipswich or Sandwich, which provided facilities for centralised regional exchange, particularly in prestige goods; but the demand for salt gave it a special importance of its own.

The significance of the *wic* is underlined by the number of neighbouring place-names which contain the same element, including the three brine-pits of Upwich, Middlewich, and Netherwich, the unidentified *Lootwic* and *Coolbeorg* of the 716–7 charter (which may be somewhere in Stoke Prior, Finburg 1972, no 206) and possibly also the *Helperic* of the Domesday survey (identifiable with Helpridge Farm, now called Old Ridgeway Farm, just above Brinepits Farm in Dodderhill (Mawer and Stenton 1927, 282). Two other significant places are named in relation to it. Wychbold, the 'building or palace by the *wic*' is on higher ground to the north-east at a decent distance from the industrial centre. Witton, the 'enclosure or farm by the *wic*', was presumably the agricultural settlement. There is an interesting parallel here with the salt-working town of Northwich in Cheshire, where the ancient parish church lay in the adjoining settlement of Witton (Mawer and Stenton 1927, 282, 285–6, 289).

Any of the central-place functions described above, or any combination of them, could in theory have prompted the beginnings of urban growth. In the particular case under discussion, however, the early estate centre (the palace of Wychbold), the various hundred centres and defensive locations and the principal ecclesiastical centre (the minster of Hanbury) all failed to serve as pre-urban nuclei. It was primarily the industrial function, and following upon that the secondary commercial function, which determined the locality upon which the town was to arise.

Direct archaeological evidence for the character of the Anglo-Saxon settlement underlying the modern town remains slight. At Friar Street (HWCM 605) a few features, stratified between late Roman and 11th and 12th century features, contained sparse and abraded Roman sherds, and one yielded grass-tempered pottery (P3ii). This would seem to imply that in the early and middle Saxon periods this formerly settled area reverted to agricultural use, though whether this was due to settlement shrinkage or settlement shift it is impossible to say at present. Further evidence comes from observation and salvage excavation by Freezer in 1974 at the Old Police Station (HWCM 601). Here part of a curvilinear ditched enclosure was traced, the fill of which also contained a few grass-tempered sherds. The plan of the enclosure and the size of the ditch seemed inappropriate for a field boundary, and this feature might be interpreted as a small, short-lived farmstead enclosure. Whether it was an isolated site or part of a more extensive pattern of dispersed settlement we do not know at present, but further archaeological evaluation of the higher ground on both sides of the Salwarpe valley might shed light on this problem. Salt extraction from the early/middle Saxon period onwards is clearly attested by evidence from the Upwich excavation (HWCM 4575). Trading patterns are suggested by the presence of Stamford, St Neots-type, Cotswold, and Stafford-type wares in the Friar Street excavation (Chapter 20).

The programme of excavations to date has not produced any evidence for settlement of distinctively urban character at Droitwich before the Norman Conquest, despite the documentary indications of significant industrial and commercial growth. What is clear, however, is that the origins of the town cannot be understood in isolation from its context. Throughout the pre-Conquest period we need to be aware of other, complementary and directly linked central-place functions operating within the same territorial unit but based in different centres. The pace of emergence of the town is likely to be very closely related to its success in acquiring or superseding those central-place functions which were originally based elsewhere.

Medieval Droitwich (Fig 120)

The urban character of Droitwich becomes more securely established after the Norman Conquest, both from written sources which establish its legal status as a borough and its economic status as a market centre, and from archaeological and topographical sources which reveal a distinctively urban plan and buildings.

The Domesday survey indicates something of the size and economic significance of Droitwich by the late 11th century. It comprised ten hides, of which two and a half were in Witton. While it is not named as a *burgus* or as a royal manor, a total of 115 burgesses and 35 houses are recorded (compared with only seven burgesses and 131 houses, of which five were waste, in Worcester). The king had only eleven houses, the bulk of his revenue coming from his substantial holding in the salt industry. The abbey of St Denis in Paris had eighteen burgesses rendering 4*s* 6*d* (undoubtedly acquired through its dependent priory at Deerhurst), Westminster Abbey had 31 burgesses rendering 15*s* 8*d*, St Guthlac's Priory in Hereford had nine burgesses rendering 30*s*, Roger de Lacy had eleven burgesses with saltpans rendering 32 mitts of salt and Herald, son of Earl Ralf of Hereford, had 20 burgesses with saltpans rendering 50 mitts of salt. In addition there were 26 contributory burgesses and 24 houses attached or belonging to surrounding settlements and estates (Round 1901, 268; Monkhouse 1954, 262–3). The fluctuating economic fortunes of the medieval town are inevitably linked very closely with the salt industry, and these aspects will be dealt with at greater length in a later volume. Our present purpose is to outline what we know of the medieval urban topography.

The borough status of Droitwich, implied by the numerous Domesday burgesses, is confirmed in the Pipe Roll of 1155–6, when the town was referred to as a *burgus* paying an aid of 100*s*. Consolidation of this status came in 1215 when King John granted a charter to the burgesses giving them the right to hold the town in fee-farm, including the king's own salt-dues, (Hunter 1844, 63) and also giving or confirming the right to a market and fair (Ballard 1913, 115, 124, 175, 190, 216, 231; Bateson 1913, 73–4; Beresford and Finberg 1973, 182). This, confirmed on several occasions, remained the governing charter of the town up to the 16th century. The bounds of the borough are described in a detailed perambulation of 1456 (quoted in full in Bateson 1913, 79–80).

The topographical expression of borough status is the presence of burgage tenements. Slater (1982, 192) has made the point that in the west midlands generally shallow burgages, ie those with a length/breadth ratio of less than 6:1 and an overall length generally less than 46m, tend to predominate. He comments further that the burgages in the central areas of towns of pre-Conquest origin, like Droitwich, Tamworth and Warwick, tend to be especially broad, presumably because lack of demand produces little mediation and may encourage the amalgamation of some plots. Unfortunately the burgage pattern in Droitwich has been badly disrupted by modern development. There are signs of shallow burgages, especially on the north side of High Street, continuing round to the west side of Queen Street and also on the south side of High Street where Tower Hill serves as a back lane. The burgage pattern appears far from uniform but metrical analysis is needed. There are also indications of burgages in Friar Street, but here the pattern is even more vestigial.

John's charter of 1215 granted to the burgesses an annual fair of eight days beginning on 9th May and a market. The dates of the fair were altered on a number of subsequent occasions (Bateson 1913, 78–9). The town plan, perhaps significantly, provides little room for the accommodation of the market, unlike some of the post-Conquest planned towns of the region. Two market-places, not necessarily in contemporary use, can be identified. The older market-place is likely to be represented by the small triangular space north of St Andrew's church, where High Street, Friar Street, and St Andrew's Street meet. This is almost certainly the site referred to by Leland, who tells us that 'There is a meane churche in the chefe strete, and in the towne is once a weke a metely celebrate market' (Smith 1964, ii, 92). Slater (1982, 188–9) has noted that triangular market-places are especially characteristic of the post-Conquest towns and those which have developed organically from villages, and the proximity of St Andrew's church suggests a possible link between early informal Sunday trading in the churchyard and the later formal establishment of the chartered market. The very limited space on this site offered little scope for market encroachment, such a familiar feature of the topography of may other medieval towns. However, an exchequer-house, first recorded in 1327, was rebuilt in timber in 1581, and a range of butcher's shambles of seven arches was added to its east end in 1628. The whole structure was swept away shortly after its description by Prattinton in 1825 and the present town hall was built on its site, on the corner of St Andrew's Street and Friar Street (Bateson 1913, 80–1; Cooper 1934; HWCM 652). A cross is also mentioned in 1629. The lack of

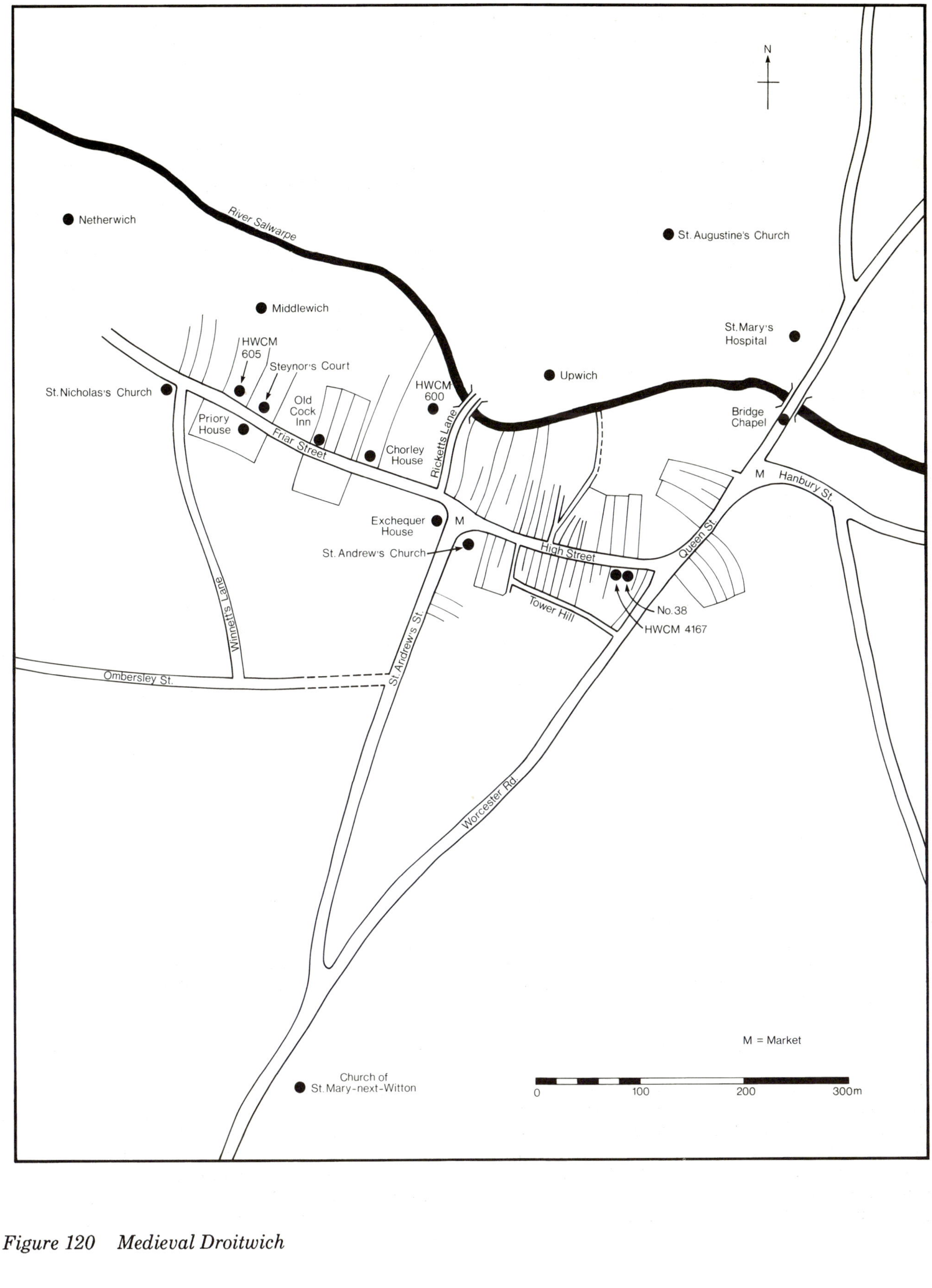

Figure 120 Medieval Droitwich

room in the St Andrew's market-place may also have led to stalls being erected on other sites. Nash (1781, 305) tells us that there was, before his time, a market-place 'near the George Inn'. The exact site of this inn cannot be identified, but the junction between Hanbury Street and Queen Street on the main Worcester–Bromsgrove road is known as St Georges Square and this probably identified the location of this second market-place. On the face of it this seems a promising site for commercial activities, at a strategic junction of routeways which had been important from the Roman period onwards. However, it is difficult to escape the feeling from the small size of both market places that the markets played a comparatively small part on the economic life of a town unusually dominated by industrial interests.

Much of the rest of the street pattern of the lower town can be dated to the medieval period (Fig 120). High Street, Friar Street, St Andrew's Street, Winnetts Lane, Tower Hill, Ricketts Lane, (also called Bagbridge Street) and Queen Street (also called Gosford Street) all existed in the Middle Ages. Some of the documented street-names do not survive, but their general position is indicated from contemporary records. *La Ruyenestret,* recorded in 1344-5 (*Cat Ancient Deeds III [1900],390, no c3621)),* appears to be identical with *Rafunestreet* (*Cal Chart R*, 1226–57, p102 1229) which ran from St Mary's Witton to *Luthbridge* and so to *Letherenebruge*, ie Leather Bridge. A grant of a void space between a tenement and shop in *Barestrete* in 1387–8 describes the site as opposite the churchyard of St Andrew (*Cat Ancient Deeds* VI (1915), p234, no C5529), and since the whole of the churchyard lies south of the church this must be an alternative name for St Andrew's Street. Vallance End was probably equivalent to or near Queen Street. Another documented medieval street, Wavenham Lane, is so far unlocated (Bateson 1913, 80).

Evidence for town defences is very tenuous. No murgage grants are recorded, and there is no archaeological or topographical evidence that Droitwich was ever surrounded by a wall. However, like many other small towns it seems to have had defences of a more minor nature, intended more to define the bounds of the town and control traffic coming into it than to serve any truly military purpose (cf Jones and Bond 1987, 100–2). The street called *Barestrete* in 1387–8 is earlier, in 1336, called *la Barrestret* (*Cat Ancient Deeds* 1 (1890), p415, no C306), clearly taking its name from one of the gates. An account of 1622 mentions the 'filling of the pit at Worcester gate' and Bateson (1913, 79) concludes that a 'ditch and tollgates were probably its sole defences'. The lack of any topographical evidence for any continuous perimeter suggests that any ditch which did exist can only have been a slight and possibly incomplete feature.

Droitwich had several churches. The Domesday survey mentions two priests of the Westminster Abbey holding in Droitwich (Round 1901, 302), and in view of their endowment with one hide it is possible that we see here a residue of a former minster (Bond 1988, 133). St Andrew's seems to have been the principal parish church of the town, standing in a prominent, if cramped central position. Its advowson was held by Deerhurst Priory apparently from before the Conquest (Bateson 1913, 87). The church of St Nicholas (HWCM 255), which stood at the junction of Friar Street and Winnetts Lane appears to have originated as a chapel-of-ease. It is first recorded in *c* 1170, when it was granted to the Cistercian nuns of Westwood by Count Matthew of Boulogne, who had probably held it as an appurtenance of his manor of Bampton, one of the Oxfordshire vills which had salt rights (Houghton 1919, 62). After 1291 it was constituted as a rectory and subsequently served as the parish church for the west end of the town, including Friar Street, Winnetts Lane and Ombersley Street, leaving St Andrew's to serve the central and eastern parts, especially High Street, St Andrew's Street and the Worcester Road. Both churches were damaged in the fire of 1297 and St Andrew's shows evidence of substantial rebuilding after that date, though the tower and fragments of the nave and chancel survive from the early 13th century (Bateson 1913, 82–4). St Nicholas's church fell into disuse after the Civil War, though its ruins survived into the early 19th century. At some stage one of its windows was removed and reused in the Cock Inn, a mid 17th century house on the north side of Friar Street (HWCM 268). The ruins were cleared, and the site redeveloped, by 1884 (OS 1:2500, 1 edn).

The parish church of St Peter-de-Witton (HWCM 253) still stands on rising ground to the south-east of the town in a somewhat isolated position, perhaps marking the original nucleus of the agricultural settlement. The present building dates from the early 12th century. St Peter's parish extended northwards to take in Hanbury Street and the bridge area of Droitwich and on the bridge it had a dependent chapel (HWCM 698). Leland records that he saw 'a bridge of four arches of stone over the broke that rennithe by the Wiche, and at the hither end of this bridge was a fayre new chaple of tymbar' (Smith 1964, ii, 94). Habington mentions ' the Chappell on the Brydge, throughe which passe the (a thinge rarely seene) the Kinge's highwaye (Amphlett 1895, i, 485), a bizarre feature confirmed by Nash (1781-2, 329), who describes how 'the public road with horses and carts passed through the chapel, the congregation sitting on one side of the road, the priest on the other'. This curious structure was pulled down in 1763. The last of the Droitwich churches was St Mary-next-Witton (HWCM 257), which stood on the eastern side of the Worcester road, beyond the southern end of the town, in an area which was perhaps affected by contraction of settlement. It was recorded in *c* 1200, but was in poor repair by 1349. In 1427–8 the parish of St Mary received exemption from the subsidy payment, as there were then less than ten inhabitants (*Feudal Aids*, 1281–1431, Vol V (1908), p315), and the church was subsequently demolished, the parish being united with St Andrew's in 1662 (Bateson 1913, 87–8). St Augustine's church in Dodderhill, technically outside the town though dominating it visually, has already been described.

Droitwich was never a monastic centre, but there were two small religious foundations. The earlier of these was the Hospital of St Mary (HWCM 689), which stood on the north side of the Salwarpe above the east end of Vines Lane, on a site now cut by the railway. This was founded in 1285 by William Dover, rector of Dodderhill, who endowed it with a bullary of salt, half a carucate of land and rents worth 26*s* 4*d* in Wych and Witton, lands granted to him for the purpose by St Peter's Abbey Gloucester. It was placed under the supervision of the prior and convent of Worcester (Locke 1906, 179–80). It was suppressed late in 1535 or early in the following year by its own patrons: a detail of 'articles to be proved against the prior and convent of Worcester' included the accusation 'That they suppressed the hospital of Doverhill [*sic*] in Droitwich, Worc., without licence of the king, and expelled the poor people, to their utter destruction... and caused the hospital to be pulled down and the building materials sold to their own use... [and that] they trouble Sir Ric. Cornewall, clerk of the said hospital, and hold the lands of the same by intrusion' (*L and P Foreign and Domestic, Henry VIII*, Vol XI, p570, no 1429).

In 1331 Thomas Alleyne of Wych granted a plot of land 300ft square for a foundation of Austin Friars (*Cal Pat Rolls* 1330-34, 44). A plan in the Prattinton manuscripts places the site in the Vines on the southern side of the river (Bateson 1913, 82). The precinct was extended on at least two occasions, in 1343 when John, son of William Dragoun, granted a plot 200ft long by 60ft broad for the enlargement of the friar's dwelling and in 1351 when John Bush and William Mercer, chaplains, granted plots five acres in extent for the further enlargement of the friary (*Cal Pat Rolls*, 1343–5, 26; *ibid*, 1350–4, 188). All three gifts were confirmed in 1385 (*ibid*, 1385–9, 93). A hermitage was built in 1388 by Thomas Beauchamp, Earl of Warwick, on the south side of the friary church for Brother Henry de Stokebrugge (Nash 1781–2, i, 332; Clay 1914, 77, 81–2). The 1531 Visitation reported to Cromwell on the great poverty of the house, with so many of its possessions sold that it was only able to support one friar, and the premises required great expenditure to be made fit for habitation (Little 1906, 173–5).

No domestic buildings are known to survive from the early Middle Ages, for which the extensive fire of 1297 may be partly to blame. However, by *c* 1500 there were a number of substantial merchants' houses in Friar Street, of which the sole survivor is Priory House (HWCM 609), on the southern side of the street, almost opposite the excavation reported here. The solar wing of this timber-framed house with its external stone stack was begun in the 15th century and it was extended or partly rebuilt some two centuries later. Another prominent late medieval and 16th century house, Steynor's Court, stood immediately west of the excavated site. It was demolished in the 19th century, but early photographs of it survive. The finest of all the Friar Street houses was Chorley House (HWCM 292), with a hall of late 14th century date flanked by additional two-storey bays at either end and side wings to the rear. Charles (1967, 55–9) has described the elaborate form of the hall with its cusped wind-braces and monumental arcade as too lavish for a normal town house, but his suggestion that it may be the house built in or just before 1388 by Thomas Beauchamp, Earl of Warwick and given to the Austin Friars seems equally unlikely, since this reference is to a anchorite's cell (see above). Regrettably this fine building was demolished in 1962. The only house of comparable status in High Street is no 38 (HWCM 632), known as Stephen's, with a timber-framed solar dating from *c* 1400 and a hall (now in separate occupation) rebuilt in the 16th century (Charles 1967, 49–51).

What can be concluded about the character and shape of Droitwich in the Middle Ages? There are no medieval maps of the town and the earliest survey to show any sort of topographical detail is the thumbnail sketch in John Ogilby's depiction of the road from Hereford to Leicester in his *Britannia*, published in 1675. The central line of Ogilby's plan is the old Roman road from Worcester to Bromsgrove, crossing the Salwarpe by the 'Wood bridg', but some of the other streets are clearly recognisable. The main axis of the town was clearly along High Street and Friar Street, while St Andrew's Street, Queen Street and Hanbury Street are also depicted as built up on both sides. If Ogilby's sketch is reliable, there has been some contraction since that date, since both the south side of Hanbury Street and the east side of Queen Street are now empty.

The High Street/Friar Street axis runs along the contour on the southern side of the Salwarpe valley. At the junction of these two streets was a significant focus of routeways and activities: the crossroads formed by St Andrew's Street and Ricketts Lane (formerly Bagbridge Lane, giving access to a minor crossing over the Salwarpe), the parish church of St Andrew, and the main market-place. Moreover, at this junction there is a marked kink in the alignment of High Street and Friar Street. On these grounds it is fair to conclude that they were not laid out in one single operation. It can be suggested that High Street appeared first, since it is directly connected to the Roman road and the focus of early routes in the St Georges Square area. A primary town plan, based on the High Street alignment, with burgage plots on either side, delimited at each end by significant road junctions and at one if not both ends by market-places, would have some topographical unity and integrity. Only one site has been excavated directly on High Street (HWCM 4167), by Peacock in the late 1960s and later by Hillelson in 1984. At this site medieval structures and deposits had been almost entirely removed by later developments (Hillelson 1985). Another excavation just off High Street at Gurneys Lane (Woodiwiss 1983; HWCM 4099) demonstrated that a very large area has been affected by subsidence and that any surviving medieval or earlier deposits exist only at a great depth. Further excavation opportunities may be offered by piecemeal developments and should be exploited wherever practicable.

Friar Street makes no topographical sense as a primary plan element, and if it is accepted as a secondary morphological unit within the town, we need to consider its dating and character. In the Friar Street excavation several post-Roman plough furrows (P3ii) ran at right angles to the medieval and later street alignment. This was in marked contrast to features of the early to mid Roman periods (P2 and P3i) which lay below the metalled medieval road at an acute angle to its alignment. We can speculate that in the post-Roman period fields were laid out in this area, their boundaries aligned with the contours of the valley, and that their orientation influenced the planning framework when urban development extended into this area. The foundation of St Nicholas's church in the 12th century at the west end of Friar Street, near the limits of the built-up area, perhaps provides a *terminus ante quem* for this development. The first substantial medieval phase of occupation recorded in the Friar Street excavation was dated to the later Saxo-Norman period, probably the later 11th or early 12th century, and consisted of a tannery, a noxious industry often banished to the urban fringes (P4ii). Since it respected the street alignment, this tannery seems likely to have appeared after the line of Friar Street had made its appearance. Subsequent occupation (P5i and P5ii, 12th and early 13th centuries) saw the building of houses along the street: initially at least one poor dwelling, succeeded by a more prestigious structure. It is possible, then, that St Nicholas's church was built to serve a residential and business development which had expanded over a former urban-fringe industrial area in the 12th century.

After the tannery, probably of later 11th or early 12th century date, was swept away to make room for residential development, the scale and character of the succeeding domestic occupation changed considerably. A very humble house of phase 5i was replaced by a large, possibly partly-ditched enclosure, presumably surrounding and draining a substantial building or complex of buildings in the early 13th century (P5ii). This enclosure in turn was subdivided later in the 13th century and in its western division a house, probably timber-framed, was built. This latter house, probably of at least middling status, was apparently destroyed by fire in the later 13th century. Whether this fire can be linked with the widespread documented fire of 1297 is a matter for speculation. After the fire which terminated phase 5ii, and possibly after an intervening lapse of time, there was a resumption of residential or commercial rebuilding which is broadly dated within the 13th or 14th centuries (P6), with a further rebuilding in the 15th to 16th centuries (P7). Thus by the later 16th century Friar Street was occupied by the substantial merchants' houses described above.

The brine-pits along the Salwarpe valley continued to provide the principal economic base of Droitwich throughout the Middle Ages. However, in 1215 King John's charter granted the brine-pits to the burgesses of the town and for the next 480 years the salt industry remained a jealously-guarded borough monopoly (Berry 1957). By 1300 Droitwich was a flourishing industrial town with municipal government and a chartered market. Hilton (1966, 176) has pointed out that much wealth was to be gained there, for the burgesses had to pay an annual fee farm of £100 to the exchequer for their privileges, compared with the £30 paid by Worcester. However, the burgesses stranglehold over the salt industry significantly inhibited the expansion of the town. Surnames of place-name origin, which provide some index of recent immigration, represent only ten percent of the total number of taxpayers' names in Droitwich compared with about forty percent in the boom town of Stratford-on-Avon (Hilton 1966, 184).

Although Droitwich was never a large town, it nonetheless possessed a significant range of urban equipment. Most importantly, it was a corporate borough with a regular market on an important route. It lacked a castle or major monastic institution, reflecting the fact that it was not a major estate centre or ecclesiastical focus before the Conquest; but it had several smaller parish churches and two minor foundations of monastic character. It remains exceptional in its economic character through the domination of the salt industry, its strength in so far as it was uniquely able to provide a commodity in much demand, but its weakness in that control of the industry could so easily fall into the restrictive hands of a monopoly.

The post-medieval town (Fig 121)

Following Ogilby, the earliest large-scale map of Droitwich, dating from the 17th century (HWRO 8000/s497.33; reproduced in Crickmore 1984b, fig 5) shows the urban area still centred on the High Street/Friar Street axis, apparently very little

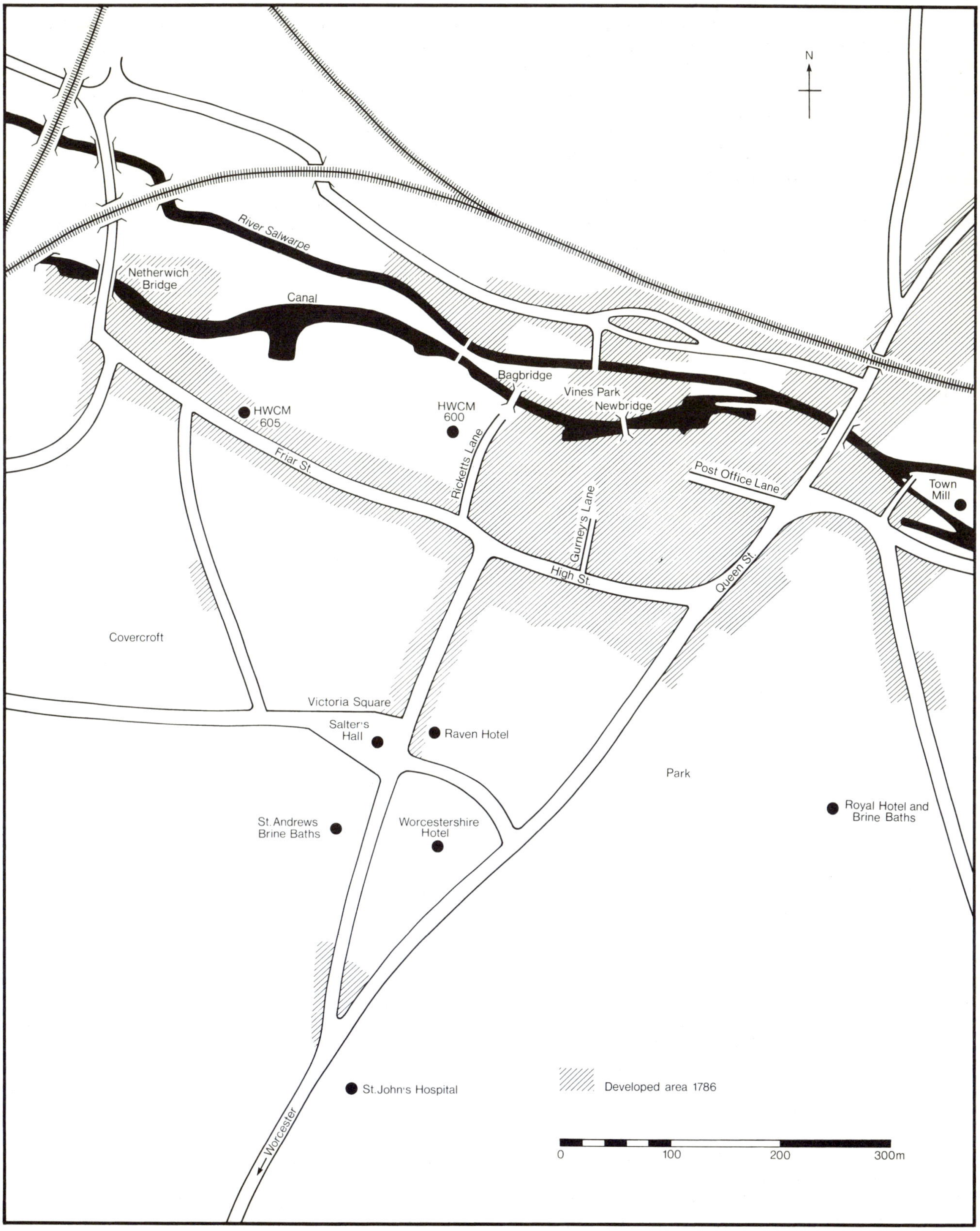

Figure 121 Post-medieval Droitwich

changed from its medieval extent. The ancient brine-pits of Upwich and Netherwich are shown in the valley bottom, though Middlewich is omitted; at this period the Salwarpe took a slightly more circuitous course around the southern side of the Upwich pit, where it was crossed by Bagbridge (at the end of what is now Ricketts Lane) and by Newbridge (by the former junction of Lechmoores Lane, now Gurney's Lane and Frog Lane, later Post Office Lane). Subsequently, in 1768–70, the river was straightened and now passes north of the site. Both pits are shown surrounded by numerous small buildings, which may be boiling-houses.

In the second quarter of the 16th century Leland had described the three salt springs and furnaces, which he estimated to be 400 in number, in some detail, and commented on the problems of fuel supply. Boiling was then taking place only six months in the year, to maintain the price level and conserve fuel, but the coppices had to produce 6000 loads of wood to enable each furnace to make four loads of salt a year. The former sources of fuel were now 'sore decayed' and fuel was having to be brought in from Worcester, Bromsgrove, Alvechurch and Alcester (Smith 1964, ii, 92–4). The Middlewich pit had fallen out of use by Habington's time, though he records two wells working in Netherwich and one in Upwich (Amphlett 1899, 296) and gives a very full account of the industry in the early 17th century. The documentation for the Upwich pit after 1695 and the problem of its co-identity with the Anglo-Saxon and medieval pit of the same name have recently been reviewed by Crickmore (1984b)

The self-perpetuating monopolistic oligarchy which had controlled the making and marketing of salt for many generations was finally broken in 1692, when Robert Steynor sunk two wells on his own land and defeated the burgesses in a lawsuit three years later, though he nearly bankrupted himself in the process. At the same time the limitations of the fuel supply were overcome by the use of coal for the first time, in 1691. A number of new salt-working enterprises then began to develop, opening many new brine wells. In 1725 Sir Richard Lane opened an experimental new boring through the layer of talc which had formed the base of the old pits and immediately tapped a much more productive brine source. The old pits fell into decay and were abandoned as new salt-producing factories spring up in the 18th and 19th centuries in the Salwarpe valley along Vines Park, below the town and in the Covercroft area (map of Droitwich, 1786: HWRO r925.99:40). A new source was discovered five kilometres to the north-east in the parish of Stoke Prior in 1828 and a major new saltworks built there by John Corbett (Anon 1910; HWCM 3348). By 1875 the production of Stoke Works exceeded that of Droitwich. Although many of the smaller concerns in the town amalgamated to form the Salt Union in 1889, their days were numbered and the last Droitwich saltworks closed in 1922 (Pillans 1906; Maddocks 1950).

The reinvigoration of the town in the 18th century was aided by improvements to its transport facilities. The old Roman road from Chapel Bridge to Worcester was the first Worcestershire road to be turnpiked and the Turnpike Act of 1713 spelled out the need: 'by the Reason of the Great and many Loads and Carriages of Salt and other goods which Daily pass through the said Road...(which was) almost impassable for the Space of Nine Months in every Year...' (12 Anne, Stat ii, Cap 3, Priv). The term of the first Act for this road was extended northwards along the Roman road to Bromsgrove and eastwards along the Roman Salt Way to Bradley Brook (22 Geo II, Cap 43). Further roads radiating from Droitwich were turnpiked under an Act of 1755, including that from Netherwich Bridge to Low Hill near Hartlebury (28 Geo II, Cap 48), and renewals and amendments to the earlier Turnpike Acts continued through to the mid-19th century.

The River Salwarpe may have been navigable in 1378, when Richard II permitted the bailiffs of Droitwich to levy tolls on it, but it did not long remain passable. There were several abortive attempts during the 17th century to open the river for transport, first in 1655 by Andrew Yarranton and Captain Wall; then in 1660–1 by Lord Windsor, who seems to have built five out of an intended six locks before his scheme failed; and then by a series of proprietors in 1662, 1673, 1693 and 1747. A London druggist named Baker promoted an alternative scheme to send brine by pipeline to the Severn at Hawford in 1755 but this also failed. Finally in 1768 the salt-makers of the town acquired an Act for making a new barge waterway from Droitwich to Hawford, to achieve their dream of exporting salt and importing coal by water. The route was surveyed by James Brindley and construction began in 1768 under the engineer John Priddey, being completed and opened in 1771 (Hadfield 1966, 59–62). It was eleven kilometres long and had eight locks, originally accommodating Severn trows and barges. The opening of the Worcester and Birmingham Canal in 1815, passing only two kilometres to the east of the town, provided a rival water link and the latter canal gained an advantage in 1830 when Stoke Works was opened up alongside it. Both waterways were soon faced with railway competition however and their companies responded by building the Droitwich Junction Canal in 1852–4, which linked the two older canals by a new line from Droitwich Town Mill to Hanbury Wharf, just over one kilometre long with seven locks. Despite considerable improvements to all three canals during the later 19th century, the income from the two Droitwich canals ultimately did not justify their maintenance and both were formally abandoned in 1939, having been derelict for many years (Hadfield 1966, 270–6).

The opening of the Birmingham and Gloucester Railway in 1845, which passed three kilometres east of the town, provided a direct service to Stoke Works and from 1847 most of the salt from this source was transported by rail. Droitwich itself was first reached by rail in 1852, when Brunel completed a branch of the Oxford, Worcester and Wolverhampton Railway (OWWR) running from Worcester to Droitwich and on to link with the Birmingham and Gloucester Railway at Stoke Works, just before his resignation as the company's engineer. Brunel's successor, John Fowler, completed the main OWWR line northwards from Droitwich on to Stourbridge very shortly afterwards. Extensive sidings serving the salt industry developed around Droitwich station (Jenkins and Quayle 1977).

Another significant development of the 19th century, which was to have a major impact upon Droitwich's character, was its development as a spa. The potential healing qualities of the brine springs were first recognised in the 1830's and in 1836 the first Royal Brine Baths (for 'rheumatic and gouty patients') and Royal Hotel were opened on a site east of Queen Street, where the former County Council offices (The Herriots) stand. There was, however, an almost intractable problem in reconciling the grimy and unsightly appearance of the old town, now riddled with subsidence and permanently overcast by a pall of smoke and steam from the saltworks, with the pleasant surroundings required by a successful health resort. It was only when John Corbett's expansion of the Stoke Works initiated the decline of the town's salt industry that its development in a new direction became possible. Corbett was himself the leading figure in the planning of what was virtually a new spa town on the plateau south of the medieval centre. The centrepiece of Corbett's town is Victoria Square (in fact an informal triangular open space), around which he grouped his principal buildings: Salter's Hall (used for public meetings and concerts), the St Andrew's Brine Baths and the Worcestershire Hotel, all built in the decade after 1881. The Raven Hotel was also rebuilt on a much grander scale at the same time, though retaining its 16th century nucleus. East of the Worcester Road, Corbett laid out a park of four hectares and built the St John's Brine Baths and Hospital in 1881 for the benefit of poor people suffering from rheumatic diseases. Droitwich was a very late spa and its development as a place of recreation, residence and retirement remained subsidiary to its primary purpose of providing specialised treatment for the sick. It is, therefore, quite different in character from Worcestershire's other major spa town, Great Malvern.

The changing fortunes and character of post-medieval Droitwich are to some extent reflected in microcosm in the sites excavated in the town. At the Friar Street site the two late or sub-medieval houses (S41 and S43, P7) had been united into one large property in the 17th century and some rebuilding or refurbishment took place (P9). By the 18th century, however, the premises had probably reverted to industrial or commercial use; the 1786 map shows that after the arrival of the canal a wharf was built nearby (HWRO r925.99:40). The clearance of the site and the building of poorly-constructed and short-lived tenements early in the 19th century (P10), represents a typically high-density solution to the need for working-class housing in a growing industrial centre. Immediately to the south-east of the tenements a large building, probably a salt factory, was in place by 1884 (OS 1:2500, 1 edn). It outlived the tenements, which were cleared by 1784, but was itself removed by the early 20th century when two halls, for church and community uses, were built (P11). These halls complete the varied pattern of chronological development on this excavated site. Its relationship with the immediately surrounding area calls for further historical and archaeological investigation, but initial impressions suggest that this western end of Friar Street had become a residential area once again by the early 20th century.

Conclusions

Inevitably this discussion has been dominated by questions and speculations rather than by solid answers, but we would hope that it may serve to stimulate discussion and further investigation. Without doubt there is a wealth of historical and archaeological potential at Droitwich. Documentary sources and vernacular buildings require research programmes of their own and there may well be other long and well-preserved archaeological sequences to be excavated within the historic urban core. Further topographical research, including particularly a close examination of property units and boundaries along the lines pioneered by Slater (1980; 1981; 1982), is likely to pay dividends. In theoretical terms we should give more attention to the early evolution of the settlement from pre-urban to pro-urban status. Again in the early periods the territorial context and the relationship with neighbouring centres, particularly Worcester, would merit closer attention. It is arguable that few small towns in the midlands of England offer so much research potential and we hope that this modest contribution, in addition to providing a somewhat sketchy overview of work to date, will point the way forward to further productive work in future.

35 Pottery fabrics; a multi-period series for the County of Hereford and Worcester

Derek Hurst and Helen Rees

The pottery fabric series has been developed since 1983 in response to a need for a single ceramic series to allow cross-referencing between different sites in the County of Hereford and Worcester. The series has been evolved primarily from the fabric analysis of ceramics from the two major Droitwich excavations reported in this volume, as these sites together provide an almost complete ceramic sequence from the Iron Age to the post-medieval period. The important contribution by Morris (1980) to ceramic studies in the region, based on the pottery from the Worcester Sidbury excavations (HWCM 117), has formed the basis of the present fabric series, especially for the medieval period. However, the Worcester fabric type series was not published with detailed fabric descriptions in the main report and, though correlation between the Morris Worcester series and the present fabric series is often possible for post-Roman fabrics, the fabric reference number may differ for any particular fabric type.

Fabric analysis

Many of the fabric types have been examined petrologically at Southampton University to confirm and further delineate their particular character. This has continued the programme of detailed fabric study commenced for the region by Dr D P S Peacock in the mid 1960s, when the potential of ceramic thin section analysis was first appreciated and the value of its application to a geologically variable area was demonstrated. Since then, extensive thin section analysis has been carried out on pottery from Beckford (Tomber pers comm, Williams pers comm) and Kenchester (Tomber 1985). The Iron Age fabrics from the Friar Street excavation and the Roman fabrics from the Old Bowling Green in particular have been the subject of petrological analysis in this volume. However, though thin section study is most important in the region for assisting fabric recognition, its pre-eminent position in ceramic analysis locally may paradoxically have tended to exclude other methods, such as elemental analysis, whenever a programme of fabric analysis has been formulated. It may, therefore, be appropriate to adopt other methods of compositional analysis, as considered by Tomber (1:F13–G2), notably to examine some of the well-known Roman wares which are less well characterised to source, such as Severn Valley ware. The correlation of stylistic and form attributes with detailed elemental analysis in a well stratified series of vessels, may yet prove useful in determining local types of Severn Valley ware and their distribution.

Pottery fabric descriptions

The fabric type series is presented in some detail with the intention that it may prove useful as a general multi-period series for development through further ceramic study in the region. Visual colour descriptions are provided, followed by Munsell colour code. Terms used for description of inclusions are as defined by Orton (1977, 28–30). The full fabric descriptions and the reference sherds are available in archive.

The Droitwich briquetage has been included in the pottery report because, although an industrial saltworking ceramic, it was also used for the storage and transportation of salt.

Fabric 1 Sandy briquetage

Manufacture	Handmade
Hardness	Soft to hard
Colour	Usually oxidized red (10R 5/8–2.5YR 6/8) throughout, though white or purplish patches sometimes evident
Surface treatment	Inner surface is often roughened, with deep finger impressions. Outer surface usually smoothed
Inclusions	Moderate to abundant quartz sand up to *c* 0.5mm with occasional larger (*c* 2.0mm) rounded quartz, moderate to sparse elongated voids of all sizes up to *c* 20.0mm, sparse rounded clay pellets (<7.0mm) and sparse hard angular dark brown iron ore fragments up to *c* 2.0mm
Source	Droitwich (Morris 1981b, 153; 1983, 171–5 and 178–82, figs 5.14 and 5.16)

Fabric 2 Organic briquetage

Manufacture	Handmade
Hardness	Soft
Colour	Orange (2.5YR 6/8) to red (5YR 6/4) often with reduced grey core (N 4/0)

Surface treatment Inner surface is often roughened, with deep finger impressions. Outer surface usually smoothed
Inclusions Burnt-out organic matter has left abundant elongated voids of all sizes up to *c* 20.0mm; sparse rounded clay pellets (*c* 2.0–7.0mm)
Source Droitwich (Morris 1981b, 153; 1983, 175–82, figs 5.15 and 5.16)

Fabric 3 Malvernian ware (Group A)

Manufacture Handmade
Hardness Soft to hard
Colour Dark grey to black (N 2.5/0); occasionally with patchy red oxidized lenses (10R 5/8) on, or beneath, one or both surfaces
Surface treatment Outer surface often smoothed. Inner surface may be smoothed. Linear tooled decoration, or pattern burnished
Inclusions Abundant angular Malvernian metamorphic rock fragments usually *c* 1.0–3.0mm in size, but ranging up to *c* 8.0mm
Source Malvern Hills area (Peacock 1968, 414–21; Morris 1981b; 1983, 112–6, figs 4.15 and 4.16)

Fabric 4.1 Palaeozoic limestone tempered ware (Group B1)

Manufacture Handmade
Hardness Soft
Colour Reduced black (N 4/0) but often with patchy beige/brown (10YR 6/4) or orange (5YR 7/6) oxidized lenses on, or beneath, one or both surfaces
Surface treatment Outer surface sometimes smoothed, or highly burnished all over
Source A source in either the Malvern, Woolhope, or Mayhill areas of Herefordshire and Gloucestershire, has been suggested (Peacock 1968, 422). However, more recent work (Morris 1983, 116–22) has argued that the Woolhope source is the most likely

Fabric 4.2 Oolitic limestone tempered ware

Manufacture Handmade
Hardness Soft
Colour Oxidized buff (7.5YR 7/4) throughout
Surface treatment None represented
Inclusions Abundant, ill-sorted oolitic limestone up to 5.0mm
Source Cotswolds area

Fabric 5.1 Sandy ware

Manufacture Handmade
Hardness Fairly hard
Colour Pink (10YR 6/6) to black (N 3/0)
Surface treatment May be smoothed or burnished on outer surface
Inclusions Abundant, medium, rounded and subangular quartz
Source Possibly local

Fabric 5.2 Fine sandstone tempered ware (Group E)

Manufacture Handmade
Hardness Soft
Colour Buff surfaces (7.5YR 7/4) and dark grey core
Surface treatment None represented
Inclusions Abundant medium (<0.5mm), well-sorted, subrounded-subangular quartz; moderate rounded clay pellets; occasional subangular medium (<0.5mm) fragments of quartzose sandstone
Source Llandovery (Silurian) deposits in Malvern Hills, Woolhope or May Hill area (Morris 1983, 135–40, figs 4.23–4.24)

Fabric 5.3 Coarse quartz tempered ware

Manufacture Handmade
Hardness Soft
Colour Oxidized red (10R 6/6) throughout
Surface treatment None represented
Inclusions Abundant medium (<0.5mm) subrounded-subangular quartz; moderate coarse quartz/quartzite; sparse-moderate clay pellets/mudstone (<2.5mm)
Source Unknown

Fabric 6 Dolerite tempered ware

Manufacture Handmade
Hardness Soft
Colour Black or buff with grey core (Gelling and Peacock 1966, 96–7)
Inclusions Quartz and dolerite rock inclusions
Source Clee Hills area, south Shropshire (Gelling and Peacock 1966)

Fabric 7 'Belgic-type' ware

Manufacture	Wheel-thrown
Hardness	Fairly hard
Colour	Beige/brown (7.5YR 7/2) to black (N 2.5/0)
Surface treatment	Outer surface often highly burnished
Inclusions	Abundant rounded and sub-rounded quartz grains up to *c* 0.8mm in size
Source	Unknown, possibly local

Fabric 8 'Belgic-type' ware

Manufacture	Wheel-thrown
Hardness	Fairly soft
Colour	Similar to fabric 7
Surface treatment	Outer surface is often highly burnished and may have a grey slip (N 4/0)
Inclusions	Moderate rounded or subangular quartz grains up to *c* 1.0mm and sparse to moderate other inclusions eg clay pellets up to *c* 1.5mm and iron ore up to *c* 0.5mm
Source	Possibly local

Fabric 9 Mudstone tempered ware (Group D)

Manufacture	Handmade
Hardness	Soft
Colour	Ranges from light brown to black
Inclusions	Vesicular as a result of the dissolution of argillaceous, sedimentary rock fragments
Source	Possibly Martley area, Worcestershire (Morris 1982)

Fabric 10 Untempered fine ware

Manufacture	Handmade
Hardness	Hard
Colour	Pink surfaces (2.5YR 6/6–6/8) with black core
Inclusions	None visible macroscopically
Source	Unknown

Fabric 12 Severn Valley ware

Manufacture	Wheel-thrown
Hardness	Soft to hard
Colour	Usually reddish orange (2.5YR 5/8) but may be brown (5YR 6/6) and sometimes with reduced grey (10R 6/1) core
Surface treatment	Outer surface often highly burnished. Simple impressed groove and cordon decoration
Inclusions	Fine fabric containing occasional limestone fragments, clay pellets or iron ore
Source	Severn Basin (Webster 1976, 18–46)

Fabric 12R Reduced Severn Valley ware

As for fabric 12 except that reduction produces a grey finish

Fabric 12.2 Severn Valley ware variant

As for fabric 12 but with sparse elongated voids usually appearing as black or dark grey streaks in fracture

Fabric 12.4 Severn Valley ware variant

Manufacture	Wheel-thrown
Hardness	Hard
Colour	Pale orange (5YR 8/3) surfaces and grey core
Surface treatment	None observed
Inclusions	Abundant shelly limestone fragments up to 0.5mm with sparse larger fragments
Source	Unknown

Fabric 13 Sandy oxidized ware

Manufacture	Wheel-thrown
Hardness	Hard
Colour	Orange throughout (2.5YR 6/8)
Surface treatment	Outer surface may be highly burnished
Inclusions	Moderate to abundant sub-angular and rounded quartz grains of all sizes up to *c* 1.0mm
Source	Unknown

Fabric 14 Fine sandy grey ware

Manufacture	Wheel-thrown
Hardness	Hard
Colour	Grey (2.5YR 5/2–N 4/0)
Surface treatment	Outer surface may be burnished
Inclusions	Moderate angular and subangular quartz grains up to *c* 0.1mm in size; sometimes micaceous.
Source	Possibly local

Fabric 15 Coarse sandy grey ware

Manufacture	Wheel-thrown
Hardness	Soft to hard
Colour	Light to dark grey (5YR 7/1–5YR 5/1)
Surface treatment	Outer surface may be burnished
Inclusions	Abundant rounded quartz grains up to *c* 3.0mm in size

Source	Possibly local

Fabric 16 Grog tempered ware

Manufacture	Wheel-thrown
Hardness	Hard
Colour	Grey (N 5/0)
Surface treatment	None represented
Inclusions	Moderate angular fragments of iron-rich grog up to *c* 0.5mm, and sparse, subangular quartz grains of a similar size
Source	Possibly local

Fabric 17 Mudstone tempered ware

Manufacture	Wheel-thrown
Hardness	Hard
Colour	Pink (5YR 6/4), usually with a grey core (N 5/0)
Surface treatment	Outer surface often has thin white, pinkish white slip or wash (7YR 8/2)
Inclusions	Abundant angular mudstone fragments up to *c* 4.0mm and occasional fragments of rounded limestone up to *c* 2.0mm
Source	Unknown; also occurs at Alcester (P Booth pers comm)

Fabric 18 Malvernian derived ware

Manufacture	Wheel-thrown
Hardness	Soft to hard
Colour	Grey to black (N 6/0–N 3/0)
Surface treatment	None represented
Inclusions	Abundant Malvernian rock fragments up to 0.5mm in size
Source	Malvern Hills area

Fabric 19 Wheel-thrown Malvernian ware

Manufacture	Wheel-thrown
Hardness	Hard
Colour	Grey in colour (5YR 5/1) with occasional oxidized, orange examples (5YR 4/6)
Surface treatment	None represented
Inclusions	Moderate to abundant, angular Malvernian rock fragments up to *c* 3.0mm in size
Source	Malvern Hills area (Peacock 1965–7, 418–28)

Fabric 20 White slipped ware

Manufacture	Wheel-thrown
Hardness	Hard
Colour	Orange red (2.5YR 6/8)
Surface treatment	Outer surface white (5YR 8/4) slipped
Inclusions	Often without macroscopically visible inclusions, but sometimes contains variable amount of rounded quartz up to *c* 0.5mm
Source	Gloucester (Rawes 1972, 18–59; C Gouge and A Vince pers comm)

Fabric 21 Micaceous ware

Manufacture	Fabric 21.1 handmade Fabric 21.2 wheel-thrown
Hardness	Fabric 21.1 soft Fabric 21.2 Soft to hard
Colour	Grey (2.5YR 4/1)
Surface treatment	Usually burnished
Inclusions	Abundant large (up to *c* 0.5mm) flakes of white mica
Source	Unknown. Also occurs at Gloucester as TF5 (C Gouge pers comm)

Fabric 22 Black Burnished ware, type 1 (BB1)

Manufacture	Handmade
Hardness	Hard
Colour	Reduced dark grey or black throughout (5YR 3/1)
Surface treatment	Highly burnished zones on outer surface, smoothed or burnished on visible inner surfaces
Inclusions	Abundant subangular quartz grains up to *c* 1.0mm but only rarely exceeding *c* 0.5mm. Some sparse white inclusions up to *c* 1.5mm, and occasional pieces of shale
Source	Dorset (Williams 1977, 163–220)

Fabric 23 Shell gritted ware

Manufacture	Wheel-thrown
Hardness	Soft to hard
Colour	Grey in colour (N 4/0) sometimes with oxidized beige brown surfaces (5YR 8/1)
Surface treatment	Surfaces are sometimes rilled
Inclusions	Abundant shell and limestone fragments, ranging in size from *c* 0.03mm to *c* 3.5mm, with many as large as *c* 1.5mm
Source	Unknown

Fabric 24 Shell and ironstone tempered ware

Manufacture	Wheel-thrown
Hardness	Hard

Colour	Mottled grey/light brown surfaces (5YR 6/1– 5YR 6/3) and dark grey core (5YR 3/1)
Surface treatment	None represented
Inclusions	Abundant large (up to *c* 5.0mm) flat calcareous inclusions, and sparse rounded iron ore up to *c* 2.0mm
Source	Unknown

Fabric 25 Terra Nigra

Manufacture	Wheel-thrown
Hardness	Hard
Colour	White or pale grey (5YR 8/1)
Surface treatment	Overall fine dark grey or black slip (N 3/0)
Inclusions	Virtually inclusion-free except for very sparse minute red flecks, and occasional flat voids up to *c* 1.0mm in length
Source	North-east Gaul/Rhineland (Rigby 1973, 7–24)

Fabric 26 Lyons ware

Manufacture	Wheel-thrown
Hardness	Hard
Colour	Pale cream, often green tinged (5YR 8/3)
Surface treatment	Dark greenish brown (10YR 3/2) colour coat
Inclusions	None visible
Source	Lyons, France (Greene 1978, 15–16)

Fabric 27 'Indixivixus'

Manufacture	Wheel-thrown
Hardness	Hard
Colour	White (N 8/0)
Surface treatment	Glossy dark grey (N 4/0) slip
Inclusions	Moderate small black specks (*c* 0.1mm) and sparse irregularly shaped black lumps up to *c* 2.0mm
Source	Nene Valley area (D F Mackreth and J R Perrin pers comm). Product of the potter 'Indixivixus' (Dannell 1973, 139–42)

Fabric 28 Nene Valley ware

Manufacture	Wheel-thrown
Hardness	Hard
Colour	(i) White or off white (10YR 8/2) (ii) Pale pink (7.5YR 7/6)
Surface treatment	Dark red (2.5YR 6/6) to dark brown (10YR 3/1) matt slip. In overfired examples the colour coat may appear metallic
Inclusions	Sparse large (up to *c* 5.0mm) iron ore
Source	Nene Valley area (Howe, Perrin and Mackreth 1980)

Fabric 29 Oxfordshire red and brown colour coated ware

Manufacture	Wheel-thrown
Hardness	Generally hard
Colour	Pale orange (5YR 7/6) to red orange (2.5YR 6/8)
Surface treatment	Orange (2.5YR 6/8) to dark brown (5YR 4/2) slip
Inclusions	Sandy fabric with sparse small black and red inclusions, and occasional lumps of chalk up to *c* 5.0mm. Frequently micaceous
Source	Oxfordshire (Young 1977, 123)

Fabric 30 Oxfordshire white colour coated ware (Young 1977, 117–22)

Manufacture	Wheel-thrown
Hardness	Generally hard
Colour	Pale orange (5YR 7/6) to red orange (2.5YR 6/8)
Surface treatment	White or off-white (7.5YR 8/2–7.5YR 8/4) slip
Inclusions	Sandy fabric with sparse small black and red inclusions and occasional lumps of chalk up to *c* 5.0mm. Frequently micaceous
Source	Oxfordshire (Young 1977, 117)

Fabric 31 Brown colour coated ware

Manufacture	Wheel-thrown
Hardness	Hard
Colour	Pale orange (5YR 7/4) often with a thick grey (N 5/0) core
Surface treatment	Dull brown (7.5YR 4/2) slip on one or both surfaces
Inclusions	Moderate to abundant very fine black and red iron ore, sparse white calcareous and non-calcareous inclusions up to *c* 2.0mm, and occasional rounded quartz grains. Frequent flat voids, up to *c* 0.7mm in length are also visible. May be micaceous
Source	Similar to Cirencester Excavation Committee fabric 105, which may have a North Wiltshire source (J Richardson pers comm). Also occurred at Andoversford (Young 1980)

Fabric 32 Mancetter / Hartshill mortarium

Manufacture	Wheel-thrown
Hardness	Hard
Colour	White (10YR 8/1)
Surface treatment	Outer surface may have a thin yellow (10YR 8/3) wash. Trituration grits are black, grey or rust brown opaque refired pottery fragments.
Inclusions	Occasional large quartz grains (up to *c* 2.0mm), and refired pottery fragments. The latter are also used for the trituration grits
Source	Mancetter/Hartshill, Warwickshire (Hartley 1973, 143–7)

Fabric 33 Oxfordshire white mortarium

Manufacture	Wheel-thrown
Hardness	Hard
Colour	White (10YR 8/3)
Surface treatment	Outer surface may have a thin yellow wash (10YR 8/4). Trituration grits are rounded white and pink quartz
Inclusions	Moderate black and red quartz
Source	Oxfordshire (Young 1977, 56)

Fabric 34 West Midlands mortarium

Manufacture	Wheel-thrown
Hardness	Hard
Colour	White (10YR 8/2)
Surface treatment	Moderate sandstone (up to *c* 4.0mm), translucent quartz/quartzite (up to *c* 3.0mm) and ironstone (up to *c* 2.0mm) trituration grits
Inclusions	Similar range of inclusions to those used for trituration gritting
Source	West Midlands, probably Wroxeter area (K Hartley pers comm)

Fabric 35 Brockley Hill / Verulamium mortarium

Manufacture	Wheel-thrown
Hardness	Hard
Colour	White (10YR 8/2)
Surface treatment	None represented
Inclusions	Abundant rounded and sub-angular quartz grains, typically *c* 0.5mm in size and sparse large (up to *c* 3.5mm) iron ore
Source	Brockley Hill and Verulamium area

Fabric 36 Kent / Continental mortarium

Manufacture	Wheel-thrown
Hardness	Hard
Colour	Pale pink (5YR 7/3) with lighter off-white (5YR 8/2) surfaces
Surface treatment	Trituration grits are small, rounded and subangular, grey and white flint fragments
Inclusions	Occasional iron ore up to *c* 2.0mm
Source	Kent and Continent (Hartley 1977, 5–17)

Fabric 37 Severn Valley mortarium

Manufacture	Wheel-thrown
Hardness	Soft
Colour	Orange (2.5YR 6/8)
Surface treatment	Self-coloured matt slip; angular white opaque quartzite trituration grits
Inclusions	Sparse brown-red flecks up to *c* 0.5mm
Source	Severn Basin (K Hartley pers comm)

Fabric 38 Oxfordshire white ware

Manufacture	Wheel-thrown
Hardness	Hard
Colour	Off-white (7.5YR 8/4)
Surface treatment	None represented
Inclusions	Moderate clear, black and red quartz grains up to *c* 0.5mm
Source	Oxfordshire (Young 1977, 93)

Fabric 39 Oxfordshire burnt white ware

Manufacture	Wheel-thrown
Hardness	Hard
Colour	Off-white (7.5YR 8/4)
Surface treatment	Burnt or fumed outer surface (7.5YR 8/2)
Inclusions	Moderate clear, black and red quartz grains up to *c* 0.5mm
Source	Oxfordshire (Young 1977, 113–16)

Fabric 40 Oxfordshire parchment ware

Manufacture	Wheel-thrown
Hardness	Hard
Colour	White (7.5YR 8/2) sometimes with pink core (7.5YR 8/4)
Surface treatment	None represented
Inclusions	Moderate black and red flecks up to *c* 0.5mm
Source	Oxfordshire (Young 1977, 81)

Fabric 41 Unprovenanced white ware

Manufacture	Wheel-thrown
Hardness	Soft to hard
Colour	White (5YR 8/2)
Surface treatment	Pale pink (5YR 8/3) or pale yellow (10YR 8/3) wash sometimes occurs on one or both surfaces
Inclusions	Sparse minute black and red flecks and rounded iron ore up to *c* 1.0mm, with sparse to moderate rounded and sub-angular quartz grains up to *c* 0.5mm, and sparse rounded or flat voids up to *c* 0.2mm
Source	Unknown

Fabric 42 Amphorae

42.1	Dressel 20 type
42.2	Dressel type 2–4
42.3	Pélichet 47 type
42.4	Rhodian type
42.5	*Camulodunum* type 186A
42.6	Unidentified *amphorae* (for fabric descriptions see 1:G8–9)

Fabric 43 Samian ware

Fabric 44 Rhenish ware

Manufacture	Wheel-thrown
Hardness	Hard
Colour	Oxidized red-orange (5YR 7/6), or laminated red and grey (7.5YR 7/7)
Surface treatment	Very fine dark brown or black glossy colour coat (7.5YR 3/2), which may have a metallic sheen
Inclusions	None represented
Source	Trier, Central Gaul (Greene 1978, 18–19)

Fabric 45

Manufacture	Wheel-thrown
Hardness	Hard
Colour	Pale orange (5YR 6/6) throughout
Surface treatment	Sparsely applied clay pellet rough casting, and matt dark grey colour coat
Inclusions	Sparse fine (<0.2mm), sub-angular quartz
Source	Possibly Rhineland

Fabric 46 Stamford ware

Manufacture	Wheel-thrown
Hardness	Hard
Colour	Usually off-white (10YR 7/8) or buff (2.5Y 7/2), but occasionally pinkish orange (5YR 7/6)
Surface treatment	Usually glazed outside yellow (2.5Y 8/8) or pale green (5YR 6/3)
Inclusions	Abundant fine quartz, and sparse coarse (<3.0mm) red inclusions
Source	Stamford, Lincolnshire (Kilmurray 1980)

Fabric 48 Stafford-type ware

Manufacture	Wheel-thrown
Hardness	Very hard
Colour	May be oxidized orange (5YR 6/8) throughout
Surface treatment	Outside surface often partially blackened/ discoloured, possibly through use
Inclusions	Abundant, medium (0.1–0.5mm), well-sorted quartz
Source	Possibly Stafford (Cane 1984, 58–61)

Fabric 49 St Neots-type ware

Manufacture	Wheel-thrown
Hardness	Hard
Colour	Colour variable from grey (5YR 5/1), through orange (5YR 6/4) to pale brown (7.5YR 6/4)
Surface treatment	Jars especially have pronounced internal throwing grooves
Inclusions	Abundant fossilized plate shell fragments 1.0–3.0mm in size
Source	Jurassic area

Fabric 50 'Grass' tempered ware

Manufacture	Handmade
Hardness	Soft to hard
Colour	Generally dense black throughout, except for greyish brown (10YR 6/2 light brownish grey) external surface
Surface treatment	Some external wiping marks
Inclusions	Organic matter has burnt out leaving abundant linear voids; occasional medium quartz; moderate fine mica
Source	Unknown

Fabric 53 Early Malvernian glazed ware

Manufacture	Handmade
Hardness	Hard

Colour	Orange brown (10YR 6/6) surfaces, and reduced grey (10YR 5/3) core
Surface treatment	Applied decoration; thin external green glaze
Inclusions	Abundant quartz; moderate Malvernian rock fragments
Source	Malvern Hills area (Vince 1977, 269)

Fabric 55 Worcester-type unglazed ware

Manufacture	Handmade; later examples wheel-thrown
Hardness	Hard
Colour	Usually reduced dark grey (N 2.5/0) throughout
Surface treatment	Occasional use of rouletting or combed line decoration
Inclusions	Abundant medium quartz
Source	Probably Worcester (Morris 1980, 224)

Fabric 56 Malvernian unglazed ware

Manufacture	Handmade, later examples wheel-thrown
Hardness	Hard
Colour	Usually reduced medium grey (10YR 5/1) throughout
Surface treatment	Occasional applied strip decoration
Inclusions	Moderate-abundant Malvernian rock inclusions usually 3.0–4.0mm; moderate medium quartz
Source	Malvern Hills area (Vince 1977, 264–6)

Fabric 57 Cotswolds unglazed ware

Manufacture	Handmade
Hardness	Soft
Colour	Dark grey (N 4/0) often with buff (10YR 7/3) margins
Surface treatment	Occasional thumbed decoration on rim, or stamp decoration
Inclusions	Abundant 0.1–3.0mm, oolitic limestone fragments
Source	Cotswolds area

Fabric 58 Sandy limestone tempered ware

Manufacture	Handmade
Hardness	Hard
Colour	Usually dark grey (N 4.0)
Surface treatment	Occasional stamp decoration
Inclusion	Moderate coarse-medium rounded quartz; variable fossiliferous limestone content
Source	Possibly Avon Valley, Somerset (Vince 1979, 28)

Fabric 62 Deritend-type ware

Manufacture	Wheel-thrown
Hardness	Hard
Colour	Generally orange (2.5YR 5/8) throughout
Surface treatment	Painted white line decoration
Inclusions	Moderate, medium quartz
Source	Possibly Deritend area, Birmingham (Sherlock 1955)

Fabric 63 Brill/Boarstall ware

Manufacture	Wheel-thrown
Hardness	Hard
Colour	Pale orange (7.5YR 7/6) throughout
Surface treatment	Horizontal grooves; external glaze with moderate (copper) green blotches
Inclusions	Abundant, medium, well-sorted quartz; occasional iron-rich (<1.0mm) inclusions
Source	Brill and Boarstall area, Buckinghamshire

Fabric 64 Glazed sandy wares

Fabric 64.1 Worcester-type sandy glazed ware

Manufacture	Wheel-thrown
Hardness	Hard to very hard
Colour	Oxidized orange-red (5YR 5/6) but often with reduced grey (N 4/0) surfaces
Surface treatment	Impressed decoration frequent; normally green (reduced iron) glazed
Inclusions	Abundant medium-coarse quartz
Source	Probably Worcester (Morris 1980, 225)

Fabric 64.2 Buff sandy ware

Manufacture	Wheel-thrown
Hardness	Hard
Colour	Usually off-white or buff (10YR 7/6) sometimes with reduced grey core
Surface treatment	Often has pinkish orange (7.5YR 6/6) wash; applied strip or stabbing decoration frequent; speckled green (copper coloured) glaze
Inclusions	Abundant medium to coarse quartz; occasional large red sandstone inclusions
Source	Possibly Warwickshire

Fabric 64.3 Green glazed white ware

Manufacture	Wheel-thrown
Hardness	Hard
Colour	White (10YR 8/1) throughout
Surface treatment	Mottled (copper) green speckled yellow glaze
Inclusions	Abundant, medium, well-sorted quartz; occasional iron-rich (usually <1.0mm) speckles
Source	Unknown, but probably a coal measures area

Fabric 65 Glazed oolitic limestone tempered ware

Manufacture	Handmade
Hardness	Hard
Colour	Dull brown-buff (7.5YR 7/4)
Surface treatment	Applied strip and comb impressed decoration. Green/brown glaze
Inclusions	Abundant oolitic limestone (usually <2.0 – 3.0mm)
Source	Cotswolds area

Fabric 69 Oxidized glazed Malvernian ware

Manufacture	Wheel-thrown
Hardness	Hard
Colour	Usually orange (5YR 5/8)
Surface treatment	Copper speckled orange glaze generally applied
Inclusions	Sparse to moderate Malvernian rock fragments; moderate medium quartz
Source	Malvern Hills area (Vince 1977, 269–70)

Fabric 70 Southern white ware

Manufacture	Wheel-thrown
Hardness	Hard
Colour	White to off-white (10YR 8/3)
Surface treatment	Overall bright green glaze
Inclusions	Moderate, usually well-sorted, fine (<0.1mm) quartz
Source	Hampshire area

Fabric 71

Manufacture	Wheel-thrown
Hardness	Hard
Colour	Red (5YR 5/6) inside surface otherwise grey (10YR 5/1)
Surface treatment	Green/brown, mottled exterior glaze
Inclusions	Sparse to moderate, well-sorted, medium quartz; abundant well-sorted fine mica
Source	Possibly Herefordshire (cf Herefordshire group A (Vince 1985, 35–6)

Fabric 72

Manufacture	Wheel-thrown
Hardness	Hard
Colour	Red (2.5YR 5/8) throughout
Surface treatment	Overall dark-brown glaze
Inclusions	Sparse to moderate, medium quartz
Source	Unknown

Fabric 75 North Devon gravel tempered ware

Manufacture	Wheel-thrown
Hardness	Very hard
Colour	Brown (7.5 YR 6/4) to grey (10YR 5/1)
Surface treatment	Overall blotchy green/yellow glaze
Inclusions	Abundant coarse (usually <3.0mm) quartz
Source	North Devon (Grant 1983)

Fabric 77 Midlands yellow ware

Manufacture	Wheel-thrown
Hardness	Hard
Colour	Pale buff (2.5Y 8/2)
Surface treatment	Bright yellow glaze usually applied overall
Inclusions	Sparse fine quartz
Source	Midlands

Fabric 78 Post-medieval red wares

Fabric 78.1 Red sandy ware

Manufacture	Wheel-thrown or press moulded
Hardness	Hard
Colour	Usually red (2.5YR 5/6) throughout
Surface treatment	Usually red slipped internally and externally; overall black glaze. Sometimes slip decorated
Inclusions	Moderate medium quartz
Source	Probably Midlands

Fabric 78.2

Same as fabric 78.1, except that fabric is purplish and very hard, and glaze often patchy and overfired

Fabric 78.3 Fine red sandy ware

Manufacture	Wheel-thrown
Hardness	Very hard
Colour	Purplish (10YR 4/1) throughout

Surface treatment	Lathe-turned decoration and sprigging used; overall even black glaze
Inclusions	Moderate fine quartz
Source	Unknown

Fabric 79 Merida ware

Manufacture	Wheel-thrown
Hardness	Hard
Colour	Oxidized pink or orange
Surface treatment	None represented
Inclusions	Abundant mica; moderate quartz
Source	Spain

Fabric 81 Stonewares

Fabric 81.2	*Westerwald*
Fabric 81.3	*Nottingham*
Fabric 81.4	*Miscellaneous late*
Fabric 81.5	*White salt glazed*
Fabric 81.6	*London*
Fabric 81.7	*Possibly Staffordshire*
Fabric 81.8	*Raeren*

Fabric 82 Tin glazed ware

Fabric 83 Porcelain

Fabric 83.1 Worcester porcelain

Manufacture	Slip cast
Hardness	Very hard
Colour	White
Surface treatment	Blue hand-painted decoration; may be distinguished by the green quality of transmitted light
Inclusions	None visible
Source	Worcester

Fabric 84 Creamware

Fabric 84.1

Manufacture	Wheel-thrown or press moulded
Hardness	Very hard
Colour	White
Surface treatment	Pale yellow glaze overall
Inclusions	None visible
Source	Mainly Staffordshire

Fabric 84.2	Same as fabric 84.1 except paler glaze used

Fabric 85 Modern stone china

Fabric 90 Post-medieval orange ware

Manufacture	Wheel-thrown
Hardness	Hard
Colour	Oxidized orange (5YR 6/6) throughout
Surface treatment	Orange glaze with moderate dark speckling
Inclusions	Occasional medium quartz
Source	Possibly Severn Valley area

Fabric 91 Post-medieval buff wares

Manufacture	Wheel-thrown or press moulded
Hardness	Hard
Colour	Buff (10YR 8/3), sometimes slightly mottled
Surface treatment	Usually slip decorated under pale yellow glaze
Inclusions	Moderate medium quartz; occasional large (<3.00mm) red and buff inclusions
Source	Mainly Staffordshire

Fabric 92

Manufacture	Handmade
Hardness	Soft
Colour	Pale brown (7.5YR 6/4) or grey (N 4/0) patchy surfaces
Surface treatment	Some surface smoothing; occasional combed wavy line decoration
Inclusions	Moderate rounded coarse (up to 3.0mm) quartz; abundant fine mica
Source	Unknown

Fabric 98 Miscellaneous Roman wares

Fabric 99 Miscellaneous medieval wares

Fabric 100 Miscellaneous post-medieval wares

Bibliography

Alexander, J A, 1975 The salt industries of Africa: their significance for European prehistory, in de Brisay and Evans 1975, 81–3

Allies, J, 1852 *On the ancient British, Roman and Saxon antiquities and folklore of Worcestershire*, 2 edn

Amphlett, J (ed), 1895 *Survey of Worcestershire*, Vol 1, Worcestershire Hist Soc

——, (ed), 1899 *Survey of Worcestershire*, Vol 2, Worcestershire Hist Soc

——, and Rea, C, 1909 *The Botany of Worcestershire: an account of the flowering plants, ferns, mosses, hepatics, lichens, fungi, and fresh-water algae, which grow or have grown spontaneously in the County of Worcester*

Anon, 1910 Salt-making in Worcestershire, *Trans Worcestershire Natur Club*, **4**, 363–6

Anon, 1976 *The English glass bottle*, unpubl exhibition catalogue 1/7/76–30/9/76, County Museum, Truro

Armitage, P L and Clutton-Brock, J, 1976 A system for the classification and description of horn cores of cattle from archaeological sites, *J Archaeol Sci*, **3**, 329–48

Atkinson, D, 1942 *Report on excavations at Wroxeter, 1923–27*

Atkinson, D R, 1975 *Tobacco pipes of Broseley, Shropshire*

Baillie, M G L, 1973 *A dendrochronological study in Ireland with reference to the dating of medieval and post-medieval timber*, unpubl PhD Thesis, Queen's Univ, Belfast

——, 1982 *Tree-ring dating and archaeology*

——, and Pilcher, J R, 1973 A simple crossdating programme for tree-ring research, *Tree-ring Bull*, **33**, 7–14

Ballard, A, 1913 *British borough charters*, **1**, 1042–216

Barnard, F P, 1916 *The casting-counter and the counting-board; a chapter in the history of early numismatics and early arithmatic*

Bassett, S R, and Dyer, C C, 1980 Hanbury Hereford and Worcester, *West Midlands Archaeol*, **23**, 88–91

Bateson, M, 1913 The borough of Droitwich, in VCH, **iii**, 72–89

Beresford, M W, and Finberg, H P R, 1973 *English medieval boroughs: a hand-list*

Berry, E K, 1957 The borough of Droitwich and its salt industry, 1215–1700, *Univ Birmingham Hist J*, **6**, 39–61

——, 1975 Medieval Droitwich and the salt trade, in de Brisay and Evans 1975, 76–80

Bestwick, J D, 1975 Romano-British inland salting at Middlewich (Salinae), Cheshire, in de Brisay and Evans 1975, 66–70

Biddle, M, 1976 Towns, in Wilson, D M, (ed), *The archaeology of Anglo-Saxon England*, 99–150

Bidwell, P T, 1979 *The legionary bath house and basilica and forum at Exeter, Exeter Archaeol Rep*, **1**

Blair, J, 1988 Minster churches in the landscape, in Hooke, D (ed), *Anglo-Saxon settlements*, 35–58

Boessneck, J, 1969 Osteological difference between sheep (*Ovis aries* Linne) and goat (*Capra Hircus* Linne), in Brothwell, D, and Higgs, E S (eds), *Science and Archaeology*, 331–58

Böhme, A, 1972 Die fibeln der Kastelle Saalburg und Zugmantel, *Saalburg Jahrbuch, Bericht des Saalburg Museums*, **29**

Bond, C J, 1988 Church and parish in Norman Worcestershire, in Blair, J (ed), *Minsters and parish churches: the local church in transition 950–1200*, Oxford Univ Comm Archaeol Monog, **17**, 119–58

Boon, G C, 1978 A silver-gilt ornament from a 'Carmarthen' trumpet-brooch, *Antiq J*, **58**, 152–3

——, and Savory, H N, 1975 A silver trumpet-brooch with relief decoration, parcel-gilt, from Carmarthen, and a note on the development of the type, *Antiq J*, **55**, 41–61

Booth, P M, and Green, S, 1989 The nature and distribution of certain pink, grog tempered vessels, *J Roman Pottery Stud*, **2**, 77-84

Bradley, R, 1975 Salt and settlement in the Hampshire/Sussex borderland, in de Brisay and Evans 1975, 20–5

——, 1984 *The social foundations of prehistoric Britain*

Brodribb, A C C, Hands, A R, and Walker, S R, 1971 *Excavation at Shakenoak Farm, near Wilcote, Oxfordshire, part IV: site C*, privately printed

Bull Hist et Scient de l'Auvergne LXII Martin, J, 1942 L'evolution des vases sigilles ornes de Lezoux, au premier siècle de notre ere, **62**, 181–209

Bushe-Fox, J P, 1913 *Excavations on the site of the Roman town at Wroxeter, Shropshire, in 1912*, Rep Res Comm Soc Antiq London, **1**

Cambell-Curtis, M, 1913 Dodderhill, in VCH, **iii**, 58–69

Cane, C B K, 1984 'Stafford ware' fabric and form: an interim statement, in Cane, C B K, Cane, J, and Carver, M O H, Saxon and medieval Stafford, new results and theories 1983, *West Midlands Archaeol*, **26**, 48–65

Carter, C O, 1975 Man's need of salt, in de Brisay and Evans 1975, 13

Carver, M O H (ed), 1980 Medieval Worcester, an archaeological framework, *Trans Worcestershire Archaeol Soc, 3 ser*, **7**

CIL *Corpus Inscriptionum Latinarum* 1863–, Berlin

Charles, F W B, 1967 *Medieval cruck building and its derivatives*, Soc Medieval Archaeol Monog ser, **2**

Charleston, R J, 1964 Medieval and later glass, in Cunliffe, B W (ed), *Winchester excavations 1947-1960, Vol 1*, 145–51

——, 1977 Some aspects of 17th century glass found in England, *Annales du 7e* Congres de l'association internationale pour l'histoire de verre, Berlin-Leipzig, 1977 [1978], Liege, 283–97

Charlesworth, D, 1966 Roman Square Bottles, *J Glass Stud*, **8**, 26–40

——, 1971 A group of vessels from the commandant's house, Housesteads, *J Glass Stud*, **13**, 34–7

——, 1972 The glass, in Frere, S S, (ed), *Verulamium Excavations, Vol 1*, Rep Res Comm Soc Antiq London, **28**, 196–215

——, 1979 Glass (including material from all other Exeter sites excavated between 1971 and 1976), in Bidwell, P T, *The legionary bath-house and basilica and forum at Exeter*, 222–31

——, 1981 The Roman glass, in Down, A, (ed), *Chichester Excavations*, **5**, 293–7

Clapham, A R, Tutin, T G, and Warburg, E F, 1962 *Flora of the British Isles*, 2 edn

Clay, R M, 1914 *The hermits and Anchorites of England*

Coope, G R, Shotton, F W, and Strachan, I, 1961 A late Pleistocene fauna and flora from Upton Warren, Worcestershire, *Phil Trans Roy Soc London*, ser B, **244**, 379–421

Cooper, W S, 1934 The Old Exchequer House, Droitwich, *Trans Worcestershire Archaeol Soc*, 3 ser, **11**, 64–74

Crickmore, J, 1984a *Romano-British urban settlements in the West Midlands*, Brit Archaeol Rep **S127**

Crickmore, J, 1984b The old pit, Upwich, Droitwich, *Trans Worcestershire Archaeol Soc*, 3 ser, **9**, 13–36

Crummy, N, 1983 *The Roman small finds from excavations in Colchester 1971–9*, Colchester Archaeol Rep, **2**

Cunliffe, B W, 1978 *Iron Age communities in Britain*, 2 edn

——, 1982 Settlement hierarchy and social change in southern Britain in the Iron Age, *Analecta Præhistorica Leidensia*

Daniels, J S, 1950 *The Woodchester glass house*

Dannell, G B, 1973 The potter Indixivixus, in Detsicas 1973, 139–42

Darby, H C, and Terrett, I B (eds), 1954 *The Domesday geography of Midland England*

Darling, M J, 1977 Pottery from early military sites in western Britain, in Dore and Greene 1977, 57–100

Darrah, R, 1982 Working unseasoned oak, in McGrail, S (ed), *Woodworking techniques before AD 1500*, Brit Archaeol Rep, **129**, 219–30

Darvill, T C, and Timby, J R, 1982 Textural analysis: a review of potentials and limitations, in Freestone, I, Johns C, and Potter, T (eds), *Current research in ceramics: Thin section studies*, British Museum Occas Pap, **32**, 73–88

de Brisay, K W, and Evans K A (eds), 1975 *Salt: the study of an ancient industry*

Detsicas, A P (ed), 1973 *Current research in Romano-British coarse pottery*, CBA Res Rep, **10**

DoE, 1975 *Principles of publication in rescue archaeology*, Ancient Monuments Board for England Committee for Rescue Archaeology, Department of the Environment

——, 1982 *The publication of archaeological excavations*, Joint Working Party of the CBA and the Department of the Environment

Dore, J and Greene, K, 1977 *Roman pottery studies in Britain and beyond*, Brit Archaeol Rep, **S30**

Dunning, G C, 1968 The pitcher imported from Saintonge at Lich Street, Worcester, *Trans Worcestershire Archaeol Soc*, 3 ser, **2**, 45–7

Earp, J R, and Hains, B A, 1971 *The Welsh borderland*, British Regional Geology

EE, *Ephemeris Epigraphica*, 1913, **9**

Ekwall, E, 1960 *The concise Oxford dictionary of English place-names*, 4 edn

Ellis, S E, 1969 The petrography and provenance of Anglo-Saxon and medieval English honestones, with notes on some other hones, *Bull Brit Mus (Natur Hist), Mineralogy*, **2**.3, 135–87

Ellison, M, Finch, M, and Harbottle, B, 1979 The excavation of a 17th century pit at the Black Gate, Newcastle-upon-Tyne, 1975, *Post-Medieval Archaeol*, **13**, 153–81

Evans, J G, 1975 *The environment of early man in the British Isles*

Finberg, H P R, 1972 *The early charters of the West Midlands*, 2 edn

Fock, J, 1966 *Metrishe untersungen an metapodien einiger Europaischer rinderassen gedruckt mit genehmigung de tierarzuchen*, Fakultat der Universitat Munchen

Freezer, D F, 1977 *From saltings to spa town, the archaeology of Droitwich*, Droitwich Archaeol Comm

Frison, E, 1961 Examen anatomique des bois du puits Romain N^{O} I de Herelbeke, *Archaeologia Belgica*, **59**, 800–5

Fulford, M G, 1975 *New Forest Roman pottery*, Brit Archaeol Rep, **17**

Fundberichte 1910 *Fundberichte aus Schwaben*, **18**

Gelling, P S, 1957 Report on excavations in Bays Meadow, Droitwich, Worcestershire, 1954–5, *Birmingham Archaeol Soc Trans Proc*, **75**, [1959], 1–23

——, and Peacock, D P S, 1966 The pottery from Caynham Camp near Ludlow, *Trans Shropshire Archaeol Soc*, **58**, 96–100

Gillam, J P, 1970 *Types of Roman coarse pottery vessels in northern Britain*, 3 edn

——, 1976 Coarse fumed ware in northern Britain, and beyond, *Glasgow Archaeol J*, **4**, 57–80
Godwin, H, 1975 *History of the British flora*, 2 edn
Goodman, W L, 1962 *Woodwork from the Stone Age to Do-it-Yourself*
Goultequer, P L, 1975 Niger, country of salt, in de Brisay and Evans 1975
Grant, A, 1975 Appendix B. The use of tooth wear as a guide to the age of domestic mammals – a brief explanation, in Cunliffe, B W, (ed), *Excavations at Portchester Castle. 1 Roman*, Rep Res Comm Soc Antiq London, **32**, 437–50
Grant, A, 1983 *North Devon pottery: the seventeenth century*
Grant, E (ed), 1986 *Central places, archaeology and history*
Greene, K, 1978 Imported fine wares in Britain to AD 250: a guide to identification, in Arthur, P, and Marsh, G (eds), *Early fine wares in Roman Britain*, Brit Archaeol Rep, **57**, 15–30
Greig, J R A, 1978–9 Appendix 4. The plant remains, in Watts, L, Birmingham Moat: its history, topography and destruction, *Trans Birmingham Warwickshire Archaeol Soc*, **89**, [1981], 66–72
——, 1988 The interpretation of some Roman well fills from the Midlands of England, in Küster, H-J (ed), *Der Prähistorische Mensch und Seine Umwelt*, Stuttgart, 367–78
Hadfield, C, 1966 *The canals of the West Midlands*
Hains, B A, and Horton, A, 1969 *Central England*, British Regional Geology, 3 edn
Harcourt, R, 1974 The dog in prehistoric and early historic Britain, *J Archaeol Sci*, **1**, 151–75
Harden, D B, 1947 The glass, in Hawkes and Hull 1947, 287–307
——, Painter, K S, Pinder-Wilson, R H, and Tait, H, 1968 *Masterpieces of glass*
——, and Price J, 1971 The glass, in Cunliffe, B W, (ed), *Fishbourne 1961–69, Vol II, the finds*, Rep Res Comm Soc Antiq London, **27**, 317–68
Hartley, K F, 1973 The kilns at Mancetter and Hartshill, Warwickshire, in Detsicas 1973, 143–7
——, 1977 Two major potteries producing *mortaria* in the first century AD, in Dore and Greene 1977, 5–17
Hawkes, C F C, and Hull, M R, 1947 *Camulodunum*, Rep Res Comm Soc Antiq London, **14**
Heighway, C, 1978 Excavations at Gloucester, fourth interim report: St Oswalds Priory, Gloucester, 1975–76, *Ant J*, **58**, 103–32
——, 1984 Anglo-Saxon Gloucestershire, in Saville, A (ed), *Archaeology in Gloucestershire*, 225–47
Henderson, J, forthcoming Glass beads, counters, *tesserae* and armlets from Castleford, West Yorkshire, in *Excavations at Castleford*, West Yorkshire County Council
Hillam, J, 1982 The dating of Roman timbers from Friar Street 1, Droitwich, *Ancient Monuments Laboratory Rep*, **3754**
——, and Morgan, R A, 1981 Dendro dates from Sheffield, *Curr Archaeol*, **7**.9, 286–7
——, ——, and Tyers, I, 1987 Sapwood estimates and the dating of short ring sequences, in Ward, R G (ed), *Applications of tree-ring studies*, Brit Archaeol Rep, **S33**, 165–85
Hillelson, D, 1985 *Excavations on the old market site, Droitwich, 1984: an interim report*, unpubl rep, Hereford and Worcester County Council
Hilton, R H, 1966 *A medieval society: the West Midlands at the end of the thirteenth century*
Hodgkinson, H R, 1925–6 Note on the Roman buildings at Droitwich, *Birmingham Archaeol Soc Trans Proc*, **51**, [1928], 35–8
——, 1931 Medieval tiles from Droitwich, *Birmingham Archaeol Soc Trans Proc*, **55**, [1933], 70
——, 1933 Roman pottery from Droitwich, *Birmingham Archaeol Soc Trans Proc*, **57**, [1935], 175–6
Hogan, D, 1970 The Du Houx and the Haughton Green glasshouse, in *Studies in glass history and design. Papers presented to the Committee B session of VIIIth international congress on glass*, 24–5
Hollister, P, 1981 Flowers which clothe the meadows in spring, the rebirth of millefiori c 1500, *Annales du 8e congrès de l'association internationale pour l'histoire du vere,* London–Liverpool 18–25/9/79, Liege
Hollstein, E, 1965 Jahrringchronologische datierung von Eichenhölzern ohne Waldkante, *Bonner Jahrbuch*, **165**, 12–27
——, 1972 Dendrochronologische datierungaus Wederath (Belginum), *Trierer Zeitschrift*, **35**, 123–5
——, 1974 Eine Romische Deichel aus Dillingen, Kreis Saarlouis, *Bericht der Staatlichen Denkmalpflege im Saarland*, **21**, 101–4
Hooke, D, 1981 The Droitwich salt industry: An examination of the West Midlands charter evidence, in Brown, D, *et al (eds), Anglo-Saxon Studies in Archaeology and History 2*, Brit Archaeol Rep, **92**, 123–69
——, 1985 *The Anglo-Saxon landscape: the kingdom of the Hwicce*
——, 1986 Territorial organisation in the Anglo-Saxon West Midlands: central places, central areas, in Grant 1986, 79–93
Hoover, H C, and Hoover, L H (trans), 1912 *Georgius Agricola de re metallica*
Houghton, F T S, 1919 The parochial and other chapels of the county of Worcester, together with some account of the development of the parochial system in the county, *Birmingham Archaeol Soc Trans Proc*, **45**, [1921], 23–114
——, 1929–30 Salt-ways, *Birmingham Archaeol Soc Trans Proc*, **54**, [1932], 1–17

Howard, H, 1985 *The petrology of the Kenchester Roman and Iron Age pottery*, unpubl, Kenchester archive, 108
Howe, M D, Perrin, J R, and Mackreth, D F, 1980 *Roman pottery from the Nene Valley: a guide*, Peterborough City Mus Occas Pap, **2**
Hughes, M K, Milson, S J, and Leggett, P A, 1981 Sapwood estimates in the interpretation of tree-ring dates, *J Archaeol Sci*, **8**, 381–90
Hüme, I N, 1961 The glass wine bottles in Colonial Virginia, *J Glass Stud*, **3**, 91–117
——, 1962 Tudor and early Stuart glasses found in London, *The Connoisseur*, **150**, 269–73
——, 1970 *Guide to the artefacts of colonial Virginia*
Hunt, A, 1975 Friar Street, Worcs, 1973–5 (SO 897635), *West Midlands Archaeol News Sheet*, **18**, 39–40
Hunter, J, (ed) 1844 *The great Rolls of the Pipe for the second, third and fourth years of the reign of Henry II,* Record Commissioners, London
Hunter, R, 1979 St Neots type ware, in Williams, J H, *St Peter's Street Northampton: excavations 1973–6*, Northampton Archaeol Monog, **2**
Hurst, D, 1991 Major Saxon discoveries at Droitwich–excavations at the Upwich brine pit, *Curr Archaeol*, **126**, 252-5
Jacobs, J, 1913 Sigillatafunde aus einem romischen Keller zu Bregenz, *Jahrbuch fur Altertumskunde*, 6te Band, 1912 [1913]
Jenkins, S C, and Quayle, H I, 1977 *The Oxfordshire, Worcester and Wolverhampton Railway*
John, E, 1964 *Land tenure in early England: a discussion of some problems*, Studies in early English history, **1**
Jones, M J, and Bond, C J, 1987 Urban defences, in Schofield, J, and Leech, R (eds), *Urban archaeology in Britain*, CBA Res Rep, **61**, 81–116
Jones, R, Wall, S M, Lockyer, A M, Coy, J, and Maltby, M, 1981 Ancient Monuments Laboratory, DoE, computer based osteometry, data capture user manual (1), *Ancient Monuments Laboratory Rep*, **3342**
Kenyon, K M, 1940 Excavations at Viroconium 1936–7, *Archaeologia*, **88**, 175–227
——, 1953 Excavations at Sutton Walls, Herefordshire 1948–51, *Archaeol J*, **110**, [1954], 1–87
Kiesewalter, 1974 in von den Dreisch, A, and Boessneck, J, Kritische Anmerkungen zur widerristhohen berechnung aus langenmassen vor-und fruhgeschichtlicher Tierknochen, *Saugetierkunglicke Mitteilungen*, **22**, 325–48
Kilby, K, 1977 *The village cooper*
Kilmurray, K, 1980 *The pottery industry of Stamford, Lincs c AD 850–1250: its manufacture, trade, and relationship with continental wares with a classification and chronology*, Brit Archaeol Rep, **84**
Knorr, R, 1912 *Die Terra-Sigillata-Gefaße von Aislingen*
——, 1919 *Topfer und Fabriken verzierter Terra-Sigillata des ersten Jahrhunderts*
Lambrick, G, 1980 Excavations in Park Street, Towcester, *Northamptonshire Archaeol*, **15**, 35–118
Leeds, E T, 1941 17th and 18th century wine bottles of Oxford Taverns, *Oxoniensia*, **6**, 44–55
Little, A G, 1906 Austin Friars, Droitwich, in *VCH*, **ii**, 173–5
Liversidge, J, 1976 Woodwork, in Strong, D, and Brown, D (eds), *Roman crafts*, 155–8
LMMC *London Museum Medieval Catalogue* [1940]
Locke, A A, 1906 The hospital of St Mary in Droitwich, in *VCH*, **ii**, 179–80
LRBC Carson, R A G, Hill, P V, and Kent, J P C, *Late Roman bronze coinage* [1960]
Lyne, M A B, and Jeffries, R S, 1979 *The Alice Holt / Farnham Roman pottery industry*, CBA Res Rep, **30**
McCarthy, M, 1979 Pottery synthesis, in Williams, J H, *St Peter's Street Northampton excavations 1973–6*, Northampton Archaeol Monog, **2**
Mackreth, D F, 1985 Brooches from Roman Derby, in Dool, J *et al*, Roman Derby: excavations 1968–83, *Derbyshire Archaeol J*, **105**, 281–99
Maddocks, P G, 1950 *The salt industry of Worcestershire*, unpubl dissertation, Birmingham Univ Geography Dept, **160**
Malam, J P, 1981 White salt-glazed stoneware manufactured at Jackfield, *West Midlands Archaeol*, **24**, 45–50
Maltolcsi, J, 1970 Historiche Erforschung der Korpergrosse des Rindes auf Grund von ungarischem Knockenmaterial, *Zeitschrift f, Tierzuchtg u Zuchtungsbiol Hamburg*, **87**, 89–137
Mason, D J P, 1985 *Excavations at Chester. 26–42 Lower Bridge Steet 1974–6. The Dark Age and Saxon periods*, Grosvenor Mus Archaeol Excavation and Survey Rep, **3**
Mathews, L G, and Green, H J M, 1969 Post-medieval pottery at the Inns of Court, *Post-Medieval Archaeol*, **3**, 1–17
Mawer, A, and Stenton, F M, 1927 *The place-names of Worcestershire*, English Place-name Soc, **4**
Megaw, J V S, and Simpson, D D A, 1979 *Introduction to British Prehistory*
Mellor, M, 1980 Late Saxon pottery from Oxfordshire: evidence and speculation, *Medieval Ceramics*, **4**, 17–27
Mitchell, G H, Pocock, R W, and Taylor, J H, 1961 *Geology of the country around Droitwich, Abberley and Kidderminster*, Memoir Geol Survey Gt Britain
Monkhouse, F J, 1954 Worcestershire, in Darby and Terrett 1954, 215–69
Moore, D T, 1983 Petrological aspects of some sharpening stones, touchstones and milling stones, in Kempe, D R C, and Harvey, A P (eds), *The Petrology of archaeological artefacts*, 277–300

Morgan, R A, and Schofield, J, 1978 Tree-rings and the archaeology of the Thames waterfront in the City of London, in Fletcher, J M (ed), *Dendrochronology in Europe*, Brit Archaeol Rep, **S51**, 223–38

Morris, C A, 1981 Early medieval separate-bladed shovels from Ireland, *J Roy Soc Antiq Ir*, **111**, 50–69

Morris, E L, 1978 Late Saxon pottery from Worcester, *Trans Worcestershire Archaeol Soc*, 3 ser, **6**, [1979], 75–80

——, 1980 Medieval and post-medieval pottery in Worcester — a type series, in Carver 1980, 221–53

——, 1981a Ceramic exchange in western Britain: a preliminary view, in Howard, H, and Morris, E L (eds), *Production and distribution: a ceramic viewpoint*, Brit Archaeol Rep, **S120**, 67–81

——, 1981b Petrological report on the Beaker and Iron Age ceramics, in Stanford, S C, *Midsummer Hill*, privately printed, 151–5

——, 1982 Iron Age pottery from western Britain: another petrological study, in Freestone, I, Johns, C, and Potter, T (eds), *Current research in ceramics: thin section studies* Brit Mus Occas Pap, **32**, 15–25

——, 1983 *Salt and ceramic exchange in western Britain during the first millennium BC*, unpubl PhD thesis, Univ Southampton

——, 1985 Prehistoric salt distributions: two case studies from western Britain, *Bull Board Celtic Stud*, **32**, 336–79

Morris, J,(ed) 1980 Nennius, British history and the Welsh Annals

Museum of London, 1986 *DUA archive report writing manual*, Department of Urban Archaeology Museum of London

Nash, T R, 1781–2 *Collections for the history of Worcestershire*, 2 vols

Nichol, J, 1811 *The history and antiquities of the county of Leicestershire*, Vol 4

Niermeyer, J F, 1954–76 *Mediae latinitatis lexicon Minus. A medieval latin — French / English dictionary*, Leiden

Northolt, A J D, and Highley, D E, 1973 *Salt*, Minerals Resourses Consultative Committee, mineral dossier no 7, HMSO

OLD *Oxford Latin Dictionary*, 1982

Orton, C, 1977 Roman pottery, in Blurton, T R (ed), Excavations at Angel Court, Walbrook, 1974, *Trans London Middlesex Archaeol Soc*, **28**, 14–100

Oswald, A, 1975 *Clay pipes for the archaeologist*, Brit Archaeol Rep, **14**

——, and Phillips, H, 1949 A restoration glass house from Gracechurch St, London, *The Connoisseur*, **123–124** (Jan–Dec), 30–6

Oswald, O, 1936–7 *Index of figure-types on Terra Sigillata 'samian ware'*

Peacock, D P S, 1965–7 Romano-British pottery production in the Malvern district of Worcestershire, *Trans Worcestershire Archaeol Soc*, 3 ser, **1**, 15–28

——, 1968 A petrological study of certain Iron Age pottery from western England, *Proc Prehist Soc*, **34**, 414–27

——, 1971 Petrology of the coarse pottery, in Cunliffe, B W (ed), *Excavations at Fishbourne 1961–1969, Vol II: the finds*, Rep Res Comm Soc Antiq London, **27**, 255–9.

——, 1977a Roman *amphorae*: typology, fabric and origins, *École Française de Rome*, **32**, 261–78

——, 1977b *Pottery and early commerce. Characterisation and trade in Roman and later ceramics*

Pillans, E B, 1906 Salt, in VCH, **ii**, 256–63

Platt, C, 1976 *The English medieval town*

Pollard, S, 1974 A late Iron Age settlement and a Romano-British villa at Holcombe, near Uplyme, Devon, *Proc Devon Archaeol Soc*, **32**, 59–161

Poole, E G, and Williams, B J, 1981 *The Keuper saliferous beds of the Droitwich area*, Inst Geol Sci Rep, **81/2**

Rackham, O, 1980 *Ancient woodland; its history, vegetation and uses in England*

——, Blair, W Y, and Munby, J T, 1978 The 13th century roof and floor of the Black Friars Priory at Gloucester, *Medieval Archaeol*, **22**, 105–22

Rastel, T, 1678 An account of the salt waters of Droytwich, *Phil Trans Roy Soc London*, A, **12**, 1059–64

Rawes, B, 1972 Roman pottery kilns at Gloucester, *Trans Bristol Gloucestershire Archaeol Soc*, **91**, [1973], 18–59

——, 1982 Gloucester Severn Valley ware — a study of the Roman pottery forms, *Trans Bristol and Gloucestershire Archaeol Soc*, **100**, 33–46

Rees, H, 1986 Ceramic salt working debris from Droitwich, *Trans Worcestershire Archaeol Soc*, 3 ser, **10**, 47–54

RIC Mattingly, H and Sydenham, E A, *The Roman imperial coinage*, **5**.2, 1923–67

Richardson, L, 1930 *Wells and springs of Worcestershire*, Memoir Geol Survey England Wales

Rigby, V, 1973 Potters Stamps on Terra Nigra and Terra Rubra found in Britain, in Detsicas 1973, 7–24

——, 1977 The Gallo-Belgic pottery from Cirencester, in Dore and Greene 1977, 37–46

Riha, E, 1979 *Die Römishcen fibeln aus Augst und Kaiseraugst*, Forschungen in Augst, Band 3

Rivet, A L F, and Smith, C, 1979 *The place-names of Roman Britain*

Rodwell, W R, 1979 Iron Age and Roman salt-winning on the Essex coast, in Burnham, B C, and Johnson, H B (eds), *Invasion and response, the case of Roman Britain*, Brit Archaeol Rep, **30**, 133–75

Roe, F E S, 1988 Report on worked stone from Beckford, Worcestershire, *Ancient Monuments Laboratory Rept*, **154**

Rogers, G B, 1974 Poterie sigillee de la Gaule centrale, *Gallia*, Suppl 28

Round, J H, 1901 Domesday Survey, in VCH, **i**, 253–323

St Joseph, J K, 1938 The Roman fort at Dodderhill, Droitwich, *Birmingham Archaeol Soc Trans Proc*, **62**, 27–31

Saville, A, 1979 *Excavations at Guiting Power Iron Age site, Gloucestershire*, Comm Rescue Archaeol Avon, Gloucestershire, Somerset, Occas Pap, **7**

Sawyer, P H, 1968 *Anglo-Saxon charters: an annotated list and bibliography*, Roy Hist Soc Guides Handb, **8**

Sellwood, L, 1984 Tribal bounderies viewed from the perspective of numismatic evidence, in Cunliffe, B W, and Miles, D (eds), *Aspects of the Iron Age in central southern Britain*, 191–204

Sherlock, R J, 1955 Excavations at Deritend, *Birmingham Archaeol Soc Trans Proc*, **73**, [1957], 109–14

Slater, T R, 1980 *The analysis of burgages in medieval towns*, Univ Birmingham Geogr Dept Working Pap, **4**

——, 1981 The analysis of burgage patterns in medieval towns, *Area*, **13**, 211–6

——, 1982 Urban genesis and medieval town plans in Warwickshire and Worcestershire, in Slater, T R, and Jarvis, P J, (eds), *Field and forest: an historical geography of Warwickshire and Worcestershire*, 173–202

Smith, L T (ed), 1964 *The Itinerary of John Leland in or about the years 1535–1545*, 5 vols

Squires, R W, 1984 *A guide to the Droitwich canals*

Stanfield, J A, and Simpson, G, 1958 *Central Gaulish potters*

Stanford, S C, 1981 *Midsummer Hill, an Iron Age Hillfort on the Malverns*, privatly printed

Stead, I M, and Rigby, V, 1986 *Baldock, the excavations of a Roman and pre-Roman settlement, 1968–72*, Britannia Monog Ser, **7**

Stewart, B H I H, and Blunt, C E, 1978 The Droitwich mint and BMC type XIV of Edward the Confessor, *Brit Numis J*, **48** 52-57

Straker, V, 1979 Macroscopic plant remains from Droitwich, Bays Meadow 1972–6, *Ancient Monuments Laboratory Rep*, **2812**

Streeten, A D F, 1979 Appendices I–II, in Freke, D J (ed), The excavation of a 16th century pottery kiln at Lower Parrock, Hartfield, East Sussex 1977, *Post-Medieval Archaeol*, **13**, 114–20

Swan, V, 1975 Oare reconsidered and the origins of Severnake Ware in Wiltshire, *Britannia*, **6**, 117–37

Symonds, R P, forthcoming *Rhenish wares: fine dark-coloured pottery from Gaul and Germany*, Oxford Univ Comm Archaeol Monog, **23**

Taylor, M, 1981 *Wood in archaeology*

Teichart, M, 1975 Osteometrische Untersuchungen zur Berechnung der Widerristhohe bei schafen, in Clason, A (ed), *Archaeozoological Studies*, Amsterdam, 51–69

Terrisse, J-R, 1968 Les ceramiques sigillees Gallo-Romaines des Martres-de-Veyre (Puy-de-Dome), *Gallia*, Suppl 19

Thompson, H, 1949 *Elementary veterinary science*, 6th ed

Thorn, F, and Thorn, C (eds), 1982 *Domesday Book: Vol 16, Worcestershire*

Tite, M S, 1972 *Methods of Physical Examination in Archaeology*

Tomber, R, 1985 III Pottery, in Wilmott, A R, and Rahtz, S P Q, An Iron Age and Roman settlement outside Kenchester, (Magnis), Herefordshire. Excavations 1977–79, *Trans Woolhope Natur Field Club*, **45**, 99–145

Vaughan, M D, 1982 *The charred plant remains from Hanbury Street, Droitwich*, unpubl MSc thesis, Univ London

VCH, Page, W (ed), 1906 *The Victoria History of the Counties of England. A history of Worcestershire*, **ii**

VCH, Page, W (ed), 1913 *The Victoria History of the Counties of England. A history of Worcestershire*, **iii**

VCH, Page, W (ed), 1924 *The Victoria History of the Counties of England. A history of Worcestershire*, **iv**

Vince, A G, 1977 The medieval and post-medieval ceramic industry of the Malvern region, the study of a ware and its distribution, in Peacock 1977, 257–305

——, 1979a The medieval pottery (Citizens House, 1970), in Cunliffe, B W (ed), *Excavations in Bath 1950–1975*, Comm Rescue Archaeol Avon, Gloucestershire, Somerset, Excav Rep, **1**

——, 1979b The pottery, in Heighway, C M, Garrod, A P, and Vince, A G, Excavations at 1 Westgate Street, Gloucester, *Medieval Archaeol*, **23**, 159–213

——, 1983 *The medieval ceramic industry of the Severn Valley*, unpubl PhD thesis, Univ Southampton

——, 1984a Grass-tempered pottery (appendix 1), in Heighway, C, Anglo-Saxon Gloucestershire, in Saville, A (ed), *Archaeology in Gloucestershire — from the earliest hunters to the industrial age*

——, 1984b Late Saxon and medieval pottery in Gloucestershire, in Saville, A (ed), *Archaeology in Gloucestershire — from the earliest hunters to the industrial age*

——, 1984c The use of petrology in the study of medieval ceramics: case studies form southern England, *Medieval Ceramics*, **8**, 31–45

——, 1985 The ceramic finds, in Shoesmith, R, *Hereford City excavations — the finds*, CBA Res Rep, **56**

——, and Whitehead, P F, 1979 An abandoned Flandrian channel at Pershore — stratigraphy, pottery and biota, *Vale Evesham Hist Soc Res Pap*, **7**, 9–24

Wade-Evans, A W, 1938 *Nennius's 'History of the Britons', together with the 'Story of the loss of Briain'*, Church Hist Soc

Walters, H B, 1908 *Catalogue of the Roman pottery in the Department of Antiquities*, British Museum

Ward, M, 1981 *Terra Nigra*-type wares from Chester, in Anderson, A C, and Anderson, A S (eds), *Roman Pottery Research in Britain and North-West Europe*, Brit Archaeol Rep, **123**, 51–68

Webster, P V, 1976 Severn Valley ware: a preliminary study, *Trans Bristol Gloucestershire Archaeol Soc*, **94**, 18–46

——, and Hobley, B, 1964 Aerial reconnaissance over the Warwickshire Avon, *Archaeol J*, **121**, [1965], 1–22

Wenham, L P, 1968 *The Romano-British cemetery at Trentholme Drive, York*, Ministry Public Buildings Works Archaeol Rep, **5**

Wheeler, R E M, 1923 *Segontium and the Roman occupation of Wales, Y Cymmrodor*, **33**

——, 1930 *London in Roman times*

Whitelock, D (ed), 1955 *English historical documents, c* 500–1042

Whithouse, D B, 1962 A note on excavation of the Roman fort at Dodderhill, Droitwich, 1961–2, *Trans Worcestershire Archaeol Soc, New Ser*, **39**, 55–8

Whiting, W, Hawley, W, and May, T, 1931 *Report on the excavation of the Roman cemetery at Ospringe, Kent*, Rep Res Comm Soc Antiq London, **7**

Whitley, D, 1923 Saltways of the Droitwich district, *Birmingham Archaeol Soc Trans Proc*, **49**, [1926], 1–15

Williams, D F, 1977 The Romano-British Black Burnished industry: an essay on characterisation by heavy mineral analysis, in Peacock 1977b, 163–220

Wills, G, 1977 *The Bottle-collector's guide*

Woodiwiss, S G, 1983 *Excavations in Droitwich 1982–3, HWCM 4099 High Street, HWCM 4574 Queen Street*, unpubl report, Hereford and Worcester County Council

Wright, R P, 1954 II Inscriptions, *J Roman Stud*, **54**, 103–11

Young, C J, 1977 *The Roman pottery industry of the Oxford region*, Brit Archaeol Rep, **43**

——, 1980 The late Roman fine wares, in Rawes, B (ed), The Romano-British site at Wycomb, Andoversford, *Trans Bristol Gloucestershire Archaeol Soc*, **98**, 11–55

Index *by Barbara Hird*

NOTE: Friar Street is abbreviated to FS, the Old Bowling Green to OBG, Hereford and Worcester to H & W. Figures in italics refer to plates and figures; there may also be textual references on these pages.